Fuzzy Mathematics
in Economics and Engineering

Studies in Fuzziness and Soft Computing

Editor-in-chief
Prof. Janusz Kacprzyk
Systems Research Institute
Polish Academy of Sciences
ul. Newelska 6
01-447 Warsaw, Poland
E-mail: kacprzyk@ibspan.waw.pl
http://www.springer.de/cgi-bin/search_book.pl?series=2941

Further volumes of this series can be found at our homepage.

Vol. 71. K. Leiviskä (Ed.)
Industrial Applications of Soft Computing, 2001
ISBN 3-7908-1388-5

Vol. 72. M. Mareš
Fuzzy Cooperative Games, 2001
ISBN 3-7908-1392-3

Vol. 73. Y. Yoshida (Ed.)
Dynamical Aspects in Fuzzy Decision, 2001
ISBN 3-7908-1397-4

Vol. 74. H.-N. Teodorescu, L.C. Jain and A. Kandel (Eds.)
Hardware Implementation of Intelligent Systems, 2001
ISBN 3-7908-1399-0

Vol. 75. V. Loia and S. Sessa (Eds.)
Soft Computing Agents, 2001
ISBN 3-7908-1404-0

Vol. 76. D. Ruan, J. Kacprzyk and M. Fedrizzi (Eds.)
Soft Computing for Risk Evaluation and Management, 2001
ISBN 3-7908-1406-7

Vol. 77. W. Liu
Propositional, Probabilistic and Evidential Reasoning, 2001
ISBN 3-7908-1414-8

Vol. 78. U. Seiffert and L.C. Jain (Eds.)
Self-Organizing Neural Networks, 2002
ISBN 3-7908-1417-2

Vol. 79. A. Osyczka
Evolutionary Algorithms for Single and Multicriteria Design Optimization, 2002
ISBN 3-7908-1418-0

Vol. 80. P. Wong, F. Aminzadeh and M. Nikravesh (Eds.)
Soft Computing for Reservoir Characterization and Modeling, 2002
ISBN 3-7908-1421-0

Vol. 81. V. Dimitrov and V. Korotkich (Eds.)
Fuzzy Logic, 2002
ISBN 3-7908-1425-3

Vol. 82. Ch. Carlsson and R. Fullér
Fuzzy Reasoning in Decision Making and Optimization, 2002
ISBN 3-7908-1428-8

Vol. 83. S. Barro and R. Marín (Eds.)
Fuzzy Logic in Medicine, 2002
ISBN 3-7908-1429-6

Vol. 84. L.C. Jain and J. Kacprzyk (Eds.)
New Learning Paradigms in Soft Computing, 2002
ISBN 3-7908-1436-9

Vol. 85. D. Rutkowska
Neuro-Fuzzy Architectures and Hybrid Learning, 2002
ISBN 3-7908-1438-5

Vol. 86. M.B. Gorzałczany
Computational Intelligence Systems and Applications, 2002
ISBN 3-7908-1439-3

Vol. 87. C. Bertoluzza, M.Á. Gil and D.A. Ralescu (Eds.)
Statistical Modeling, Analysis and Management of Fuzzy Data, 2002
ISBN 3-7908-1440-7

Vol. 88. R.P. Srivastava and T.J. Mock (Eds.)
Belief Functions in Business Decisions, 2002
ISBN 3-7908-1451-2

Vol. 89. B. Bouchon-Meunier, J. Gutiérrez-Ríos, L. Magdalena and R.R. Yager (Eds.)
Technologies for Constructing Intelligent Systems 1, 2002
ISBN 3-7908-1454-7

Vol. 90. B. Bouchon-Meunier, J. Gutiérrez-Ríos, L. Magdalena and R.R. Yager (Eds.)
Technologies for Constructing Intelligent Systems 2, 2002
ISBN 3-7908-1455-5

James J. Buckley
Esfandiar Eslami
Thomas Feuring

Fuzzy Mathematics in Economics and Engineering

With 69 Figures
and 27 Tables

Physica-Verlag
A Springer-Verlag Company

Professor James J. Buckley
University of Alabama at Birmingham
Mathematics Department
Birmingham, AL 35294
USA
buckley@math.uab.edu

Professor Esfandiar Eslami [1]
Shahid Bahonar University
Department of Mathematics
Kerman
Iran
eslami@arg3.uk.ac.ir
eslami@math.uab.edu

Dr. Thomas Feuring
University of Siegen
Electrical Engineering and Computer Science
Hölderlinstraße 3
57068 Siegen
Germany

[1] Thanks to the University of Shahid Bahonar, Kerman, Iran, for financial support during my sabbatical leave at UAB. Thanks to UAB for producing a good atmosphere to do research and teaching. Special thanks to Prof. James J. Buckley for his kind cooperation that made all possible.

ISSN 1434-9922
ISBN 3-7908-1456-3 Physica-Verlag Heidelberg New York

Cataloging-in-Publication Data applied for
Die Deutsche Bibliothek – CIP-Einheitsaufnahme
Buckley, James J.: Fuzzy mathematics in economics and engineering: with 27 tables / James J. Buckley; Esfandiar Eslami; Thomas Feuring. – Heidelberg; New York: Physica-Verl., 2002
(Studies in fuzziness and soft computing; Vol. 91)
ISBN 3-7908-1456-3

Physica-Verlag Heidelberg New York
a member of BertelsmannSpringer Science+Business Media GmbH

Printed in Germany

Hardcover Design: Erich Kirchner, Heidelberg

SPIN 10858007 88/2202-5 4 3 2 1 0 – Printed on acid-free paper

To Julianne, Birgit and Mehra

Helen, Lioba, Jason, Pooya, Peyman and Payam

Contents

Chapter 1

Introduction

This book surveys certain applications of fuzzy sets to economics and engineering based on the authors' research over the last 16 years. We cover many of the topics addressed in the literature, and if we omit an area of application, we may state this within the pertinent chapter or it will be found in the references.

No previous knowledge of fuzzy sets is needed because in Chapter 2 we survey the basic ideas needed for the rest of the book. The basic prerequisite is elementary differential calculus because we use derivatives, and partial derivatives, from time to time. The chapters on differential equations are all elementary and should be understandable from basic calculus.

We sometimes define new concepts in the chapters. Usually, when we do this the concept is useful only for that chapter and therefore was not included in Chapter 2.

Our policy on "theorems" is always to state the theorem and present the proof whenever the proof is short and elementary. Longer, and more complicated proofs, are relegated to the references.

We do not give a complete list of references. We, of course, give all our references since the book is based on these papers. For other references we: (1) give the recent (last couple of years) references; and (2) for older references we only give a few "key" citations from which the reader can find other relevant papers.

An overview of the book can be seen from the table of contents. A more detailed overview is in Chapter 12. So, if you want a quick reading about what is in the book, please turn to the summary section of Chapter 12.

To achieve a uniform notation in a book with lots of mathematics is always difficult. What we have done is introduce the basic notation, to be uniform throughout the book, in Chapter 2. Other notation is chapter dependent. By chapter dependent we mean some symbols may change their meaning from chapter to chapter. For example, the letters "a" and "b" may be used as $[a, b]$ to represent a closed interval in one chapter but they can be parameters in a differential equation in another chapter.

Our recommendation reading for applications to economics is

$$1 \rightarrow 2 \rightarrow \{3,4,5,6,8,10\} \rightarrow 12 ,$$

where "1" stands for Chapter 1, etc. For engineering we suggest

$$1 \rightarrow 2 \rightarrow \{3,7,8,9,10,11\} \rightarrow 12 .$$

This book is basically about solving fuzzy equations (linear, eigenvalue, differential, etc.) and fuzzy optimization. What is new is that we introduce three solution concepts for fuzzy equations: (1) the classical solution; (2) the extension principle solution; and (3) the α-cut and interval arithmetic solution. Therefore, it is most important that the reader understand the ideas involved in these three methods of solving fuzzy equations before proceeding to the rest of the book. The easiest introduction to the solution methods is in Sections 3.1 and 3.2 of Chapter 3 where we look at solving a simple fuzzy linear equation.

Also what is new is that we use an evolutionary algorithm to solve fuzzy problems, especially the fuzzy optimization problems. Usually the classical methods (calculus, etc.) do not apply to solving fuzzy optimization problems so we employ a direct search algorithm to generate good (approximate) solutions to these fuzzy problems.

Our general purpose evolutionary algorithm (abbreviated EA) is described in Chapter 13 at the end of this book. It has to be adopted to the different fuzzy problems and this is usually discussed in the chapters where it is used. We have not included our evolutionary algorithm in the book since you can download this software (genetic, evolutionary) from the internet. Use the terms "genetic", or "evolutionary" algorithm in your search engine and then you can solve your own fuzzy optimization problems.

Some of the figures in the book are difficult to obtain so they were created using different methods. First, many were made using the graphics package in $LaTeX2_\epsilon$. For example, Figures 3.6, 3.11, many in Chapters 6 and 10, and Figures 7.1, 7.3, were done this way. Some others, impossible to do within the graphics package in $LaTeX2_\epsilon$, were first drawn in Maple [1] and then exported to $LaTeX2_\epsilon$. We did those in Chapter 7 (not Figures 7.1 and 7.3) and some in Chapter 3 this way. There are some other figures that we considered easier to do in Maple and then export to $LaTeX2_\epsilon$ (all those in Chapter 2 plus some in Chapter 3). These figures are defined as $x = f(y)$, for $0 \leq y \leq 1$, which is backwards from the usual y a function of x, and the "implicitplot" command in Maple made it easy to graph x a function of y (y axis vertical and the x axis horizontal). Finally, there were figures done using the graphics package in $LaTeX2_\epsilon$, but the data for the graph was obtained from our evolutionary algorithm or from Maple. The graphs in Chapter 4 plus some in Chapter 3, 6 and Figure 10.9 were done this way.

Bibliography

[1] Maple 6, Waterloo Maple Inc., Waterloo, Canada.

Chapter 2

Fuzzy Sets

In this chapter we have collected together the basic ideas from fuzzy sets and fuzzy functions needed for the book. Any reader familiar with fuzzy sets, fuzzy numbers, the extension principle, α-cuts, interval arithmetic, possibility theory and fuzzy functions may go on to the rest of the book. A good general reference for fuzzy sets and fuzzy logic is [1], [6].

Our notation specifying a fuzzy set is to place a "bar" over a letter. So $\overline{A}$, $\overline{B}, \ldots$, $\overline{X}$, $\overline{Y}, \ldots$, $\overline{\alpha}$, $\overline{\beta}, \ldots$, will all denote fuzzy sets.

2.1 Fuzzy Sets

If Ω is some set, then a fuzzy subset $\overline{A}$ of Ω is defined by its membership function, written $\overline{A}(x)$, which produces values in $[0,1]$ for all x in Ω. So, $\overline{A}(x)$ is a function mapping Ω into $[0,1]$. If $\overline{A}(x_0) = 1$, then we say x_0 belongs to $\overline{A}$, if $\overline{A}(x_1) = 0$ we say x_1 does not belong to $\overline{A}$, and if $\overline{A}(x_2) = 0.6$ we say the membership value of x_2 in $\overline{A}$ is 0.6. When $\overline{A}(x)$ is always equal to one or zero we obtain a crisp (non–fuzzy) subset of Ω. For all fuzzy sets $\overline{B}$, $\overline{C}, \ldots$ we use $\overline{B}(x)$, $\overline{C}(x), \ldots$ for the value of their membership function at x. Most of the fuzzy sets we will be using will be fuzzy numbers.

The term "crisp" will mean not fuzzy. A crisp set is a regular set. A crisp number is just a real number. A crisp matrix (vector) has real numbers as its elements. A crisp function maps real numbers (or real vectors) into real numbers. A crisp solution to a problem is a solution involving crisp sets, crisp numbers, crisp functions, etc.

2.1.1 Fuzzy Numbers

A general definition of fuzzy number may be found in [1],[6], however our fuzzy numbers will be almost always triangular (shaped), or trapezoidal (shaped), fuzzy numbers. A triangular fuzzy number $\overline{N}$ is defined by three numbers $a < b < c$ where the base of the triangle is the interval $[a, c]$ and its vertex is at $x = b$. Triangular fuzzy numbers will be written as $\overline{N} = (a/b/c)$. A triangular fuzzy number $\overline{N} = (1.2/2/2.4)$ is shown in Figure 2.1. We see that $\overline{N}(2) = 1$, $\overline{N}(1.6) = 0.5$, etc.

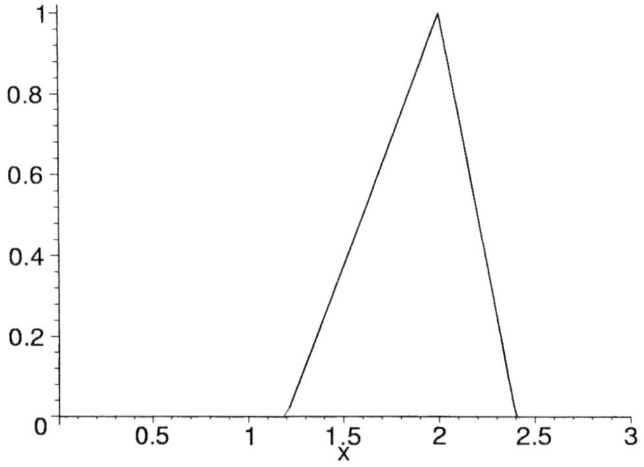

Figure 2.1: Triangular Fuzzy Number $\overline{N}$

A trapezoidal fuzzy number $\overline{M}$ is defined by four numbers $a < b < c < d$ where the base of the trapezoid is the interval $[a, d]$ and its top (where the membership equal one) is over $[b, c]$. We write $\overline{M} = (a/b, c/d)$ for trapezoidal fuzzy numbers. Figure 2.2 shows $\overline{M} = (1.2/2, 2.4/2.7)$.

A triangular shaped fuzzy number $\overline{P}$ is given in Figure 2.3. $\overline{P}$ is only partially specified by the three numbers 1.2, 2, 2.4 since the graph on $[1.2, 2]$, and $[2, 2.4]$, is not a straight line segment. To be a triangular shaped fuzzy number we require the graph to be continuous and: (1) monotonically increasing on $[1.2, 2]$; and (2) monotonically decreasing on $[2, 2.4]$. For triangular shaped fuzzy number $\overline{P}$ we use the notation $\overline{P} \approx (1.2/2/2.4)$ to show that it is partially defined by the three numbers 1.2, 2, and 2.4. If $\overline{P} \approx (1.2/2/2.4)$ we know its base is on the interval $[1.2, 2.4]$ with vertex (membership value one) at $x = 2$. Similarly we define trapezoidal shaped fuzzy number $\overline{Q} \approx (1.2/2, 2.4/2.7)$ whose base is $[1.2, 2.7]$ and top is over the interval $[2, 2.4]$. The graph of $\overline{Q}$ is similar to $\overline{M}$ in Figure 2.2 but it has continuous curves for its sides.

Although we will be using triangular (shaped) and trapezoidal (shaped) fuzzy numbers throughout the book, many results can be extended to more

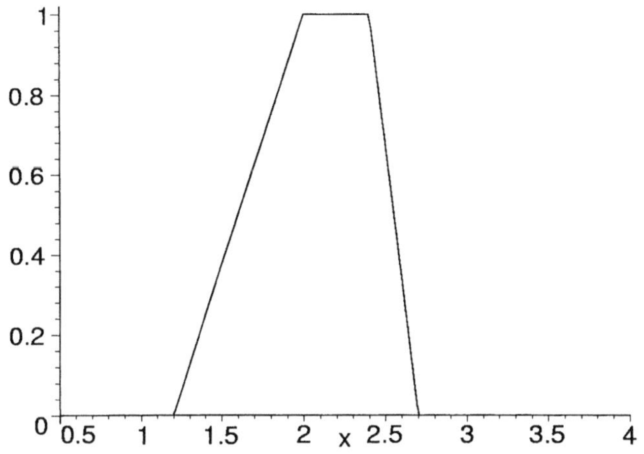

Figure 2.2: Trapezoidal Fuzzy Number $\overline{M}$

general fuzzy numbers, but we shall be content to work with only these special fuzzy numbers.

We will be using fuzzy numbers in this book to describe uncertainty. For example, in Chapter 4 if a future interest rate is uncertain but believed to be between 7 and 8%, we could model it using a trapezoidal fuzzy number (0.065/0.07,0.08/0.085). Also, in Chapter 7, if the initial value of the derivative is uncertain but known to be around 5, we may model it as $y'(0) = (4/5/6)$.

2.1.2 Alpha–Cuts

Alpha–cuts are slices through a fuzzy set producing regular (non-fuzzy) sets. If $\overline{A}$ is a fuzzy subset of some set Ω, then an α–cut of $\overline{A}$, written $\overline{A}[\alpha]$ is defined as

$$\overline{A}[\alpha] = \{x \in \Omega | \overline{A}(x) \geq \alpha\} , \tag{2.1}$$

for all α, $0 < \alpha \leq 1$. The $\alpha = 0$ cut, or $\overline{A}[0]$, must be defined separately.

Let $\overline{N}$ be the fuzzy number in Figure 2.1. Then $\overline{N}[0] = [1.2, 2.4]$. Using equation (2.1) to define $\overline{N}[0]$ would give $\overline{N}[0] =$ all the real numbers. Similarly, $\overline{M}[0] = [1.2, 2.7]$ from Figure 2.2 and in Figure 2.3 $\overline{P}[0] = [1.2, 2.4]$. For any fuzzy set $\overline{A}$, $\overline{A}[0]$ is called the support, or base, of $\overline{A}$. Many authors call the support of a fuzzy number the open interval (a, b) like the support

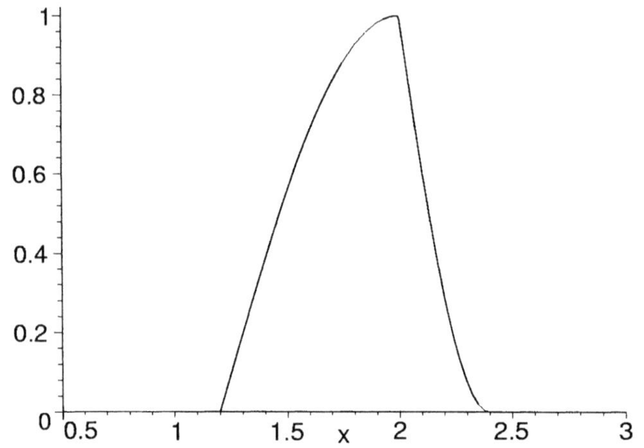

Figure 2.3: Triangular Shaped Fuzzy Number $\overline{P}$

of $\overline{N}$ in Figure 2.1 would then be $(1.2, 2.4)$. However in this book we use the closed interval $[a, b]$ for the support (base) of the fuzzy number.

The core of a fuzzy number is the set of values where the membership value equals one. If $\overline{N} = (a/b/c)$, or $\overline{N} \approx (a/b/c)$, then the core of $\overline{N}$ is the single point b. However, if $\overline{M} = (a/b, c/d)$, or $\overline{M} \approx (a/b, c/d)$, then the core of $\overline{M} = [b, c]$.

For any fuzzy number $\overline{Q}$ we know that $\overline{Q}[\alpha]$ is a closed, bounded, interval for $0 \leq \alpha \leq 1$. We will write this as

$$\overline{Q}[\alpha] = [q_1(\alpha), q_2(\alpha)] \ , \tag{2.2}$$

where $q_1(\alpha)$ ($q_2(\alpha)$) will be an increasing (decreasing) function of α with $q_1(1) \leq q_2(1)$. If $\overline{Q}$ is a triangular shaped or a trapezoidal shaped fuzzy number then: (1) $q_1(\alpha)$ will be a continuous, monotonically increasing function of α in $[0, 1]$; (2) $q_2(\alpha)$ will be a continuous, monotonically decreasing function of α, $0 \leq \alpha \leq 1$; and (3) $q_1(1) = q_2(1)$ ($q_1(1) < q_2(1)$ for trapezoids). We sometimes check monotone increasing (decreasing) by showing that $dq_1(\alpha)/d\alpha > 0$ ($dq_2(\alpha)/d\alpha < 0$) holds.

For the $\overline{N}$ in Figure 2.1 we obtain $\overline{N}[\alpha] = [n_1(\alpha), n_2(\alpha)]$, $n_1(\alpha) = 1.2 + 0.8\alpha$ and $n_2(\alpha) = 2.4 - 0.4\alpha$, $0 \leq \alpha \leq 1$. Similarly, $\overline{M}$ in Figure 2.2 has $\overline{M}[\alpha] = [m_1(\alpha), m_2(\alpha)]$, $m_1(\alpha) = 1.2 + 0.8\alpha$ and $m_2(\alpha) = 2.7 - 0.3\alpha$, $0 \leq \alpha \leq 1$. The equations for $n_i(\alpha)$ and $m_i(\alpha)$ are backwards. With the y–axis vertical and the x–axis horizontal the equation $n_1(\alpha) = 1.2 + 0.8\alpha$ means

$x = 1.2 + 0.8y$, $0 \leq y \leq 1$. That is, the straight line segment from $(1.2, 0)$ to $(2, 1)$ in Figure 2.1 is given as x a function of y whereas it is usually stated as y a function of x. This is how it will be done for all α–cuts of fuzzy numbers.

2.1.3 Inequalities

Let $\overline{N} = (a/b/c)$. We write $\overline{N} \geq \delta$, δ some real number, if $a \geq \delta$, $\overline{N} > \delta$ when $a > \delta$, $\overline{N} \leq \delta$ for $c \leq \delta$ and $\overline{N} < \delta$ if $c < \delta$. We use the same notation for triangular shaped and trapezoidal (shaped) fuzzy numbers whose support is the interval $[a, c]$.

If $\overline{A}$ and $\overline{B}$ are two fuzzy subsets of a set Ω, then $\overline{A} \leq \overline{B}$ means $\overline{A}(x) \leq \overline{B}(x)$ for all x in Ω, or $\overline{A}$ is a fuzzy subset of $\overline{B}$. $\overline{A} < \overline{B}$ holds when $\overline{A}(x) < \overline{B}(x)$, for all x. There is a potential problem with the symbol $\leq$. In some places in the book (Chapters 4, 6 and 10) $\overline{M} \leq \overline{N}$, for fuzzy numbers $\overline{M}$ and $\overline{N}$, means that $\overline{M}$ is less than or equal to $\overline{N}$. It should be clear on how we use "$\leq$" as to which meaning is correct.

2.1.4 Discrete Fuzzy Sets

Let $\overline{A}$ be a fuzzy subset of Ω. If $\overline{A}(x)$ is not zero only at a finite number of x values in Ω, then $\overline{A}$ is called a discrete fuzzy set. Suppose $\overline{A}(x)$ is not zero only at x_1, x_2, x_3 and x_4 in Ω. Then we write the fuzzy set as

$$\overline{A} = \{\frac{\mu_1}{x_1}, \cdots, \frac{\mu_4}{x_4}\}, \tag{2.3}$$

where the μ_i are the membership values. That is, $\overline{A}(x_i) = \mu_i$, $1 \leq i \leq 4$, and $\overline{A}(x) = 0$ otherwise. We can have discrete fuzzy subsets of any space Ω. Notice that α-cuts of discrete fuzzy sets of $\mathbb{R}$, the set of real numbers, do not produce closed, bounded, intervals.

2.2 Fuzzy Arithmetic

If $\overline{A}$ and $\overline{B}$ are two fuzzy numbers we will need to add, subtract, multiply and divide them. There are two basic methods of computing $\overline{A} + \overline{B}$, $\overline{A} - \overline{B}$, etc. which are: (1) extension principle; and (2) α–cuts and interval arithmetic.

2.2.1 Extension Principle

Let $\overline{A}$ and $\overline{B}$ be two fuzzy numbers. If $\overline{A} + \overline{B} = \overline{C}$, then the membership function for $\overline{C}$ is defined as

$$\overline{C}(z) = \sup_{x,y}\{\min(\overline{A}(x), \overline{B}(y)) | x + y = z\} . \tag{2.4}$$

If we set $\overline{C} = \overline{A} - \overline{B}$, then

$$\overline{C}(z) = \sup_{x,y}\{\min(\overline{A}(x), \overline{B}(y)) | x - y = z\} \ . \tag{2.5}$$

Similarly, $\overline{C} = \overline{A} \cdot \overline{B}$, then

$$\overline{C}(z) = \sup_{x,y}\{\min(\overline{A}(x), \overline{B}(y)) | x \cdot y = z\} \ , \tag{2.6}$$

and if $\overline{C} = \overline{A}/\overline{B}$,

$$\overline{C}(z) = \sup_{x,y}\{\min(\overline{A}(x), \overline{B}(y)) | x / y = z\} \ . \tag{2.7}$$

In all cases $\overline{C}$ is also a fuzzy number [6]. We assume that zero does not belong to the support of $\overline{B}$ in $\overline{C} = \overline{A}/\overline{B}$. If $\overline{A}$ and $\overline{B}$ are triangular (trapezoidal) fuzzy numbers then so are $\overline{A} + \overline{B}$ and $\overline{A} - \overline{B}$, but $\overline{A} \cdot \overline{B}$ and $\overline{A}/\overline{B}$ will be triangular (trapezoidal) shaped fuzzy numbers.

We should mention something about the operator "sup" in equations (2.4) – (2.7). If Ω is a set of real numbers bounded above (there is a M so that $x \leq M$, for all x in Ω), then $\sup(\Omega)$ = the least upper bound for Ω. If Ω has a maximum member, then $\sup(\Omega) = \max(\Omega)$. For example, if $\Omega = [0, 1)$, $\sup(\Omega) = 1$ but if $\Omega = [0, 1]$, then $\sup(\Omega) = \max(\Omega) = 1$. The dual operator to "sup" is "inf". If Ω is bounded below (there is a M so that $M \leq x$ for all $x \in \Omega$), then $\inf(\Omega)$ = the greatest lower bound. For example, for $\Omega = (0, 1]$ $\inf(\Omega) = 0$ but if $\Omega = [0, 1]$, then $\inf(\Omega) = \min(\Omega) = 0$.

Obviously, given $\overline{A}$ and $\overline{B}$, equations (2.4) – (2.7) appear quite complicated to compute $\overline{A} + \overline{B}$, $\overline{A} - \overline{B}$, etc. So, we now present an equivalent procedure based on α–cuts and interval arithmetic. First, we present the basics of interval arithmetic.

2.2.2 Interval Arithmetic

We only give a brief introduction to interval arithmetic. For more information the reader is referred to ([7],[8]). Let $[a_1, b_1]$ and $[a_2, b_2]$ be two closed, bounded, intervals of real numbers. If $*$ denotes addition, subtraction, multiplication, or division, then $[a_1, b_1] * [a_2, b_2] = [\alpha, \beta]$ where

$$[\alpha, \beta] = \{a * b | a_1 \leq a \leq b_1, a_2 \leq b \leq b_2\} \ . \tag{2.8}$$

If $*$ is division, we must assume that zero does not belong to $[a_2, b_2]$. We may simplify equation (2.8) as follows:

$$[a_1, b_1] + [a_2, b_2] = [a_1 + a_2, b_1 + b_2] \ , \tag{2.9}$$

$$[a_1, b_1] - [a_2, b_2] = [a_1 - b_2, b_1 - a_2] \ , \tag{2.10}$$

$$[a_1, b_1] / [a_2, b_2] = [a_1, b_1] \cdot \left[\frac{1}{b_2}, \frac{1}{a_2}\right] \ , \tag{2.11}$$

and

$$[a_1, b_1] \cdot [a_2, b_2] = [\alpha, \beta] , \tag{2.12}$$

where

$$\alpha = \min\{a_1 a_2, a_1 b_2, b_1 a_2, b_1 b_2\} , \tag{2.13}$$
$$\beta = \max\{a_1 a_2, a_1 b_2, b_1 a_2, b_1 b_2\} . \tag{2.14}$$

Multiplication and division may be further simplified if we know that $a_1 > 0$ and $b_2 < 0$, or $b_1 > 0$ and $b_2 < 0$, etc. For example, if $a_1 \geq 0$ and $a_2 \geq 0$, then

$$[a_1, b_1] \cdot [a_2, b_2] = [a_1 a_2, b_1 b_2] , \tag{2.15}$$

and if $b_1 < 0$ but $a_2 \geq 0$, we see that

$$[a_1, b_1] \cdot [a_2, b_2] = [a_1 b_2, a_2 b_1] . \tag{2.16}$$

Also, assuming $b_1 < 0$ and $b_2 < 0$ we get

$$[a_1, b_1] \cdot [a_2, b_2] = [b_1 b_2, a_1 a_2] , \tag{2.17}$$

but $a_1 \geq 0$, $b_2 < 0$ produces

$$[a_1, b_1] \cdot [a_2, b_2] = [a_2 b_1, b_2 a_1] . \tag{2.18}$$

2.2.3 Fuzzy Arithmetic

Again we have two fuzzy numbers $\overline{A}$ and $\overline{B}$. We know α–cuts are closed, bounded, intervals so let $\overline{A}[\alpha] = [a_1(\alpha), a_2(\alpha)]$, $\overline{B}[\alpha] = [b_1(\alpha), b_2(\alpha)]$. Then if $\overline{C} = \overline{A} + \overline{B}$ we have

$$\overline{C}[\alpha] = \overline{A}[\alpha] + \overline{B}[\alpha] . \tag{2.19}$$

We add the intervals using equation (2.9). Setting $\overline{C} = \overline{A} - \overline{B}$ we get

$$\overline{C}[\alpha] = \overline{A}[\alpha] - \overline{B}[\alpha] , \tag{2.20}$$

for all α in $[0, 1]$. Also

$$\overline{C}[\alpha] = \overline{A}[\alpha] \cdot \overline{B}[\alpha] , \tag{2.21}$$

for $\overline{C} = \overline{A} \cdot \overline{B}$ and

$$\overline{C}[\alpha] = \overline{A}[\alpha] / \overline{B}[\alpha] , \tag{2.22}$$

when $\overline{C} = \overline{A}/\overline{B}$. This method is equivalent to the extension principle method of fuzzy arithmetic [6]. Obviously, this procedure, of α–cuts plus interval arithmetic, is more user (and computer) friendly.

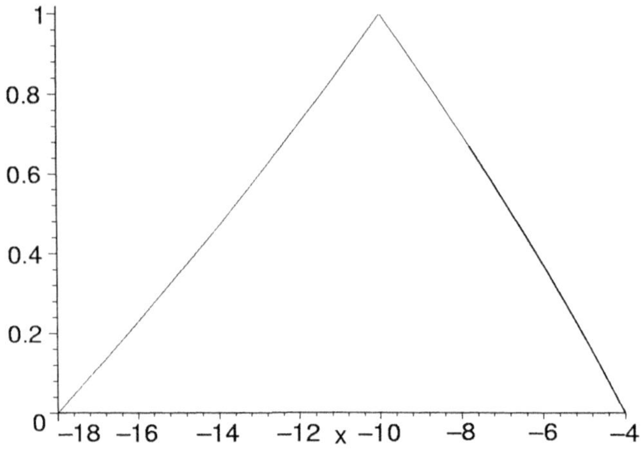

Figure 2.4: The Fuzzy Number $\overline{C} = \overline{A} \cdot \overline{B}$

Example 2.2.3.1

Let $\overline{A} = (-3/-2/-1)$ and $\overline{B} = (4/5/6)$. We determine $\overline{A} \cdot \overline{B}$ using α–cuts and interval arithmetic. We compute $\overline{A}[\alpha] = [-3+\alpha, -1-\alpha]$ and $\overline{B}[\alpha] = [4+\alpha, 6-\alpha]$. So, if $\overline{C} = \overline{A} \cdot \overline{B}$ we obtain $\overline{C}[\alpha] = [(\alpha-3)(6-\alpha), (-1-\alpha)(4+\alpha)]$, $0 \leq \alpha \leq 1$. The graph of $\overline{C}$ is shown in Figure 2.4.

2.3 Fuzzy Functions

In this book a fuzzy function is a mapping from fuzzy numbers into fuzzy numbers. We write $H(\overline{X}) = \overline{Z}$ for a fuzzy function with one independent variable $\overline{X}$. Usually $\overline{X}$ will be a triangular (trapezoidal) fuzzy number and then we usually obtain $\overline{Z}$ as a triangular (trapezoidal) shaped fuzzy number. For two independent variables we have $H(\overline{X}, \overline{Y}) = \overline{Z}$.

Where do these fuzzy functions come from? They are usually extensions of real–valued functions. Let $h : [a, b] \to \mathbb{R}$. This notation means $z = h(x)$ for x in $[a, b]$ and z a real number. One extends $h : [a, b] \to \mathbb{R}$ to $H(\overline{X}) = \overline{Z}$ in two ways: (1) the extension principle; or (2) using α–cuts and interval arithmetic.

2.3.1 Extension Principle

Any $h : [a, b] \to \mathbb{R}$ may be extended to $H(\overline{X}) = \overline{Z}$ as follows

$$\overline{Z}(z) = \sup_{x} \{ \overline{X}(x) \mid h(x) = z, \ a \leq x \leq b \} \ . \tag{2.23}$$

Equation (2.23) defines the membership function of $\overline{Z}$ for any triangular (trapezoidal) fuzzy number $\overline{X}$ in $[a, b]$.

If h is continuous, then we have a way to find α–cuts of $\overline{Z}$. Let $\overline{Z}[\alpha] = [z_1(\alpha), z_2(\alpha)]$. Then [3]

$$z_1(\alpha) = \min\{ h(x) \mid x \in \overline{X}[\alpha] \} \ , \tag{2.24}$$

$$z_2(\alpha) = \max\{ h(x) \mid x \in \overline{X}[\alpha] \} \ , \tag{2.25}$$

for $0 \leq \alpha \leq 1$.

If we have two independent variables, then let $z = h(x, y)$ for x in $[a_1, b_1]$, y in $[a_2, b_2]$. We extend h to $H(\overline{X}, \overline{Y}) = \overline{Z}$ as

$$\overline{Z}(z) = \sup_{x,y} \left\{ \min\left(\overline{X}(x), \overline{Y}(y)\right) \mid h(x, y) = z \right\} \ , \tag{2.26}$$

for $\overline{X}$ ($\overline{Y}$) a triangular or trapezoidal fuzzy number in $[a_1, b_1]$ ($[a_2, b_2]$). For α–cuts of $\overline{Z}$, assuming h is continuous, we have

$$z_1(\alpha) = \min\{ h(x, y) \mid x \in \overline{X}[\alpha], \ y \in \overline{Y}[\alpha] \} \ , \tag{2.27}$$

$$z_2(\alpha) = \max\{ h(x, y) \mid x \in \overline{X}[\alpha], \ y \in \overline{Y}[\alpha] \} \ , \tag{2.28}$$

$0 \leq \alpha \leq 1$. We use equations (2.24) – (2.25) and (2.27) – (2.28) throughout this book.

2.3.2 Alpha–Cuts and Interval Arithmetic

All the functions we usually use in engineering and science have a computer algorithm which, using a finite number of additions, subtractions, multiplications and divisions, can evaluate the function to required accuracy ([2]). Such functions can be extended, using α–cuts and interval arithmetic, to fuzzy functions. Let $h : [a, b] \to \mathbb{R}$ be such a function. Then its extension $H(\overline{X}) = \overline{Z}$, $\overline{X}$ in $[a, b]$ is done, via interval arithmetic, in computing $h(\overline{X}[\alpha]) = \overline{Z}[\alpha]$, α in $[0, 1]$. We input the interval $\overline{X}[\alpha]$, perform the arithmetic operations needed to evaluate h on this interval, and obtain the interval $\overline{Z}[\alpha]$. Then put these α–cuts together to obtain the value $\overline{Z}$. The extension to more independent variables is straightforward.

For example, consider the fuzzy function

$$\overline{Z} = H(\overline{X}) = \frac{\overline{A}\,\overline{X} + \overline{B}}{\overline{C}\,\overline{X} + \overline{D}} \ , \tag{2.29}$$

for triangular fuzzy numbers $\overline{A}$, $\overline{B}$, $\overline{C}$, $\overline{D}$ and triangular fuzzy number $\overline{X}$ in $[0, 10]$. We assume that $\overline{C} \geq 0$, $\overline{D} > 0$ so that $\overline{C}\ \overline{X} + \overline{D} > 0$. This would be the extension of

$$h(x_1, x_2, x_3, x_4, x) = \frac{x_1 x + x_2}{x_3 x + x_4} \ . \tag{2.30}$$

We would substitute the intervals $\overline{A}[\alpha]$ for x_1, $\overline{B}[\alpha]$ for x_2, $\overline{C}[\alpha]$ for x_3, $\overline{D}[\alpha]$ for x_4 and $\overline{X}[\alpha]$ for x, do interval arithmetic, to obtain interval $\overline{Z}[\alpha]$ for $\overline{Z}$. Alternatively, the fuzzy function

$$\overline{Z} = H(\overline{X}) = \frac{2\overline{X} + 10}{3\overline{X} + 4} \ , \tag{2.31}$$

would be the extension of

$$h(x) = \frac{2x + 10}{3x + 4} \ . \tag{2.32}$$

2.3.3 Differences

Let $h : [a, b] \rightarrow \mathbb{R}$. Just for this subsection let us write $\overline{Z}^* = H(\overline{X})$ for the extension principle method of extending h to H for $\overline{X}$ in $[a, b]$. We denote $\overline{Z} = H(\overline{X})$ for the α–cut and interval arithmetic extension of h.

We know that $\overline{Z}$ can be different from $\overline{Z}^*$. But for basic fuzzy arithmetic in Section 2.2 the two methods give the same results. In the example below we show that for $h(x) = x(1 - x)$, x in $[0, 1]$, we can get $\overline{Z}^* \neq \overline{Z}$ for some $\overline{X}$ in $[0, 1]$. What is known ([3],[7]) is that for usual functions in science and engineering $\overline{Z}^* \leq \overline{Z}$. Otherwise, there is no known necessary and sufficient conditions on h so that $\overline{Z}^* = \overline{Z}$ for all $\overline{X}$ in $[a, b]$.

There is nothing wrong in using α–cuts and interval arithmetic to evaluate fuzzy functions. Surely, it is user, and computer, friendly. However, we should be aware that whenever we use α–cuts plus interval arithmetic to compute $\overline{Z} = H(\overline{X})$ we may be getting something larger than that obtained from the extension principle. The same results hold for functions of two or more independent variables.

Example 2.3.3.1

The example is the simple fuzzy expression

$$\overline{Z} = (1 - \overline{X})\ \overline{X} \ , \tag{2.33}$$

for $\overline{X}$ a triangular fuzzy number in $[0, 1]$. Let $\overline{X}[\alpha] = [x_1(\alpha), x_2(\alpha)]$. Using interval arithmetic we obtain

$$z_1(\alpha) = (1 - x_2(\alpha))x_1(\alpha) \ , \tag{2.34}$$

$$z_2(\alpha) = (1 - x_1(\alpha))x_2(\alpha) \ , \tag{2.35}$$

for $\overline{Z}[\alpha] = [z_1(\alpha), z_2(\alpha)]$, α in $[0,1]$.

The extension principle extends the regular equation $z = (1-x)x$, $0 \leq x \leq 1$, to fuzzy numbers as follows

$$\overline{Z}^*(z) = \sup_x \left\{ \overline{X}(x) | (1-x)x = z,\ 0 \leq x \leq 1 \right\} . \tag{2.36}$$

Let $\overline{Z}^*[\alpha] = [z_1^*(\alpha), z_2^*(\alpha)]$. Then

$$z_1^*(\alpha) = \min\{(1-x)x | x \in \overline{X}[\alpha]\} , \tag{2.37}$$
$$z_2^*(\alpha) = \max\{(1-x)x | x \in \overline{X}[\alpha]\} , \tag{2.38}$$

for all $0 \leq \alpha \leq 1$. Now let $\overline{X} = (0/0.25/0.5)$, then $x_1(\alpha) = 0.25\alpha$ and $x_2(\alpha) = 0.50 - 0.25\alpha$. Equations (2.34) and (2.35) give $\overline{Z}[0.50] = [5/64, 21/64]$ but equations (2.37) and (2.38) produce $\overline{Z}^*[0.50] = [7/64, 15/64]$. Therefore, $\overline{Z}^* \neq \overline{Z}$. We do know that if each fuzzy number appears only once in the fuzzy expression, the two methods produce the same results ([3],[7]). However, if a fuzzy number is used more than once, as in equation (2.33), the two procedures can give different results.

2.4 Possibility Theory

We will be using some of possibility theory only in Section 6.2 (Fuzzy PERT) and Section 6.4 (Fuzzy Queuing Theory). For a general introduction to possibility theory we suggest ([4],[5]), and a brief review of the parts of possibility theory needed in Chapter 6 is in this section.

Let $\overline{X}$ be a fuzzy variable whose values are restricted by a possibility distribution $\overline{A}$, where $\overline{A}$ is a fuzzy subset of $\mathbb{R}$. Possibility distributions need to be normalized which means that $\overline{A}(x) = 1$ for some x in $\mathbb{R}$. If E is any subset of $\mathbb{R}$, not a fuzzy set, we compute the possibility that $\overline{X}$ takes its values in E as follows

$$\text{Poss}[\overline{X} \in E] = \sup \left\{ \overline{A}(x) \mid x \in E \right\} . \tag{2.39}$$

This is analogous to probability theory where we use in fuzzy set theory sup (or max) in place of summing (addition) in probability and min in place of multiplication. If X is a random variable with probability density $f(x)$, then the probability that X takes its values in E is

$$\text{Prob}[X \in E] = \int_E f(x)dx . \tag{2.40}$$

In place of integration (summing) we use "sup" in fuzzy set theory.

Now consider $\overline{X}_1, \ldots, \overline{X}_n$ fuzzy variables with associated possibility distributions $\overline{A}_1, \ldots, \overline{A}_n$, respectively. Let $\overline{X} = (\overline{X}_1, \ldots, \overline{X}_n)$ and $x =$

$(x_1, \ldots, x_n) \in \mathbb{R}^n$. Assuming that the fuzzy variables are non–interactive (analogous to independent in probability theory) we form their joint π possibility distribution as

$$\pi(x_1, \ldots, x_n) = \min \{ \overline{A}_1(x_1), \ldots, \overline{A}_n(x_n) \} , \tag{2.41}$$

all $x \in \mathbb{R}^n$. For any crisp (non–fuzzy) $E \subset \mathbb{R}^n$ we compute

$$\text{Poss}[\overline{X} \in E] = \sup_x \{ \pi(x) \mid x \in E \} . \tag{2.42}$$

Finally, given two E_1 and E_2, subsets of $\mathbb{R}^n$, we need to find the possibility that $\overline{X}$ takes its values in $E_1 \cup E_2$. It is

$$\text{Poss}[\overline{X} \in E_1 \cup E_2] = \max \{ \text{Poss}[\overline{X} \in E_1], \text{Poss}[\overline{X} \in E_2] \} . \tag{2.43}$$

The possibility of a union is the maximum of the individual possibilities.

Bibliography

[1] J.J.Buckley and E.Eslami: Introduction to Fuzzy Logic and Fuzzy Sets, Physica-Verlag, Heidelberg, Germany, 2001.

[2] J.J. Buckley and Y. Hayashi: Can Neural Nets be Universal Approximators for Fuzzy Functions? Fuzzy Sets and Systems, 101 (1999), pp.323–330.

[3] J.J. Buckley and Y. Qu: On Using α–cuts to Evaluate Fuzzy Equations, Fuzzy Sets and Systems, 38 (1990), pp. 309–312.

[4] D. Dubois and H. Prade: Possibility Theory: An Approach to Computerized Processing of Uncertainty, Plenum Press, N.Y., 1988.

[5] D. Dubois and H. Prade (eds.): Fundamentals of Fuzzy Sets, Kluwer, The Netherlands, 2000.

[6] G.J. Klir and B. Yuan: Fuzzy Sets and Fuzzy Logic: Theory and Applications, Prentice Hall, Upper Saddle River, N.J., 1995.

[7] R.E. Moore: Methods and Applications of Interval Analysis, SIAM Studies in Applied Mathematics, Philadelphia, 1979.

[8] A. Neumaier: Interval Methods for Systems of Equations, Cambridge University Press, Cambridge, U.K., 1990.

Chapter 3

Solving Fuzzy Equations

In this chapter we first look at different types of solutions to the simple fuzzy linear equation $\overline{A} \cdot \overline{X} + \overline{B} = \overline{C}$ and then systems of fuzzy linear equations. Solving fuzzy differential equations, fuzzy difference equations and fuzzy integral equations, come later on in the book. In the applications section we also look at solving $\overline{A}\ \overline{X}^2 + \overline{B}\ \overline{X} = \overline{C}$, the fuzzy quadratic. At the end of the chapter we discuss fuzzy input-output analysis. Solutions to more complicated fuzzy equations are discussed in ([2], [3], [6]). Throughout this chapter, except in Section 3.5, we use triangular, and triangular shaped, fuzzy numbers. In Section 3.5 we use trapezoidal (shaped) fuzzy numbers.

3.1 $\overline{A}\,\overline{X} + \overline{B} = \overline{C}$

$\overline{A}$, $\overline{B}$ and $\overline{C}$ will be triangular fuzzy numbers so let $\overline{A} = (a_1/a_2/a_3)$, $\overline{B} = (b_1/b_2/b_3)$ and $\overline{C} = (c_1/c_2/c_3)$. $\overline{X}$, if it exists, will be a triangular shaped fuzzy number so let $\overline{X} \approx (x_1/x_2/x_3)$. In the crisp equation

$$ax + b = c\ , \tag{3.1}$$

we immediately obtain $X = (c - b)/a$, if $a \neq 0$. We used the important facts $b - b = 0$ and $(1/a)a = 1$ from real numbers to get the solution.

We try this same approach with the fuzzy equation

$$\overline{A}\,\overline{X} + \overline{B} = \overline{C}\ , \tag{3.2}$$

we get

$$(1/\overline{A})(\overline{A}\,\overline{X} + (\overline{B} - \overline{B})) = (1/\overline{A})(\overline{C} - \overline{B})\ . \tag{3.3}$$

But the left side of the equation (3.3) does not equal $\overline{X}$ since $\overline{B} - \overline{B} \neq 0$ and $(1/\overline{A})(\overline{A}) \neq 1$. For example, if $\overline{B} = (1/2/3)$, then $\overline{B} - \overline{B} = (-2/0/2)$ not zero. Also, if $\overline{A} = (1/2/3)$, $(1/\overline{A})(\overline{A}) \approx ((1/3)/1/3)$, a triangular shaped fuzzy number, not one.

This shows a major problem in solving fuzzy equations: some basic operations we used to solve crisp equations do not hold for fuzzy equations. Actually, this comes as no great surprises because this also happens in probability theory. If X is a random variable with positive variance, then $X - X \neq 0$ and $X/X \neq 1$ since both $X - X$ and X/X will have positive variance.

We now introduce our first solution method, called the classical method, producing solution $\overline{X}_c$ (when it exists). This procedure employs α–cuts and interval arithmetic to solve for $\overline{X}_c$. Let $\overline{A}[\alpha] = [a_1(\alpha), a_2(\alpha)]$, $\overline{B}[\alpha] = [b_1(\alpha), b_2(\alpha)]$, $\overline{C}[\alpha] = [c_1(\alpha), c_2(\alpha)]$ and $\overline{X}_c[\alpha] = [x_1(\alpha), x_2(\alpha)]$, $0 \leq \alpha \leq 1$. Substitute these into equation (3.2) producing

$$[a_1(\alpha), a_2(\alpha)][x_1(\alpha), x_2(\alpha)] + [b_1(\alpha), b_2(\alpha)] = [c_1(\alpha), c_2(\alpha)] \ . \tag{3.4}$$

We now use interval arithmetic to solve equation (3.4) for $x_1(\alpha)$ and $x_2(\alpha)$. We say that this method defines solution $\overline{X}_c$ when $[x_1(\alpha), x_2(\alpha)]$ defines the α–cuts of a fuzzy number. For the $x_1(\alpha)$, $x_2(\alpha)$ to specify a fuzzy number we need:

1. $x_1(\alpha)$ monotonically increasing, $0 \leq \alpha \leq 1$;

2. $x_2(\alpha)$ monotonically decreasing, $0 \leq \alpha \leq 1$; and

3. $x_1(1) \leq x_2(1)$.

We did not mention anything about the $x_i(\alpha)$ being continuous because throughout this book $x_1(\alpha)$, $x_2(\alpha)$ will be continuous. If $x_1(1) < x_2(1)$ we obtain $\overline{X}_c$ a trapezoidal shaped fuzzy number.

Example 3.1.1

Let $\overline{A} = (1/2/3)$, $\overline{B} = (-3/-2/-1)$ and $\overline{C} = (3/4/5)$. Then $\overline{A}[\alpha] = [1+\alpha, 3-\alpha]$, $\overline{B}[\alpha] = [-3+\alpha, -1-\alpha]$, $\overline{C}[\alpha] = [3+\alpha, 5-\alpha]$. Since $\overline{A} > 0$ and $\overline{C} > 0$, we must have $\overline{X}_c > 0$, and equation (3.4) gives

$$[a_1(\alpha)x_1(\alpha) + b_1(\alpha), a_2(\alpha)x_2(\alpha) + b_2(\alpha)] = [c_1(\alpha), c_2(\alpha)] \ , \tag{3.5}$$

or

$$x_1(\alpha) = \frac{6}{1+\alpha} \ , \tag{3.6}$$

$$x_2(\alpha) = \frac{6}{3-\alpha} \ , \tag{3.7}$$

after substituting for $a_1(\alpha), \ldots, c_2(\alpha)$ and solving for $x_i(\alpha)$. But $x_1(\alpha)$ is decreasing and $x_1(\alpha)$ is increasing. So, $\overline{X}_c$ does not exist.

Example 3.1.2

Now set $\overline{A} = (8/9/10)$, $\overline{B} = (-3/-2/-1)$ and $\overline{C} = (3/5/7)$. So $\overline{A}[\alpha] = [8+\alpha, 10-\alpha]$, $\overline{B}[\alpha] = [-3+\alpha, -1-\alpha]$, $\overline{C}[\alpha] = [3+2\alpha, 7-2\alpha]$. Again we must have $\overline{X}_c > 0$ so we obtain

$$x_1(\alpha) = \frac{6+\alpha}{8+\alpha}, \tag{3.8}$$

$$x_2(\alpha) = \frac{8-\alpha}{10-\alpha}. \tag{3.9}$$

We see that $x_1(\alpha)$ is increasing (its derivative is positive), $x_2(\alpha)$ is decreasing (derivative is negative) and $x_1(1) = 7/9 = x_2(1)$. The solution $\overline{X}_c$ exists, with α–cuts $[x_1(\alpha), x_2(\alpha)]$, shown in Figure 3.1.

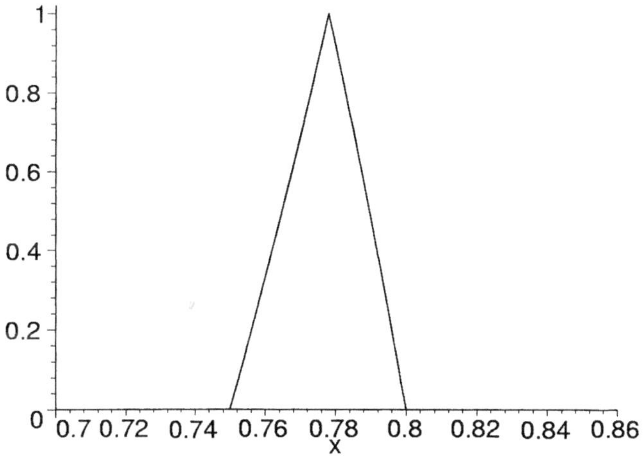

Figure 3.1: Solution to Example 3.1.2

Working more examples, like Examples 3.1.1 and 3.1.2 above, we conclude that too often fuzzy equations have no solution $(\overline{X}_c)$. This motivated the authors in ([2], [3], [6], [7]) to propose new solutions for fuzzy equations. These new solutions will be introduced in the next section. The classical solution, plus the new solutions, will be used throughout this book.

3.2 New Solutions

We continue working with the fuzzy equation $\overline{A}\ \overline{X} + \overline{B} = \overline{C}$. The new solutions simply fuzzify the crisp solution $(c-b)/a$, $a \neq 0$. The fuzzified crisp solution is

$$(\overline{C} - \overline{B})/\overline{A} \ , \tag{3.10}$$

where we assume zero does not belong to the support of $\overline{A}$. There are two ways to evaluate equations (3.10). The first method is the extension principle. If $\overline{X}_e$ is the value of equation (3.10) using the extension principle, then

$$\overline{X}_e(x) = \max\{\pi(a,b,c) \mid (c-b)/a = x\} \ , \tag{3.11}$$

where

$$\pi(a,b,c) = \min\{\overline{A}(a), \overline{B}(b), \overline{C}(c)\} \ . \tag{3.12}$$

Since the expression $(c-b)/a$, $a \neq 0$, is continuous in a, b, c we know how to find α–cuts of $\overline{X}_e$ [5]

$$x_{e1}(\alpha) = \min\{(c-b)/a \mid a \in \overline{A}[\alpha], b \in \overline{B}[\alpha], c \in \overline{C}[\alpha]\} \ , \tag{3.13}$$

$$x_{e2}(\alpha) = \max\{(c-b)/a \mid a \in \overline{A}[\alpha], b \in \overline{B}[\alpha], c \in \overline{C}[\alpha]\} \ , \tag{3.14}$$

where

$$\overline{X}_e[\alpha] = [x_{e1}(\alpha), x_{e2}(\alpha)] \ , \tag{3.15}$$

$0 \leq \alpha \leq 1$. $\overline{X}_e$ will be a triangular shaped fuzzy number when $\overline{A}$, $\overline{B}$, $\overline{C}$ are triangular fuzzy numbers.

Theorem 3.1 *If $\overline{X}_c$ exists, $\overline{X}_c \leq \overline{X}_e$.*

Proof 3.1 *It suffices to show $\overline{X}_c[\alpha] \subseteq \overline{X}_e[\alpha]$, $0 \leq \alpha \leq 1$. Choose $\alpha \in [0,1]$ and $x \in \overline{X}_c[\alpha]$. From equation (3.4) we see there is an $a \in \overline{A}[\alpha]$, $b \in \overline{B}[\alpha]$, $c \in \overline{C}[\alpha]$ so that $ax+b=c$. Or, $x = (c-b)/a$ for $a \in \overline{A}[\alpha]$, $b \in \overline{B}[\alpha]$, $c \in \overline{C}[\alpha]$. This implies, from equations (3.13) and (3.14), that $x_{e1}(\alpha) \leq x \leq x_{e2}(\alpha)$, or $x \in \overline{X}_e[\alpha]$.*

An important fact about $\overline{X}_c$ is that it will satisfy the fuzzy equation. That is $\overline{A}\ \overline{X}_c + \overline{B} = \overline{C}$ holds using α–cuts and interval arithmetic. However, $\overline{X}_e$ may, or may not, satisfy the fuzzy equation. However, $\overline{X}_e$ will always exist but $\overline{X}_c$ may fail to exist.

The second way to evaluate equation (3.10) is to use α–cuts and interval arithmetic. If the result is $\overline{X}_I$, we have

$$\overline{X}_I[\alpha] = \frac{\overline{C}[\alpha] - \overline{B}[\alpha]}{\overline{A}[\alpha]} \ , \tag{3.16}$$

to be simplified by interval arithmetic, $0 \leq \alpha \leq 1$.

Theorem 3.2 $\overline{X}_e \leq \overline{X}_I$.

Proof 3.2 *Theorem 3.1 in [14].*

$\overline{X}_I$ may or may not satisfy the fuzzy equation. $\overline{X}_I$ will be a triangular shaped fuzzy number when $\overline{A}$, $\overline{B}$, $\overline{C}$ are all triangular fuzzy numbers. We summarize these results as:

1. If $\overline{X}_c$ exists, then $\overline{X}_c \leq \overline{X}_e \leq \overline{X}_I$;

2. $\overline{X}_c$ always satisfies the fuzzy equation;

3. $\overline{X}_e \leq \overline{X}_I$.

Our general strategy for solving fuzzy equations will be:

1. the solution is $\overline{X}_c$ when it exists;

2. if $\overline{X}_c$ fails to exist, the solution is $\overline{X}_e$; and

3. if $\overline{X}_c$ fails to exist and $\overline{X}_e$ is difficult to construct, use $\overline{X}_I$ as the (approximate) solution.

For more complicated fuzzy equations $\overline{X}_e$ will be difficult to compute. However, $\overline{X}_I$ is usually easily constructed, since it uses only max, min and the arithmetic of real numbers. For this reason we suggest approximating $\overline{X}_e$ by $\overline{X}_I$ when we do not have $\overline{X}_e$.

Example 3.2.1

This continues Example 3.1.1 where $\overline{X}_c$ does not exist. To calculate $\overline{X}_e$ we need to evaluate equations (3.13) and (3.14). But this is easily done since $(c-b)/a$ is increasing in c and decreasing in both b and a. So

$$x_{e1}(\alpha) = \frac{c_1(\alpha) - b_2(\alpha)}{a_2(\alpha)} = \frac{4+2\alpha}{3-\alpha}, \tag{3.17}$$

$$x_{e2}(\alpha) = \frac{c_2(\alpha) - b_1(\alpha)}{a_1(\alpha)} = \frac{8-2\alpha}{1+\alpha}. \tag{3.18}$$

$\overline{X}_e$ is shown in Figure 3.2.

In calculating $\overline{X}_I[\alpha]$ we get

$$\overline{X}_I[\alpha] = [c_1(\alpha) - b_2(\alpha), c_2(\alpha) - b_1(\alpha)] \left[\frac{1}{a_2(\alpha)}, \frac{1}{a_1(\alpha)}\right], \tag{3.19}$$

which is the same as $\overline{X}_e[\alpha]$ because intervals in equation (3.19) are positive. In this example, we get $\overline{X}_e = \overline{X}_I$.

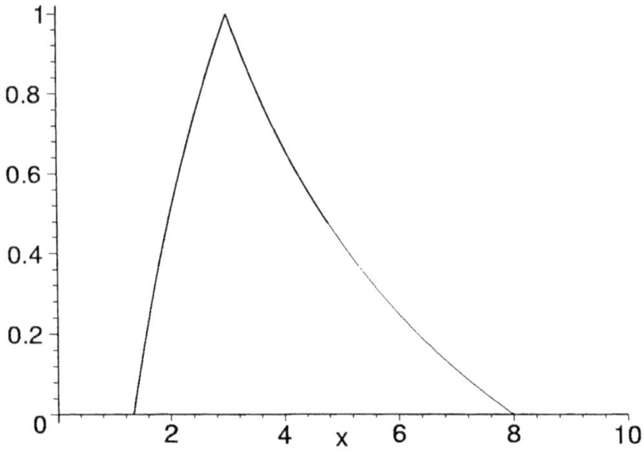

Figure 3.2: Solution to Example 3.2.1

Example 3.2.2

This continues Example 3.1.2. We notice that, since $\overline{A} > 0$ and $\overline{C} - \overline{B} > 0$, $\frac{\partial}{\partial c}[\frac{c-b}{a}] = \frac{1}{a} > 0$, $\frac{\partial}{\partial b}[\frac{c-b}{a}] = -\frac{1}{a} < 0$ and $\frac{\partial}{\partial a}[\frac{c-b}{a}] = \frac{b-c}{a^2} < 0$. This means that the expression $\frac{c-b}{a}$ is increasing in c but decreasing in a and b. Then equations (3.13) and (3.14) become

$$x_{e1}(\alpha) = \frac{c_1(\alpha) - b_2(\alpha)}{a_2(\alpha)} = \frac{4+3\alpha}{10-\alpha}, \quad (3.20)$$

$$x_{e2}(\alpha) = \frac{c_2(\alpha) - b_1(\alpha)}{a_1(\alpha)} = \frac{10-3\alpha}{8+\alpha}. \quad (3.21)$$

As in Example 3.2.1 we obtain $\overline{X}_I = \overline{X}_e$. $\overline{X}_c$ and $\overline{X}_e$ are shown in Figure 3.3.

Let us next look at solutions to systems of fuzzy linear equations.

3.3 Systems of Fuzzy Linear Equations

This section is based on [7], see also [15], [16] and [17]. Let $\overline{A} = [\overline{a}_{ij}]$ be a $n \times n$ matrix of triangular fuzzy numbers $\overline{a}_{ij}$, $\overline{B}^t = (\overline{b}_1, \ldots, \overline{b}_n)$ a $n \times 1$ vector of triangular fuzzy numbers $\overline{b}_i$ and $\overline{X}^t = (\overline{x}_1, \ldots, \overline{x}_n)$ a $n \times 1$ vector

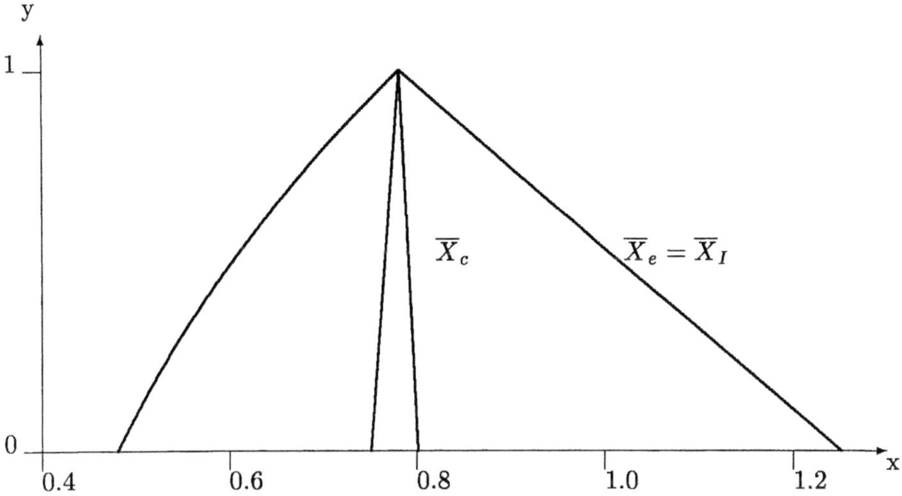

Figure 3.3: Solution to Example 3.2.2

of unknown triangular shaped fuzzy numbers $\overline{x}_j$. Set $\overline{a}_{ij} = (a_{ij1}/a_{ij2}/a_{ij3})$, $\overline{b}_i = (b_{i1}/b_{i2}/b_{i3})$, and $\overline{x}_j \approx (x_{j1}/x_{j2}/x_{j3})$. We wish to solve

$$\overline{A}\,\overline{X} = \overline{B} \ , \tag{3.22}$$

for $\overline{X}$.

We need to introduce some more notation. Define

$$a[\alpha] = \prod_{i,j=1}^{n} a_{ij}[\alpha] \ , \tag{3.23}$$

$$b[\alpha] = \prod_{i=1}^{n} b_i[\alpha] \ , \tag{3.24}$$

for $0 \leq \alpha \leq 1$. Let $v = (a_{11}, a_{12}, \ldots, a_{nn}) \in \mathbb{R}^k$, $k = n^2$, be a vector in $a[0]$. Each $v \in a[0]$ determines a crisp $n \times n$ matrix $A = [a_{ij}]$. Also, $b^t = (b_1, \ldots, b_n) \in \mathbb{R}^n$ is a vector in $b[0]$. As in ([3], [7]) we assume A^{-1} exists for all v in $a[0]$. The existence of A^{-1} over $a[0]$ simplifies the discussion of the joint solution to be introduced below.

The joint solution $\overline{X}_J$, a fuzzy subset of $\mathbb{R}^n$, is based on the extension principle

$$\overline{X}_J(x) = \max\left\{ \ \pi(v,b) \mid x = A^{-1}b \ \right\} \ , \tag{3.25}$$

where

$$\pi(v,b) = \min\left\{ \ \overline{a}_{ij}(a_{ij}), \overline{b}_i(b_i) \mid \text{all } i,j \ \right\} \ . \tag{3.26}$$

The vertex of $\overline{X}_J(x)$, where the membership value is equal to one, is at $x = A^{-1}b$ for $v = (a_{112}, a_{122}, \ldots, a_{nn2})$, $b^t = (b_{12}, \ldots, b_{n2})$. In the crisp case the solution to $Ax = b$ is a vector $x = A^{-1}b$ in $\mathbb{R}^n$, so for the fuzzy case $\overline{A}\,\overline{X} = \overline{B}$, the (joint) solution is a fuzzy vector about the crisp solution $A^{-1}b$, for v and b at the vertex values of all the triangular fuzzy numbers.

In the crisp case the marginals, the x_i, are just the components of the vector $x = A^{-1}b$. In the fuzzy case we obtain the marginals $\overline{X}_{Ji}$ by projecting $\overline{X}_J$ onto the coordinate axes. Then

$$\overline{X}_{Ji}(w) = \max\left\{\overline{X}_J(x) \mid x \in \mathbb{R}^n, x_i = w\right\}, \tag{3.27}$$

for $1 \le i \le n$. Obviously, it will be difficult to compute $\overline{X}_J$ and $\overline{X}_{Ji}$, $1 \le i \le n$, for $n \ge 4$. We will determine the joint solution, and its marginals, in two examples at the end of this section for $n = 2$.

Since $\overline{X}_J$ is difficult to determine we now turn to methods of finding the marginals directly without first computing the joint solution. As in the previous section there will be three solutions $\overline{X}_{ci}$, $\overline{X}_{ei}$ and $\overline{X}_{Ii}$, $1 \le i \le n$.

The classical solution is determined by substituting the intervals $\overline{a}_{ij}[\alpha]$, $\overline{b}_i[\alpha]$ and $\overline{X}_i[\alpha] = [x_{i1}(\alpha), x_{i2}(\alpha)]$ into $\overline{A}\,\overline{X} = \overline{B}$ and solving for the $x_{i1}(\alpha)$, $x_{i2}(\alpha)$, $1 \le i \le n$. The resulting equations are evaluated using interval arithmetic. If the intervals $[x_{i1}(\alpha), x_{i2}(\alpha)]$ define a triangular shaped fuzzy number $\overline{X}_i$ for $0 \le \alpha \le 1$, $1 \le i \le n$, then this solution is called the classical solution and we write $\overline{X}_{ci} = \overline{X}_i$, $1 \le i \le n$. The conditions for $[x_{i1}(\alpha), x_{i2}(\alpha)]$ to define $\overline{X}_{ci}$ are given in the previous section. The equations to solve for $x_{i1}(\alpha)$ and $x_{i2}(\alpha)$ are

$$\sum_{j=1}^{n}[a_{ij1}(\alpha), a_{ij2}(\alpha)][x_{j1}(\alpha), x_{j2}(\alpha)] = [b_{i1}(\alpha), b_{i2}(\alpha)], \tag{3.28}$$

for $1 \le i \le n$, where $\overline{a}_{ij}[\alpha] = [a_{ij1}(\alpha), a_{ij2}(\alpha)]$, $\overline{b}_i[\alpha] = [b_{i1}(\alpha), b_{i2}(\alpha)]$. After using interval arithmetic we obtain a $(2n) \times (2n)$ system to solve for $x_{i1}(\alpha)$, $x_{i2}(\alpha)$, $0 \le \alpha \le 1$.

As in the previous section too often the $\overline{X}_{ci}$ fail to exist. We only need $\overline{X}_{ci}$, for one value of i, to fail to exist for the classical solution to not exist. When the classical solution does not exist we turn to $\overline{X}_{ei}$, $1 \le i \le n$.

We will use Cramer's rule ([12, p. 218]) on $Ax = b$ to solve for each x_i. A comes from $v \in a[0]$ and let $b \in b[0]$. Let A_j be A with its j–th column replaced by b. Then

$$x_j = \frac{|A_j|}{|A|}, \tag{3.29}$$

$1 \le j \le n$, where $|\cdot|$ denotes the determinant. We fuzzify equation (3.29), using the extension principle, to get

$$\overline{X}_{ej}(x_j) = \max\{\pi(v, b) \mid x_j = |A_j|/|A|\}, \tag{3.30}$$

$1 \le j \le n$. If $\overline{X}_{ej}[\alpha] = [x_{ej1}(\alpha), x_{ej2}(\alpha)]$, we may find the α–cuts of $\overline{X}_{ej}$ as [5]

$$x_{ej1}[\alpha] = \min\left\{\frac{|A_j|}{|A|} \mid v \in a[\alpha], b \in b[\alpha]\right\}, \tag{3.31}$$

$$x_{ej2}[\alpha] = \max\left\{ \frac{|A_j|}{|A|} \mid v \in a[\alpha], b \in b[\alpha] \right\}, \tag{3.32}$$

To get the $\overline{X}_{Ij}$ we evaluate equation (3.29) using α–cuts and interval arithmetic. Substitute intervals $\overline{a}_{ij}[\alpha]$ and $\overline{b}_i[\alpha]$ for a_{ij} and b_i in $|A_j|/|A|$, evaluate using interval arithmetic, and the result is $\overline{X}_{Ij}[\alpha]$, $0 \leq \alpha \leq 1$, $1 \leq j \leq n$.

Theorem 3.3 *If the $\overline{X}_{ci}$ exist, $1 \leq i \leq n$, then $\overline{X}_{ci} \leq \overline{X}_{Ji} \leq \overline{X}_{ei} \leq \overline{X}_{Ii}$, $1 \leq i \leq n$.*

Proof 3.3 *1. We show $\overline{X}_{ci} \leq \overline{X}_{Ji}$ for all i by showing $\overline{X}_{ci}[\alpha] \subset \overline{X}_{Ji}[\alpha]$. Choose $\alpha \in [0,1]$ and $x_i \in \overline{X}_{ci}[\alpha]$, $1 \leq i \leq n$.Then $x_i \in [x_{ci1}(\alpha), x_{ci2}(\alpha)]$, the α–cuts of $\overline{X}_{ci}$, $1 \leq i \leq n$. From equation (3.28) there are $a_{ij} \in \overline{a}_{ij}[\alpha]$ and $b_i \in \overline{b}_i[\alpha]$ so that*

$$\sum_{j=1}^{n} a_{ij}x_j = b_i, \tag{3.33}$$

$1 \leq i \leq n$. Let $x^t = (x_1, \ldots, x_n)$. Let $v = (a_{11}, \ldots, a_{nn})$ in $a[\alpha]$, $b^t = (b_1, \ldots, b_n) \in b[\alpha]$. For this v and b we have $x = A^{-1}b$. This implies, from the definition of $\overline{X}_J$, that $\overline{X}_J(x) \geq \alpha$. Hence $\overline{X}_{Ji}(x_i) \geq \alpha$ and $x_i \in \overline{X}_{Ji}[\alpha]$, all i.

2. We argue that $\overline{X}_{Ji} \leq \overline{X}_{ei}$ by showing $\overline{X}_{Ji}[\alpha] \subset \overline{X}_{ei}[\alpha]$, $1 \leq i \leq n$. Choose $\alpha \in [0,1]$ and $x_i \in \overline{X}_{Ji}[\alpha]$, $1 \leq i \leq n$. Let $x^t = (x_1, \ldots, x_n)$.

We first show $\overline{X}_J[\alpha] \subset \prod_{i=1}^{n} \overline{X}_{ei}[\alpha]$. Let $w \in \overline{X}_J[\alpha]$. From the definition of $\overline{X}_J$ there are $a_{ij} \in \overline{a}_{ij}[\alpha]$, $b_i \in \overline{b}_i[\alpha]$ so that if $v = (a_{11}, \ldots, a_{nn})$, $b^t = (b_1, \ldots, b_n)$, $w = A^{-1}b$. This means $w_j = |A_j|/|A|$ for $v \in a[\alpha]$, $b \in b[\alpha]$ and $w_j \in \overline{X}_{ej}[\alpha]$, $1 \leq j \leq n$.

Now we argue that $x \in \prod_{i=1}^{n} \overline{X}_{ei}[\alpha]$. If $x \in \overline{X}_J[\alpha]$, then this follows from above. So assume $x \notin \overline{X}_J[\alpha]$. But x must then be in $\prod_{i=1}^{n} \overline{X}_{ei}[\alpha]$ or else there is a value of i, say $i = k$, so that $\overline{X}_{Jk}(x_k) < \alpha$, a contradiction.

But $x \in \prod_{i=1}^{n} \overline{X}_{ei}[\alpha]$ means that $x_i \in \overline{X}_{ei}[\alpha]$ all i.

3. $\overline{X}_{ei} \leq \overline{X}_{Ii}$ follows from Theorem 3.1 of [14].

Our solution strategy is: (1) use $\overline{X}_{ci}$, $1 \leq i \leq n$, if it exists; (2) if the classical solution does not exist use $\overline{X}_{Ji}$, $1 \leq i \leq n$. However, if the joint solution is too difficult to compute use $\overline{X}_{ei}$, $1 \leq i \leq n$. Equations (3.31) and (3.32) may be hard to evaluate to get the $\overline{X}_{ei}$. We have used evolutionary algorithms in the past ([8] – [11]) to do this. One can always use the $\overline{X}_{Ii}$ because they are the easiest to calculate. Notice how the fuzziness grows (the supports do not decrease) as we go from $\overline{X}_{ci}$ to $\overline{X}_{Ii}$. The only solution guaranteed to satisfy the fuzzy equations in the classical solution.

In the following two examples we only consider 2×2 fuzzy matrices since then we can easily see pictures of α–cuts of the joint solution. It is known that, in general, α–cuts of the joint solution need not be convex [13]. For example, in two dimensions $\overline{X}_J[\alpha]$ need not be a rectangle.

Example 3.3.1

Let

$$\overline{A} = \begin{pmatrix} \overline{a}_{11} & 0 \\ 0 & \overline{a}_{22} \end{pmatrix} , \tag{3.34}$$

and $\overline{b}^t = (\overline{b}_1, \overline{b}_2)$, where $\overline{a}_{11} = (4/5/7)$, $\overline{a}_{22} = (6/8/12)$, $\overline{b}_1 = (1/2/3)$ and $\overline{b}_2 = (2/5/8)$.

Then $\overline{a}_{11}[\alpha] = [4+\alpha, 7-2\alpha]$, $\overline{a}_{22}[\alpha] = [6+2\alpha, 12-4\alpha]$, $\overline{b}_1[\alpha] = [1+\alpha, 3-\alpha]$ and $\overline{b}_2[\alpha] = [2+3\alpha, 8-3\alpha]$. Since all the fuzzy numbers are positive we will solve for $\overline{X}_{ci} > 0$, $i = 1, 2$. The equations are

$$[4+\alpha, 7-2\alpha] \cdot [x_{c11}(\alpha), x_{c12}(\alpha)] = [1+\alpha, 3-\alpha] , \tag{3.35}$$

$$[6+2\alpha, 12-4\alpha] \cdot [x_{c21}(\alpha), x_{c22}(\alpha)] = [2+3\alpha, 8-3\alpha] , \tag{3.36}$$

which define triangular shaped fuzzy numbers.

$$\overline{X}_{c1}[\alpha] = \left[\frac{1+\alpha}{4+\alpha}, \frac{3-\alpha}{7-2\alpha}\right] , \tag{3.37}$$

$$\overline{X}_{c2}[\alpha] = \left[\frac{2+3\alpha}{6+2\alpha}, \frac{8-3\alpha}{12-4\alpha}\right] , \tag{3.38}$$

$0 \leq \alpha \leq 1$.

It was shown in [7] that a way to find α–cuts of $\overline{X}_J$ is

$$\overline{X}_J[\alpha] = \{A^{-1}b \mid v \in a[\alpha], b \in b[\alpha] \} . \tag{3.39}$$

Now $A^{-1}b = (b_1/a_{11}, b_2/a_{22})^t$ so that

$$\overline{X}_J[\alpha] = \frac{\overline{b}_1[\alpha]}{\overline{a}_{11}[\alpha]} \times \frac{\overline{b}_2[\alpha]}{\overline{a}_{22}[\alpha]} , \tag{3.40}$$

which is a rectangle in $\mathbf{R}^2$ for all $0 \leq \alpha \leq 1$. The first part of $\overline{X}_J[\alpha]$ is $\overline{X}_{J1}[\alpha]$ and $\overline{X}_{J2}[\alpha]$ is the second part. Then

$$\overline{X}_{J1}[\alpha] = \left[\frac{1+\alpha}{7-2\alpha}, \frac{3-\alpha}{4+\alpha}\right] , \tag{3.41}$$

$$\overline{X}_{J2}[\alpha] = \left[\frac{2+3\alpha}{12-4\alpha}, \frac{8-3\alpha}{6+2\alpha}\right] , \tag{3.42}$$

$0 \leq \alpha \leq 1$.

Next we find that $|A_1|/|A| = b_1/a_{11}$ and $|A_2|/|A| = b_2/a_{22}$. From equations (3.31) and (3.32) we obtain $\overline{X}_{ei} = \overline{X}_{Ji}$, $i = 1, 2$.

Finally, we substitute the intervals $\overline{a}_{11}[\alpha]$, $\overline{a}_{22}[\alpha]$, $\overline{b}_1[\alpha]$ and $\overline{b}_2[\alpha]$ into $x_1 = b_1/a_{11}$ and $x_2 = b_2/a_{22}$ and we see that $\overline{X}_{Ii} = \overline{X}_{ei}$, $i = 1, 2$.

For this 2×2 fuzzy diagonal matrix $\overline{A}$ we get

$$\overline{X}_{ci} \subsetneq \overline{X}_{Ji} = \overline{X}_{ei} = \overline{X}_{Ii} \ , \tag{3.43}$$

$i = 1, 2$. The graphs of $\overline{X}_{ci}$ and $\overline{X}_{Ji}$, $i = 1, 2$ are in Figures 3.4 and 3.5.

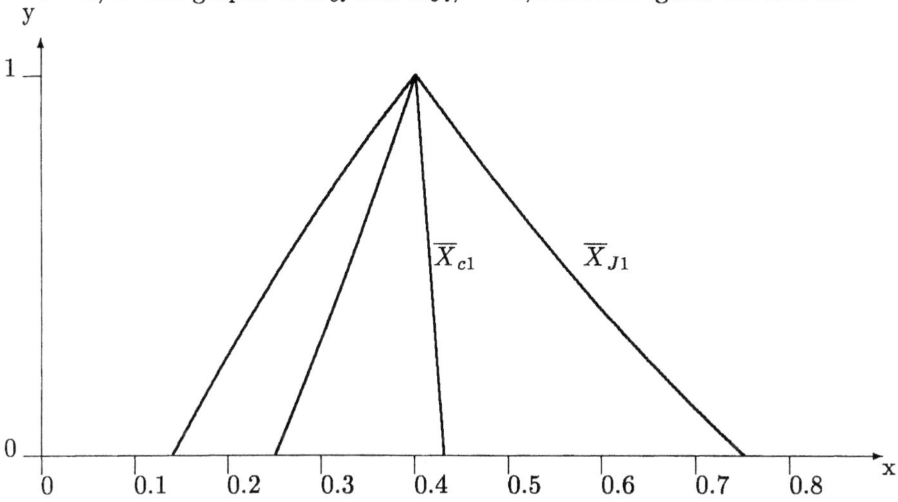

Figure 3.4: $\overline{X}_{c1}$ and $\overline{X}_{J1}$ in Example 3.3.1

Example 3.3.2

Let

$$\overline{A} = \begin{pmatrix} \overline{a}_{11} & 0 \\ 1 & \overline{a}_{22} \end{pmatrix} , \tag{3.44}$$

and $\overline{b}^t = (\overline{b}_1, \overline{b}_2)$, where $\overline{a}_{11} = (1/2/3)$, $\overline{a}_{22} = (2/5/8)$, $\overline{b}_1 = (4/5/7)$ and $\overline{b}_2 = (6/8/12)$. Then $\overline{a}_{11}[\alpha] = [1 + \alpha, 3 - \alpha]$, $\overline{a}_{22}[\alpha] = [2 + 3\alpha, 8 - 3\alpha]$, $\overline{b}_1[\alpha] = [4 + \alpha, 7 - 2\alpha]$ and $\overline{b}_2[\alpha] = [6 + 2\alpha, 12 - 4\alpha]$.

As in Example 3.3.1 we solve for $\overline{X}_{c1} > 0$ and obtain

$$\left[\frac{4+\alpha}{1+\alpha}, \frac{7-2\alpha}{3-\alpha} \right] , \tag{3.45}$$

which does not define a fuzzy number since $\partial/\partial\alpha[(4 + \alpha)/(1 + \alpha)] < 0$, or $(4 + \alpha)/(1 + \alpha)$ is a decreasing function of α in $[0, 1]$. The classical solution does not exist.

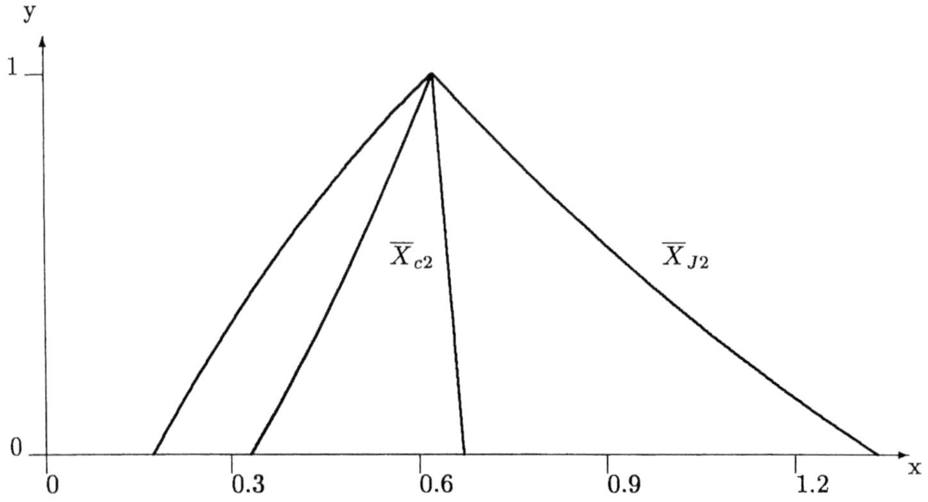

Figure 3.5: $\overline{X}_{c2}$ and $\overline{X}_{J2}$ in Example 3.3.1

We find α–cuts of $\overline{X}_J$ using equation (3.39). We only go through the details for $\alpha = 0$ and $\alpha = 1$. An equivalent expression to equation (3.39) is

$$\overline{X}_J[\alpha] = \{\, x \in \mathbb{R}^n \mid Ax = b, v \in a[\alpha], b \in b[\alpha]\} \ . \tag{3.46}$$

For $\alpha = 1$ we get $x = (2.5, 1.1)$. For $\alpha = 0$ first assume $x_1 \geq 0$, $x_2 \geq 0$. Then we want all solutions for x_1 and x_2 so that

$$([1,3]x_1 + [0,0]x_2) \cap [4,7] \neq \emptyset \ , \tag{3.47}$$
$$([1,1]x_1 + [2,8]x_2) \cap [6,12] \neq \emptyset \ . \tag{3.48}$$

We have used the $\alpha = 0$ cuts of $\overline{a}_{11}$, $\overline{a}_{22}$, $\overline{b}_1$ and $\overline{b}_2$. This means

$$x_1 \leq 7 \ , \tag{3.49}$$
$$3x_1 \geq 4 \ , \tag{3.50}$$
$$x_1 + 2x_2 \leq 12 \ , \tag{3.51}$$
$$x_1 + 8x_2 \geq 6 \ , \tag{3.52}$$

for $x_1 \geq 0$, $x_2 \geq 0$ in the first quadrant. Now x_1 must be non–negative so we can now only consider the fourth quadrant.

Assume $x_1 \geq 0$ and $x_2 \leq 0$. Then the equations become

$$x_1 \leq 7 \ , \tag{3.53}$$
$$3x_1 \geq 4 \ , \tag{3.54}$$
$$x_1 + 8x_2 \leq 12 \ , \tag{3.55}$$
$$x_1 + 2x_2 \geq 6 \ , \tag{3.56}$$

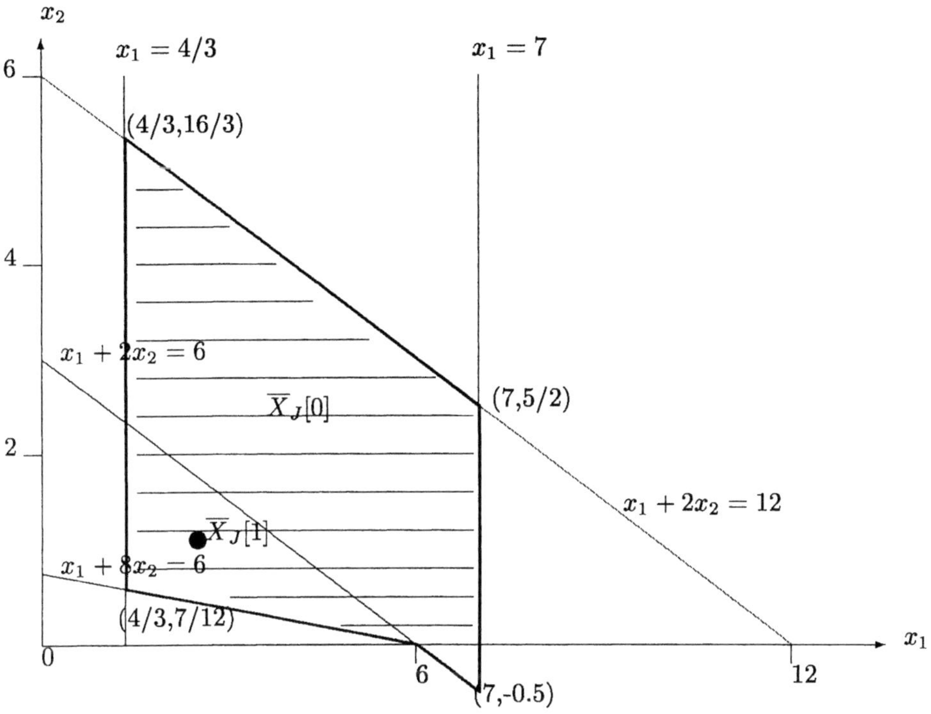

Figure 3.6: Support of the Joint Solution in Example 3.3.2

for $x_1 \geq 0$, $x_2 \leq 0$. The solution $\overline{X}_J[0]$ is shown in Figure 3.6. It is not convex since the line joining $(4/3, 7/12)$ and $(7, -1/2)$ is not entirely in $\overline{X}_J[0]$. Projecting $\overline{X}_J$ onto the x_i–axes produces $\overline{X}_{Ji}$, $i = 1, 2$. These marginals are in Figures 3.7 and 3.8.

We find the α-cuts of the $\overline{X}_{Ji}$ as follows: (1) we first construct a diagram like Figure 3.6 for $\overline{X}_J[\alpha]$ for each $0 \leq \alpha \leq 1$; (2) project the diagram onto the x_1-axis to get $\overline{X}_{J1}[\alpha]$; and (3) project the picture onto the x_2-axis to obtain $\overline{X}_{J2}[\alpha]$. It turns out, for this example, that $\overline{X}_{J1} = \overline{X}_{e1}$ and $\overline{X}_{J2} = \overline{X}_{e2}$.

Using equations (3.31) and (3.32) we find α–cuts of $\overline{X}_{ej}$, $j = 1, 2$. It is easy to see that

$$\overline{X}_{e1}[\alpha] = \left[\frac{4+\alpha}{3-\alpha}, \frac{7-2\alpha}{1+\alpha}\right] , \tag{3.57}$$

$0 \leq \alpha \leq 1$. However, $\overline{X}_{e2}[\alpha]$ is a little more difficult since we need to find the max and min of

$$\frac{a_{11}b_2 - b_1}{a_{11}a_{22}} , \tag{3.58}$$

for $a_{11} \in \overline{a}_{11}[\alpha]$, $a_{22} \in \overline{a}_{22}[\alpha]$, $b_1 \in \overline{b}_1[\alpha]$ and $b_2 \in \overline{b}_2[\alpha]$. We did this and $\overline{X}_{e2}$ is shown in Figure 3.8.

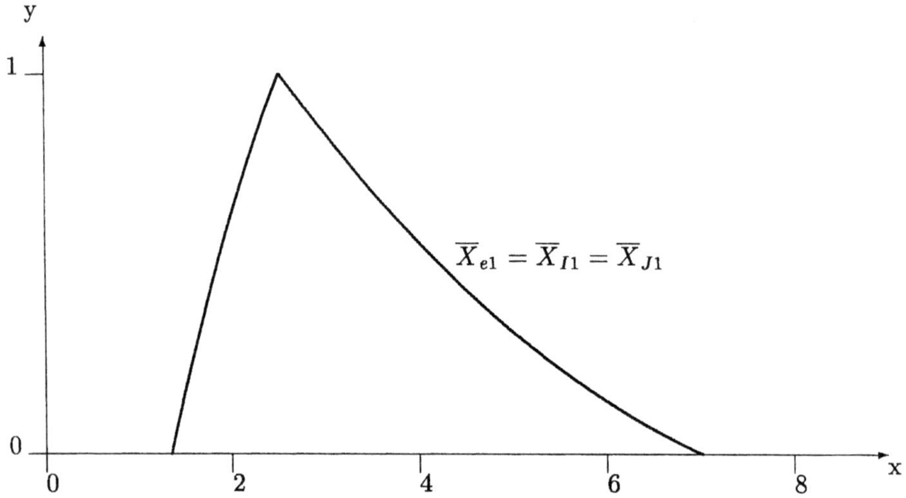

Figure 3.7: $\overline{X}_{J1}, \overline{X}_{e1}, \overline{X}_{I1}$ in Example 3.3.2

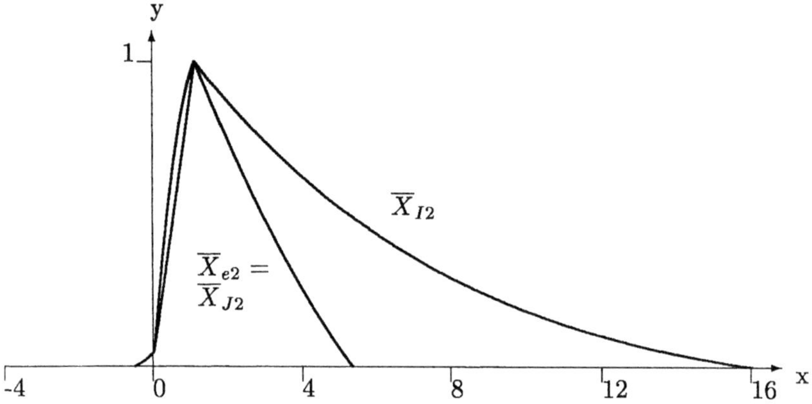

Figure 3.8: $\overline{X}_{J2}, \overline{X}_{e2}, \overline{X}_{I2}$ in Example 3.3.2

Lastly, we see that $\overline{X}_{I1} = \overline{X}_{e1}$ and

$$\overline{X}_{I2}[\alpha] = [x_{I21}(\alpha), x_{I22}(\alpha)] \ , \tag{3.59}$$

with $x_{I21}(\alpha) = N_1(\alpha)/D_2(\alpha)$, $0 \leq \alpha \leq 0.0981 = (\sqrt{108} - 10)/4$, $x_{I21}(\alpha) = N_1(\alpha)/D_1(\alpha)$ for $0.0981 \leq \alpha \leq 1$ and $x_{I22}(\alpha) = N_2(\alpha)/D_2(\alpha)$ for all α and

$$N_1(\alpha) = (1+\alpha)(6+2\alpha) - (7-2\alpha), \tag{3.60}$$

$$N_2(\alpha) = (3-\alpha)(12-4\alpha) - (4+\alpha), \tag{3.61}$$

$$D_1(\alpha) = (3-\alpha)(8-3\alpha), \tag{3.62}$$

$$D_2(\alpha) = (1+\alpha)(2+3\alpha). \tag{3.63}$$

The reason for the change in the denominator for $x_{I21}(\alpha)$ is that $N_1(\alpha)$ is negative for $0 \leq \alpha \leq 0.0981$. We used the fact that $[a,b][c,d] = [ad, bd]$ if $a < 0 < b$ and $0 < c < d$ but $[a,b][c,d] = [ac, bd]$ when $0 < a$ and $0 < c$.

$\overline{X}_{ei}$ and $\overline{X}_{Ii}$ are in Figures 3.7 and 3.8 for $i = 1, 2$.

3.4 Applications

We consider three applications: (1) the first involves solving a fuzzy linear equation; (2) the second has a fuzzy quadratic to solve; and (3) the last application is a 2×2 system of fuzzy linear equations.

3.4.1 Fuzzy Linear Equation

A dealer bought 1000 items at \$ 150 each. He wishes to sell around 40% ($\overline{A} = (0.30/0.40/0.50)$) of them for a profit of approximately 25% ($\overline{B} = (0.15/0.25/0.35)$). What price $\overline{P}$ should he sell the rest if he wants his average profit for all the items to be about 40% ($\overline{C} = (0.35/0.40/0.45)$)? The fuzzy equation to solve is

$$(1000\overline{A})(150\overline{B}) + (1000 - 1000\overline{A})(\overline{P} - 150) = (1000)(150)\overline{C}. \tag{3.64}$$

Let $\overline{A}[\alpha] = [0.30 + 0.10\alpha, 0.50 - 0.10\alpha]$, $\overline{B}[\alpha] = [0.15 + 0.10\alpha, 0.35 - 0.10\alpha]$ and $\overline{C}[\alpha] = [0.35 + 0.05\alpha, 0.45 - 0.05\alpha]$. Set $\overline{P}[\alpha] = [p_1(\alpha), p_2(\alpha)]$, and we assume that $\overline{P} > 150$.

We first look at the classical solution. Solving for $p_1(\alpha)$ we get

$$p_1(\alpha) = 150\left(1 + \frac{(0.35 + 0.05\alpha) - (0.30 + 0.10\alpha)(0.15 + 0.10\alpha)}{0.50 + 0.10\alpha}\right). \tag{3.65}$$

But $\partial p_1/\partial\alpha < 0$, and p_1 is a decreasing function of α. The classical solution does not exist.

The crisp solution is $(a, b, c \in (0,1))$

$$p = 150\left(1 + \frac{c - ab}{1 - a}\right). \tag{3.66}$$

We fuzzify this crisp solution to obtain $\overline{P}_e$. The α–cuts of $\overline{P}_e$ are

$$p_{e1}(\alpha) = \min\{\, p \mid a \in \overline{A}[\alpha], b \in \overline{B}[\alpha], c \in \overline{C}[\alpha] \,\}, \tag{3.67}$$

$$p_{e2}(\alpha) = \max\{\, p \mid a \in \overline{A}[\alpha], b \in \overline{B}[\alpha], c \in \overline{C}[\alpha] \,\}, \tag{3.68}$$

for p in equation (3.66). We see that $\partial p/\partial a > 0$, $\partial p/\partial b < 0$, $\partial p/\partial c > 0$ so $\overline{P}_e[\alpha] = 150 + 150I[\alpha]$,

$$I[\alpha] = [i_1(\alpha), i_2(\alpha)], \tag{3.69}$$

with

$$i_1(\alpha) = \frac{(0.35+0.05\alpha)-(0.30+0.10\alpha)(0.35-0.10\alpha)}{0.70-0.10\alpha}, \quad (3.70)$$

$$i_2(\alpha) = \frac{(0.45-0.05\alpha)-(0.50-0.10\alpha)(0.15+0.10\alpha)}{0.50+0.10\alpha}. \quad (3.71)$$

Figure 3.7 shows $\overline{P}_e$. We do not compute $\overline{P}_I$.

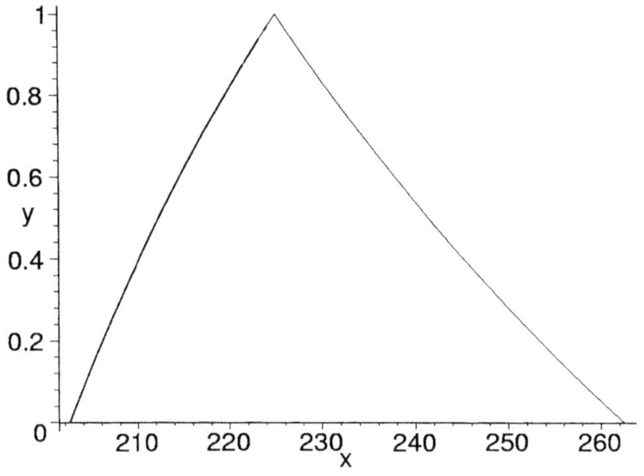

Figure 3.9: Solution $\overline{P}_e$ in the First Application

3.4.2 Fuzzy Quadratic Equation

Suppose an investment firm wishes to set aside around one million dollars ($\overline{A} = (0.8/1.0/1.2)$) to be invested at interest rate $\overline{r}$, a fuzzy number whose support will be in $[0,1]$, so that after one year they may withdraw approximately \$250,000 ($\overline{B} = (0.20/0.25/0.30)$). And after two years the amount that is left will accumulate to about \$900,000 ($\overline{C} = (0.6/0.9/1.2)$). Solve for $\overline{r}$.

After one year the amount will be $\overline{A} + \overline{A}\,\overline{r}$. Now withdraw $\overline{B}$ dollars and we have $(\overline{A} - \overline{B}) + \overline{A}\,\overline{r}$ to start off the second year. At the end of the second year we have

$$[(\overline{A} - \overline{B}) + \overline{A}\,\overline{r}] + [(\overline{A} - \overline{B}) + \overline{A}\,\overline{r}]\,\overline{r}\,. \quad (3.72)$$

For positive fuzzy numbers multiplication distributes over additions and equation (3.72) becomes

$$\overline{A}(\overline{r})^2 + \overline{D}\,\overline{r} + \overline{E}\ , \tag{3.73}$$

where $\overline{D} = 2\overline{A} - \overline{B}$, $\overline{E} = \overline{A} - \overline{B}$. So we must solve the fuzzy quadratic

$$\overline{A}(\overline{r})^2 + \overline{D}\,\overline{r} + \overline{E} = \overline{C}\ , \tag{3.74}$$

for $\overline{r}$. Let $\overline{r}[\alpha] = [r_1(\alpha), r_2(\alpha)]$ and $\overline{A}[\alpha] = [0.8 + 0.2\alpha, 1.2 - 0.2\alpha]$, $\overline{B}[\alpha] = [0.20 + 0.05\alpha, 0.30 - 0.05\alpha]$, $\overline{C}[\alpha] = [0.6 + 0.3\alpha, 1.2 - 0.3\alpha]$. Then $\overline{D}[\alpha] = [1.3 + 0.45\alpha, 2.2 - 0.45\alpha]$ and $\overline{E}[\alpha] = [0.5 + 0.25\alpha, 1 - 0.25\alpha]$. We take α–cuts of equation (3.74), use interval arithmetic, and solve for $r_1(\alpha)$ and $r_2(\alpha)$. If $[r_1(\alpha), r_2(\alpha)]$, $0 \leq \alpha \leq 1$, defines a fuzzy number, it is the classical solution (only one solution is in $[0, 1]$).

The equation for $r_1(\alpha)$ is

$$(0.8 + 0.2\alpha)r_1^2(\alpha) + (1.3 + 0.45\alpha)r_1(\alpha) + (-0.1 - 0.05\alpha) = 0\ . \tag{3.75}$$

Then

$$r_1(\alpha) = \frac{-(1.3 + 0.45\alpha) + \sqrt{W(\alpha)}}{2(0.8 + 0.2\alpha)}\ , \tag{3.76}$$

where

$$W(\alpha) = (1.3 + 0.45\alpha)^2 + 4(0.8 + 0.2\alpha)(0.1 + 0.05\alpha)\ . \tag{3.77}$$

Similarly, we solve for $r_2(\alpha)$.

The hypotheses of Theorem 13 in [4] are satisfied which means $\partial/\partial\alpha(r_1(\alpha)) > 0$, $\partial/\partial\alpha(r_2(\alpha)) < 0$. Also $r_1(1) = r_2(1) = 0.0819$. Hence, $\overline{r}$ exists and is the solution. The graph of $\overline{r}$ is in Figure 3.10.

Since the classical solution exists, we omit $\overline{r}_e$ and $\overline{r}_I$.

3.4.3 System of Linear Equations

A winery wishes to mix white wine with brandy to obtain sherry. The white wine has an alcohol content of round 10% by volume ($\overline{A} = (0.05/0.10/0.15)$) and the brandy has an alcohol content of approximately 35% ($\overline{B} = (0.30/0.35/0.40)$) by volume and they wish to produce about 10000 liters ($\overline{L} = (9990/10000/10010)$) of sherry of about 15% ($\overline{C} = (0.14/0.15/0.16)$) alcohol. How many liters of wine and brandy should be mixed together to produce the desired result?

Let $\overline{x}_1$ and $\overline{x}_2$ be fuzzy numbers representing the amount of wine and brandy, respectively, to mix to get sherry. Then

$$\overline{A}\,\overline{x}_1 + \overline{B}\,\overline{x}_2 = \overline{C}\,\overline{L}\ , \tag{3.78}$$

$$\overline{x}_1 + \overline{x}_2 = \overline{L}\ . \tag{3.79}$$

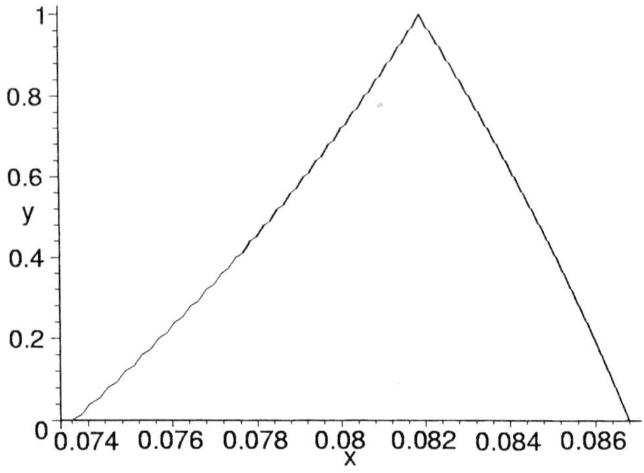

Figure 3.10: Fuzzy Interest Rate in the Second Application

Taking α–cuts we wish to solve for $[x_{11}(\alpha), x_{12}(\alpha)] = \overline{x}_1[\alpha]$ and $[x_{21}(\alpha), x_{22}(\alpha)] = \overline{x}_2[\alpha]$. The system of equations to solve is

$$a_1(\alpha)x_{11}(\alpha) + b_1(\alpha)x_{21}(\alpha) = c_1(\alpha)l_1(\alpha) , \tag{3.80}$$

$$a_2(\alpha)x_{12}(\alpha) + b_2(\alpha)x_{22}(\alpha) = c_2(\alpha)l_2(\alpha) , \tag{3.81}$$

$$x_{11}(\alpha) + x_{21}(\alpha) = l_1(\alpha) , \tag{3.82}$$

$$x_{12}(\alpha) + x_{22}(\alpha) = l_2(\alpha) , \tag{3.83}$$

$0 \leq \alpha \leq 1$, where $\overline{A}[\alpha] = [a_1(\alpha), a_2(\alpha)]$, $\overline{B}[\alpha] = [b_1(\alpha), b_2(\alpha)]$, $\overline{C}[\alpha] = [c_1(\alpha), c_2(\alpha)]$ and $\overline{L}[\alpha] = [l_1(\alpha), l_2(\alpha)]$. The classical solution does not exist. For example, solve for $x_{21}(\alpha)$ and we obtain

$$x_{21}(\alpha) = \frac{l_1(\alpha)(a_1(\alpha) - c_1(\alpha))}{a_1(\alpha) - b_1(\alpha)} . \tag{3.84}$$

Now substitute $a_1(\alpha) = 0.05 + 0.05\alpha$, $\cdots$, $l_1(\alpha) = 9990 + 10\alpha$ and we then compute $\partial/\partial\alpha(x_{21}(\alpha)) < 0$. That is, $x_{21}(\alpha)$ is a decreasing function of α where it should be an increasing function of α.

So, we now look to the joint solution $\overline{X}_J$ and its marginals. As in Example 3.3.2 we construct inequalities as in equations (3.49) – (3.52), for $\overline{X}_J[0]$. Notice that now we must have $x_1 \geq 0$, $x_2 \geq 0$. The $\alpha = 0$ cut gives

$$([0.05, 0.15]x_1 + [0.3, 0.4]x_2) \cap [1398.6, 1601.6] \neq \emptyset , \tag{3.85}$$

$$([1, 1]x_1 + [1, 1]x_2) \cap [9990, 10010] \neq \emptyset . \tag{3.86}$$

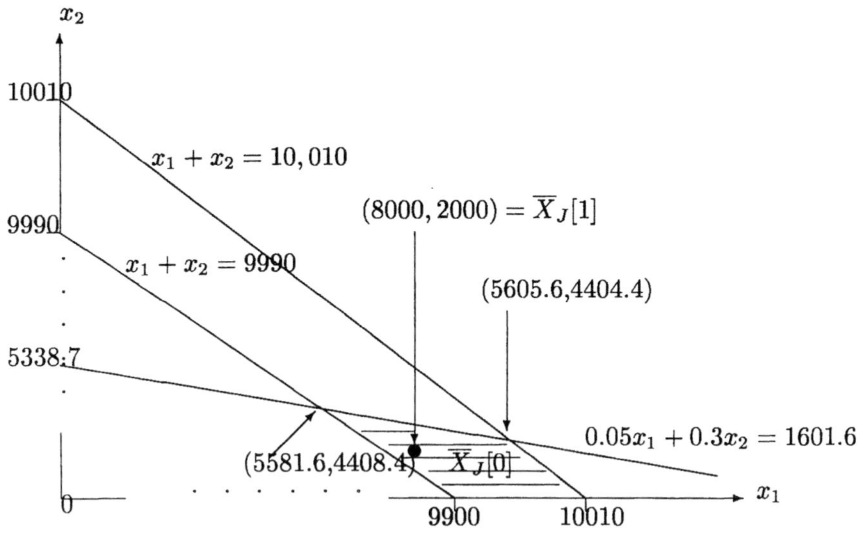

Figure 3.11: Support for the Joint Solution in the Third Application

The inequalities are

$$0.05x_1 + 0.3x_2 \leq 1601.6 \,, \tag{3.87}$$
$$0.15x_1 + 0.4x_2 \geq 1398.6 \,, \tag{3.88}$$
$$x_1 + x_2 \leq 10010 \,, \tag{3.89}$$
$$x_1 + x_2 \geq 9990 \,. \tag{3.90}$$

$\overline{X}_J[0]$ is shown in Figure 3.11. The marginals are in Figures 3.12 and 3.13. We find the α-cuts as in Example 3.2.2. We do not compute the $\overline{X}_{ei}$ or $\overline{X}_{Ii}$, $i = 1, 2$.

The graphs in Figures 3.12 and 3.13 look like triangular fuzzy numbers, but they are not triangular fuzzy numbers. In Figure 3.12 the left side of $\overline{X}_{J1}$ is the graph of $x = 4[(0.3+0.05\alpha)(9990+10\alpha)-(0.16-0.01\alpha)(10010-10\alpha)]$. The right side of $\overline{X}_{J1}$ is the graph of two functions connected at the "dot" in Figure 3.12. The right side is the graph of $x = 4[(0.4 - 0.05\alpha)(10010 - 10\alpha) - (0.14 + 0.01\alpha)(9990 + 10\alpha)]$ for $0.1706 \leq \alpha \leq 1$ and the graph of $x = 10010 - 10\alpha$ for $0 \leq \alpha \leq 0.1706$. We obtain similar results for the graph of $\overline{X}_{J2}$ in Figure 3.13 except that the graph gets cut off at zero since x_2 must be non-negative.

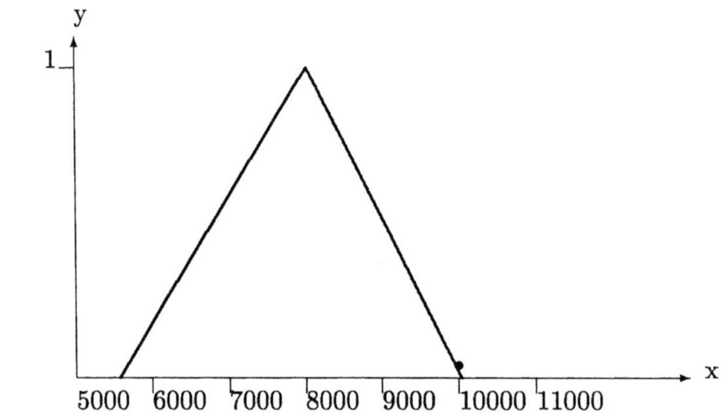

Figure 3.12: $\overline{X}_{J1}$ for the Third Application

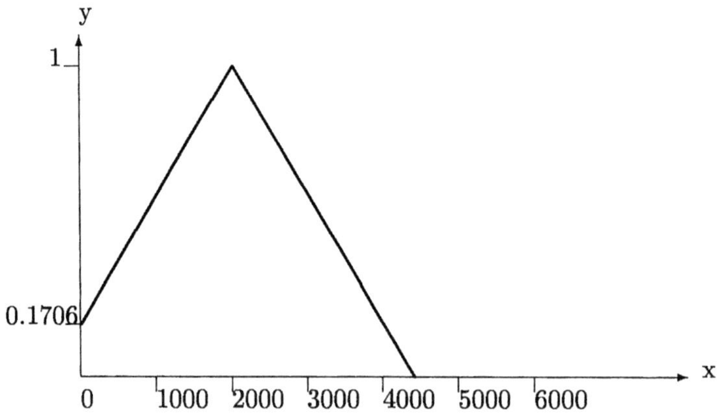

Figure 3.13: $\overline{X}_{J2}$ for the Third Application

	Industry I	Industry II	Final Demands	Gross Output
Industry I	a_{11}	a_{12}	$F_1(B_1)$	$T_1(X_1)$
Industry II	a_{21}	a_{22}	$F_2(B_2)$	$T_2(X_2)$
Outside Inputs	o_1	o_2		
Total	1.00	1.00		

Table 3.1: Input-Output Table

3.5 Fuzzy Input-Output Analysis

This section is based on [1] and another fuzzy input-output model (the "closed" model) is discussed in Section 10.2 in Chapter 10.

We will first briefly review Leontief's open input-output model for an economy. Then we generalize the model to allow for fuzzy numbers. We also present a general sufficient condition for the existence of fuzzy (open) input-output models for an economy. We then present two simple two industry input-output models showing: (1) the sufficient condition holds and the fuzzy input-output models exists; and (2) the sufficient condition is violated and the fuzzy input-output model fails to exist.

3.5.1 The Open Model

The open input-output model for a two industry economy is shown in Table 3.1. The constraints on the numbers in this table are that the a_{ij} and the o_j are in the interval $[0, 1]$, the F_i and T_i are non-negative numbers, and

$$a_{11}T_1 + a_{12}T_2 + F_1 = T_1, \tag{3.91}$$

$$a_{21}T_1 + a_{22}T_2 + F_2 = T_2, \tag{3.92}$$

$$a_{11} + a_{21} + o_1 = 1.00, \tag{3.93}$$

$$a_{12} + a_{22} + o_2 = 1.00. \tag{3.94}$$

We now explain the meaning of all the numbers in Table 3.1. T_1 (T_2) is the total annual output, measured possibly in dollars, of industry I (II). F_1 (F_2) is the amount of T_1 (T_2) consumed within the economy (government, individuals, etc.), excluding industries I and II, or is available for export. In Table 3.1 the B_i will be new values of the F_i and the X_i are to be new values of the T_i, $1 \leq i \leq 2$, to be introduced below. The a_{11} and a_{21} represent the percent of the total input to industry I which comes from industry I and II, respectively. Industry I may have other inputs, besides from industries I and II, and this percentage is given by o_1. Usually labor comes under o_j. Similarly, a_{12} (a_{22}) is the percent of the total input of II arising from I (II) and o_2 is the percent originating outside industries I and II.

We therefore see that the first two columns in Table 3.1 must sum to one (equations (3.93) and (3.94)). In equation (3.91) $a_{11}T_1$ ($a_{12}T_2$) is the total dollar input to I (II) coming from I. The left hand side of equation (3.91) shows how the total output of I is distributed between I, II and other consumption and export. Equation (3.92) has a similar interpretation for the distribution of the total output of industry II.

If the model does not allow for outside inputs, it is called a closed input-output model. We will consider only the open input-output model in this section.

Let us write equations (3.91) and (3.92) in matrix form. Set $A = [a_{ij}]$, $F = [F_i]$ (the vector of final demands) and $T = [T_i]$ (the vector of total outputs). Then we have

$$A\,T + F = T. \tag{3.95}$$

The matrix A is called the input-output , or technological, matrix for the economy. One use of the model is to predict total outputs for given different final demands. We assume that the technological matrix A is known, or has been estimated, and does not change over the planning horizon. Let B be a vector of new final demands (we hope to export more) and let X be the new vector of total outputs. Then we must have

$$A\,X + B = X, \tag{3.96}$$

which needs to be solved for X. Solving for X we obtain

$$X = (I - A)^{-1}B, \tag{3.97}$$

assuming that $I - A$ is nonsingular (has an inverse) and the X in equation (3.97) turns out to be non-negative ($X \geq 0$).

We have been considering a two industry economy so let us now extend these ideas to an m industry economy. Then $A = [a_{ij}]$ is a $m \times m$ matrix of percentages so that $0 \leq a_{ij} \leq 1$, $O = [o_j]$ is a $1 \times m$ vector of outside inputs so that $0 \leq o_j \leq 1$, and F, B, T and X are all $m \times 1$ vectors of non-negative numbers. Equations (3.91) to (3.94) still hold for the m industry economy.

The purpose of this section is now to allow all the numbers in the input-output table to be fuzzy numbers. In practice all the a_{ij} and B must be known exactly. We now assume that some of these numbers are not precisely known and may then be modeled by an appropriate fuzzy number. For example, the a_{11} in Table 3.1 can now be around 30%, or between 20% and 25%. A fuzzy number can capture the vagueness in these statements. Therefore, A will be a $m \times m$ matrix of fuzzy percentages and B will be a $m \times 1$ vector of non-negative fuzzy final demands. The main problem now is to solve equation (3.96) for the unknown $m \times 1$ vector X of non-negative fuzzy total outputs. A further restriction is that the equality in equation (3.96) must be exact (crisp) and not some fuzzy equality (approximately equal). When equation (3.96) has a solution for fuzzy $X \geq 0$ we will say that the fuzzy input-output model exists for this economy.

3.5.2 Fuzzy Model

We first introduce the notation needed to describe the fuzzy input-output model. $\overline{A} = [\overline{a}_{ij}]$ is a $m \times m$ matrix of fuzzy percentages where $\overline{a}_{ij}$ is a trapezoidal fuzzy number in $[0,1]$ all i,j. Next $\overline{B}$ is a $m \times 1$ fuzzy vector whose components $\overline{B}_i$ are non-negative trapezoidal fuzzy numbers, $1 \leq i \leq m$. Also $\overline{X}$ is a $m \times 1$ fuzzy vector with unknown components $\overline{X}_i$, $1 \leq i \leq m$, which are non-negative trapezoidal shaped fuzzy numbers. $\overline{A}$ is the fuzzy technological matrix for the economy, $\overline{B}$ is the $m \times 1$ fuzzy vector of new final demands, and $\overline{X}$ is a $m \times 1$ fuzzy vector of total outputs for the industries in this economy. We therefore must have

$$\overline{A}\,\overline{X} + \overline{B} = \overline{X}, \tag{3.98}$$

where we employ α-cuts and interval arithmetic to evaluate equation (3.98). Let $\overline{O} = [\overline{O}_j]$, where $\overline{O}_j = (o_{j1}/o_{j2}, o_{j3}/o_{j4})$ and $0 \leq o_{j1} \leq o_{j2} \leq o_{j3} \leq o_{j4} \leq 1$, $1 \leq j \leq m$, be the $1 \times m$ fuzzy vector of outside inputs. Define

$$\overline{C}_j = \sum_{i=1}^{m} \overline{a}_{ij} + \overline{O}_j, \tag{3.99}$$

where $\overline{C}_j = (c_{j1}/c_{j2}, c_{j3}/c_{j4})$. To satisfy the constraint that the column sums in the fuzzy input-output table equal one (equations (3.93) and (3.94)) we now require in the fuzzy case that $c_{j2} \leq 1 \leq c_{j3}$, or $\overline{C}_j \approx 1$ all j. The column sums in the fuzzy input-output table are approximately equal to 1.00 (100%).

We will use α-cuts and interval arithmetic to solve for $\overline{X}$. So let $\overline{a}_{ij}[\alpha] = [a_{ij1}(\alpha), a_{ij2}(\alpha)]$, $\overline{B}_i[\alpha] = [b_{i1}(\alpha), b_{i2}(\alpha)]$, $\overline{X}_i[\alpha] = [x_{i1}(\alpha), x_{i2}(\alpha)]$ all i,j and $\alpha \in [0,1]$. Define $m \times m$ matrices $A_1[\alpha] = [a_{ij1}(\alpha)]$, $A_2[\alpha] = [a_{ij2}(\alpha)]$ and $m \times 1$ vectors $B_1[\alpha] = [b_{i1}(\alpha)]$, $B_2[\alpha] = [b_{i2}(\alpha)]$, $X_1[\alpha] = [x_{i1}(\alpha)]$ and $X_2[\alpha] = [x_{i2}(\alpha)]$. Substitute α-cuts of $\overline{a}_{ij}$, $\overline{B}_i$ and $\overline{X}_i$ into equation (3.96) and then using interval arithmetic (all intervals are non-negative) we obtain

$$X_1[\alpha] = (I - A_1[\alpha])^{-1} \cdot B_1[\alpha], \tag{3.100}$$

$$X_2[\alpha] = (I - A_2[\alpha])^{-1} \cdot B_2[\alpha]. \tag{3.101}$$

for $0 \leq \alpha \leq 1$.

Even if the inverses exist for all α much more is required to obtain a solution for $\overline{X}$. We say a solution exists for $\overline{X}$ whenever: (1) $(I - A_1[\alpha])^{-1}$ and $(I - A_2[\alpha])^{-1}$ exist for all $\alpha \in [0,1]$; (2) $0 \leq x_{i1}(\alpha)$ is a monotonically increasing function of α for all i; (3) $x_{i2}(\alpha)$ is a monotonically decreasing function of $\alpha \in [0,1]$ all i; and (4) $x_{i1}(1) \leq x_{i2}(1)$ all i. If all of this is true, then we can put the α-cuts $[x_{i1}(\alpha), x_{i2}(\alpha)]$ together to obtain trapezoidal shaped fuzzy numbers $\overline{X}_i$, $1 \leq i \leq m$.

We say $A_1[\alpha]$ ($A_2[\alpha]$) is semi-positive if all the elements in the matrix are non-negative and each row (column) has at least one positive element. We now assume that both $A_1[\alpha]$ and $A_2[\alpha]$ are semi-positive for all α. If a whole row in $\overline{A}$ is zero, then we would delete this industry from consideration. We

	Industry I	Industry II	Final Demands	Gross Output
Industry I	(0.25/0.3/0.35)	(0.3/0.4/0.5)	(60/65,75/80)	$\overline{X}_1$
Industry II	(0.4/0.45,0.55/0.6)	(0.2/0.25,0.35/0.4)	(50/55,65/70)	$\overline{X}_2$
Outside Inputs	(0.1/0.2/0.3)	(0.2/0.3/0.4)		
Total	(0.75/0.95,1.05/1.25)	(0.7/0.95,1.05/1.3)		

Table 3.2: Data for Example 3.5.1

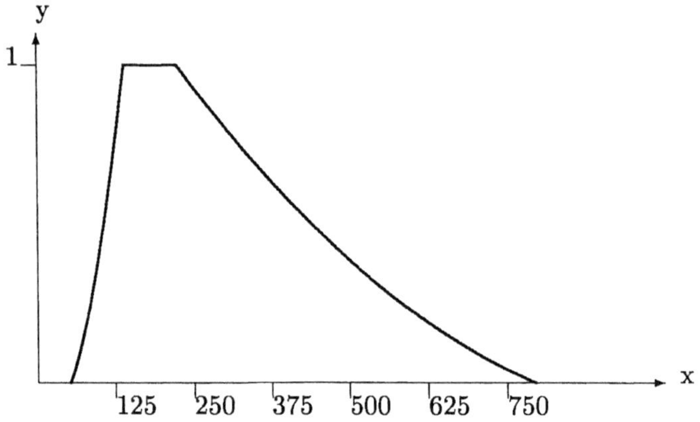

Figure 3.14: Fuzzy Total Output for Industry I in Example 3.5.1

would then take the inputs to this industry (if any) and add them to the final demands column. If a whole column in $\overline{A}$ is zero we may delete this industry and place its inputs (if any) into the outside input row.

Theorem 3.4 *Let $\overline{a}_{ij} = (a_{ij1}/a_{ij2}, a_{ij3}/a_{ij4})$ all i, j. If $\sum_{i=1}^{m} a_{ij4} < 1$ all j, then the fuzzy input-output model exists for the economy. $\overline{X}$ exists for any non-negative final demand vector $\overline{B}$.*

Proof 3.4 *The proof in given in [1].*

Example 3.5.1

The data for the fuzzy input-output model is presented in Table 3.2. We notice that $a_{114} + a_{214} < 1$ and $a_{124} + a_{224} < 1$ so Theorem 3.4 guarantees a solution for $\overline{X}_1$ and $\overline{X}_2$. We now solve equations (3.100) and (3.101) for α=0.0,0.1,...,0.9,1.0 producing the fuzzy numbers $\overline{X}_1$ and $\overline{X}_2$ for the needed gross output. The graphs of $\overline{X}_1$ and $\overline{X}_2$ are shown in Figures 3.14 and 3.15, respectively. To obtain $\overline{X}_1$ and $\overline{X}_2$, in general, for selected values of α, we simply solve a system of linear equations for their unique solution.

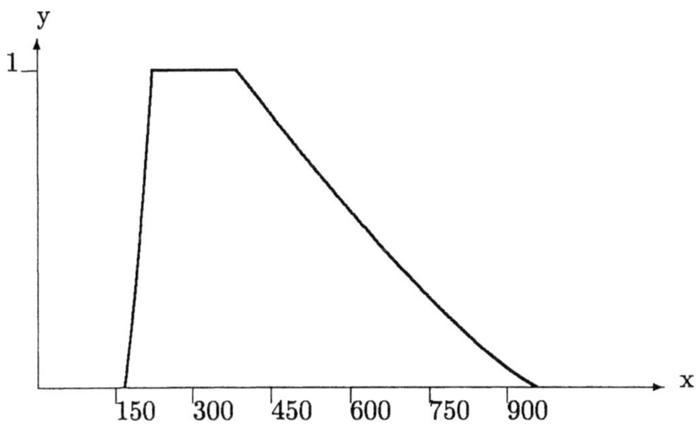

Figure 3.15: Fuzzy Total Output for Industry II in Example 3.5.1

Example 3.5.2

In Table 3.2 let $\overline{a}_{11} = (0.3/0.4/0.5)$, $\overline{a}_{12} = (0.5/0,6/0.7)$, $\overline{a}_{21} = (0.4/0.5/0.6)$ and $\overline{a}_{22} = (0.2/0.3/0.4)$. Also, let $\overline{o}_1 = \overline{o}_2 = (0/0.1/0.2)$ but keep the same $\overline{B}_i$ as in Example 3.5.1. We see that $a_{114} + a_{214} = 1.1$ and $a_{124} + a_{224} = 1.1$ so Theorem 3.4 does not apply to say that $\overline{X}_i$ exists for $i = 1, 2$. We find that : (1) $I - A_2[\alpha]$ is singular (no inverse) for α=0.5 ; (2) $x_{i2}(\alpha) < 0$ for $i = 1, 2$ and $0 \leq \alpha < 0.5$; and (3) $x_{i2}(\alpha)$ is not a decreasing function of α for $i = 1, 2$. Obviously, $\overline{X}_i$ will not be a fuzzy number for $i = 1, 2$.

Example 3.5.2 shows that if the conditions of Theorem 3.4 are not satisfied, then the economy may not have a fuzzy input-output model. The conditions of Theorem 3.4 are only sufficient and not necessary for a fuzzy input-output model to exist. For example, the sum $a_{1j4} + \cdots + a_{mj4}$ may equal one for some j and be less than one otherwise and the fuzzy input-output model can exist.

3.6 Summary and Conclusions

We introduced our three solution methods for fuzzy equations that will be used throughout this book. We applied these solution techniques first to the simple fuzzy linear equation $\overline{A}\,\overline{X} + \overline{B} = \overline{C}$ and then to systems of fuzzy linear equations. Three applications were presented: (1) the first required solving a fuzzy linear equation; (2) the second had a fuzzy quadratic equation that we needed to solve; and (3) the third had a 2×2 fuzzy linear system to solve.

We finished the chapter with the fuzzification of Leontief's open input-output model for an economy. We gave a sufficient condition for the fuzzy input-output model to exist, so what is needed is a necessary and sufficient condition for the existence of the fuzzy input-output model.

Bibliography

[1] J.J. Buckley: Fuzzy Input-Output Analysis, European J. Operational Research, 39 (1989), pp. 54–60.

[2] J.J. Buckley: Solving Fuzzy Equations in Economics and Finance, Fuzzy Sets and Systems, 48 (1992), pp. 289–296.

[3] J.J. Buckley: Solving Fuzzy Equations, Fuzzy Sets and Systems, 50 (1992), pp. 1–14.

[4] J.J. Buckley and Y. Qu: Solving Linear and Quadratic Fuzzy Equations, Fuzzy Sets and Systems, 38 (1990), pp. 43–59.

[5] J.J. Buckley and Y. Qu: On Using Alpha–Cuts to Evaluate Fuzzy Equations, Fuzzy Sets and Systems, 38 (1990), pp. 309–312.

[6] J.J. Buckley and Y. Qu: Solving Fuzzy Equations: A New Solution Concept, Fuzzy Sets and Systems, 39 (1991), pp. 291–301.

[7] J.J. Buckley and Y. Qu: Solving Systems of Fuzzy Linear Equations, Fuzzy Sets and Systems, 43 (1991), pp. 33–43.

[8] J.J. Buckley, Th. Feuring and Y. Hayashi: Fuzzy Hierarchical Analysis Revisited, Proc. IFSA, Taipei, Taiwan, Aug. 17–20, 1999, Vol. 1, pp. 1–5.

[9] J.J. Buckley, Th. Feuring and Y. Hayashi: Fuzzy Hierarchical Analysis, Proc. FUZZ-IEEE, Seoul, Korea, Aug. 22–25, 1999, Vol. 2, pp. 1009–1013.

[10] J.J. Buckley, Th. Feuring and Y. Hayashi: Fuzzy Hierarchical Analysis Revisited, European J. Operational Research, 129 (2001), pp. 48–64.

[11] J.J. Buckley, Th. Feuring and Y. Hayashi: Fuzzy Eigenvalues, Fuzzy Sets and Systems. Under revision.

[12] S.I. Grossman: Elementary Linear Algebra, Fifth Edition, Saunders, Forth Worth, Texas, 1994.

[13] E. Hansen: On the Solution of Linear Equations with Interval Coefficients, Linear Algebra and its Applications, 2 (1969), pp. 153–165.

[14] R.E. Moore: Methods and Applications of Interval Analysis, SIAM Studies in Applied Mathematics, Philadelphia, 1979.

[15] M. Wagenknecht, R. Hampel and V. Schneider: Solving Linear Fuzzy Equation Systems (LFES) by Inclusion, Proc. IPUM, Madrid, Spain, July 3–7, 2000, pp. 1758–1763.

[16] X. Wang and M. Ha: Solving a System of Fuzzy Linear Equations, in: M.Delgado, J.Kacprzyk, J.-L. Verdegay and M.A.Vila (eds.), Fuzzy Optimization, Physica-Verlag, Heidelberg, 1994, pp. 102–108.

[17] W. Wang, Z. Zhong and M. Ha: Iteration Algorithm for Solving a System of Fuzzy Linear Equations, Fuzzy Sets and Systems, 119 (2001), pp. 121–128.

Chapter 4

Fuzzy Mathematics in Finance

In this chapter we first consider the elementary concepts, in the mathematics of finance, future value, present value and regular annuities. In all cases the cash amounts, interest rates and number of compoundings may all be fuzzy. Then we look at two methods of comparing fuzzy net cash flows in order to rank fuzzy investment alternatives from best to worst. For other discussions of the mathematics of finance we refer the reader to ([1], [2], [10], [11], [14] – [25]). This chapter is based on ([5],[6],[7],[8]), and we will be using both triangular and trapezoidal (shaped) fuzzy numbers.

Let us explain our procedure for fuzzifying the elementary mathematics of finance. We first write down the mathematical expression for the problem in finance. Then we substitute fuzzy numbers for some, or all, the parameters in the expression. If in the fuzzy equation we need to solve for the value of some variable, we then solve using α–cuts and interval arithmetic producing the classical solution (if it exists). For example in the fuzzy equation $\overline{A}\,\overline{X} = \overline{B}$ we first solve for $\overline{X}$, given $\overline{A}$ and $\overline{B}$, as in Chapter 3, giving $\overline{X}_c$ for the classical solution. If $\overline{X}_c$ fails to exist, we then fuzzify the crisp solution giving solutions $\overline{X}_e$ and $\overline{X}_I$. In $\overline{A}\,\overline{X} = \overline{B}$, when $\overline{X}_c$ does not exist, we fuzzify $x = b/a$ to get $\overline{X} = \overline{B}/\overline{A}$. If we evaluate $\overline{B}/\overline{A}$ using the extension principle we obtain $\overline{X}_e$ and $\overline{X}_I$ computes $\overline{B}/\overline{A}$ using α–cuts and interval arithmetic.

The other possibility, after fuzzifying the original financial expression, is that all we need to do is evaluate it. Consider $\overline{X} = \overline{A}(1 + \overline{B})^n$ for fuzzy numbers $\overline{A}$, $\overline{B}$ and positive integer n. Given $\overline{A}$, $\overline{B}$ and n all we need to do is compute $\overline{X}$. This can be done in two ways: (1) using the extension principle producing $\overline{X}_e$; or (2) by α–cuts and interval arithmetic giving $\overline{X}_I$.

In the first case we can get $\overline{X}_c$, $\overline{X}_e$ and $\overline{X}_I$ and we expect $\overline{X}_c \leq \overline{X}_e \leq \overline{X}_I$. In the second case we have $\overline{X}_e$ and $\overline{X}_I$ with usually $\overline{X}_e \leq \overline{X}_I$. We prefer $\overline{X}_e$ to $\overline{X}_I$ and only use $\overline{X}_I$ to approximate $\overline{X}_e$ when it is very difficult to obtain

$\overline{X}_e$. If we need to compute $\overline{X}_c$, it is the preferred answer, when it exists.

4.1 Future Value

Assume an amount A is invested today at rate r per period for n periods. If S is the amount in the account after n periods, then $S = A(1+r)^n$.

Throughout this chapter we are dealing with compound interest. Interest rates are usually quoted as some percentage per year and then converted to the correct decimal rate per interest period. For example, 9% per year compounded monthly becomes $(0.09)/12 = 0.0075$ per month. We will assume the r is the interest rate, as a decimal, per interest period. So, $A = \$1000$ at 9% compounded monthly for 4 years produces $S = 1000(1+0.0075)^{48}$ since $n = 48 = 12 \cdot 4$ is the number of compoundings in 4 years.

We first fuzzify the compound interest formula by substituting $\overline{A}$ for A and $\overline{r}$ for r. We know $\overline{A} \geq 0$ and the support of $\overline{r}$ will be in $[0,1]$. The interest rate r may or may not be known exactly over the n periods so we can model it as a trapezoidal fuzzy number. For example, $\overline{r} = (0.0059/0.0067, 0.0075/0.0083)$ means the rate is approximately between 8 and 9% compounded monthly. The amount invested is usually known so $\overline{A}$ could be a crisp number. However, we will use a trapezoidal (or triangular) fuzzy number for $\overline{A}$. We wish to compute $\overline{S}$ where

$$\overline{S} = \overline{A}(1+\overline{r})^n \ . \tag{4.1}$$

Let $\overline{A} = (a_1/a_2, a_3/a_4)$, $\overline{r} = (r_1/r_2, r_3/r_4)$ with $\overline{A}[\alpha] = [a_1(\alpha), a_2(\alpha)]$, $\overline{r}[\alpha] = [r_1(\alpha], r_2(\alpha)]$. Before we compute $\overline{S}$ in equation (4.1) let us justify the fuzzy compound interest expression in (4.1).

At the end of the first period we have $\overline{A} + \overline{A}\ \overline{r} = \overline{A}(1+\overline{r})$ because, for positive fuzzy numbers, multiplication distributes over addition. After two periods

$$\overline{S} = \overline{A}(1+\overline{r}) + \overline{A}(1+\overline{r})\overline{r} \ , \tag{4.2}$$

or after factoring

$$\overline{S} = \overline{A}(1+\overline{r})^2 \ . \tag{4.3}$$

Hence, equation (4.1) is correct for fuzzy numbers.

There are two methods to evaluate equation (4.1). Using extension principle we obtain $\overline{S}_e$ whose α–cuts are

$$s_{e1}(\alpha) = \min\{a(1+r)^n \mid a \in \overline{A}[\alpha], r \in \overline{r}[\alpha]\} \ , \tag{4.4}$$

$$s_{e2}(\alpha) = \max\{a(1+r)^n \mid a \in \overline{A}[\alpha], r \in \overline{r}[\alpha]\} \ . \tag{4.5}$$

Clearly, the expression $a(1+r)^n$ is increasing in both a and r so that

$$s_{e1}(\alpha) = a_1(\alpha)(1+r_1(\alpha))^n \ , \tag{4.6}$$

$$s_{e2}(\alpha) = a_2(\alpha)(1+r_2(\alpha))^n \ , \tag{4.7}$$

$0 \leq \alpha \leq 1$.

The second method is to use α–cuts and interval arithmetic defining $\overline{S}_I$. So

$$\overline{S}_I[\alpha] = \overline{A}[\alpha](1 + \overline{r}[\alpha])^n \ , \tag{4.8}$$

and $\overline{S}_I = \overline{S}_e$ because all intervals are non–negative.

Now we fuzzify by substituting $\overline{A}$ for A, $\overline{r}$ for r and also $\overline{n}$ for n. The fuzzy number of periods $\overline{n}$ will be a non–negative discrete fuzzy set. That is, there are positive integers n_i, $1 \leq i \leq K$, for some positive integer K, and $\lambda_i \in (0, 1]$, $1 \leq i \leq K$, so that

$$\overline{n}(x) = \begin{cases} \lambda_i & : \ \text{if } x = n_i \ , \\ 0 & : \ \text{otherwise} \ . \end{cases} \tag{4.9}$$

The termination of the investment is uncertain and is modeled by $\overline{n}$. An interpretation of λ_i is "the possibility that $n = n_i$ is λ_i", $1 \leq i \leq K$. Using this "possibility" interpretation we then need $\lambda_i = 1$ for some i.

Use the extension principle to compute $\overline{S}_e$ in

$$\overline{S}_e = \overline{A}(1 + \overline{r})^{\overline{n}} \ . \tag{4.10}$$

We will not attempt to calculate $\overline{S}_I$ now because α-cuts of $\overline{n}$ will not be intervals but are subsets of $\{\lambda_1, \cdots, \lambda_K\}$. In this case we may not get $\overline{S}_e \leq \overline{S}_I$ and $\overline{S}_I$ is therefore not an approximation to $\overline{S}_e$. Since $\overline{S}_e$ is not difficult to compute (see the next theorem) we will only find $\overline{S}_e$ in this case.

We only use equations like (4.4) and (4.5) when the expression to be evaluated is continuous in all its variables. The expression $a(1 + r)^n$ is now discrete in n. The membership function for $\overline{S}_e$ is

$$\overline{S}_e(x) = max\{\pi(a, r, n) \mid a(1 + r)^n = x\} \ , \tag{4.11}$$

where

$$\pi(a, r, n) = \min\{\overline{A}(a), \overline{r}(r), \overline{n}(n)\} \ . \tag{4.12}$$

There is an easier way to get $\overline{S}_e$. Let $\overline{S}_{n_i} = \overline{A}(1 + \overline{r})^{n_i}$, $1 \leq i \leq K$.

Theorem 4.1 $\overline{S}_e(x) = \max_{1 \leq i \leq K} \{\min\{\overline{S}_{n_i}(x), \lambda_i\}\}$.

Proof 4.1 *Let the right hand side of the above equation be $\Gamma(x)$.*

1. *We first show $\overline{S}_e(x) \leq \Gamma(x)$. Assume $\overline{S}_e(x) = \gamma$. There exists a, r, $n = n_i$ so that $\pi(a, r, n_i) = \gamma$, $a(1 + r)^{n_i} = x$ and $\overline{A}(a) \geq \gamma$, $\overline{r}(r) \geq \gamma$ and $\lambda_i \geq \gamma$. Using $n = n_i$ let us call $\overline{S}_e$ by the name $\overline{S}_{n_i}$. From equations (4.4) and (4.5) we get $x \in \overline{S}_{n_i}[\gamma]$. So the minimum of $\overline{S}_{n_i}(x)$ and λ_i is greater than, or equal to γ. Hence, $\Gamma(x) \geq \gamma = \overline{S}_e(x)$.*

2. *Now show $\Gamma(x) \leq \overline{S}_e(x)$. Let $\Gamma(x) = \gamma$. There is an i between 1 and K so that $\min\{\overline{S}_{n_i}(x), \lambda_i\} = \gamma$. Hence $\overline{S}_{n_i}(x) \geq \gamma$ and $\lambda_i \geq \gamma$. So $x \in \overline{S}_{n_i}[\gamma]$ and there is an $a \in \overline{A}[\gamma]$, $r \in \overline{r}[\gamma]$ so that $x = a(1 + r)^{n_i}$. This means $\pi(a, r, n_i) \geq \gamma$ and $a(1 + r)^{n_i} = x$. Hence $\overline{S}_e(x) \geq \gamma$.*

Theorem 4.1 allows us to easily compute $\overline{A}(1+\overline{r})^{\overline{n}}$. We first find the $\overline{S}_{n_i}$ as in equations (4.6), (4.7) or equation (4.8), $1 \leq i \leq K$. Cut each $\overline{S}_{n_i}$ off at height λ_i, and then take the max of the resulting fuzzy sets.

Example 4.1.1

Let $\overline{A} = (90/100, 100/110)$ and $\overline{r} = (0.07/0.08, 0.09/0.10)$ with $\overline{n}$ given as follows: (1) $n_1 = 4$, $\lambda_1 = 0.6$; (2) $n_2 = 5$, $\lambda_2 = 1.0$; and (3) $n_3 = 6$, $\lambda_2 = 0.5$. $\overline{S}_e$ is shown in Figure 4.1. The individual trapezoidal shaped fuzzy numbers $\overline{S}_{n_i}$, $i = 1, 2, 3$, disappear when aggregated into $\overline{S}_e$. $\overline{S}_e$ may not be a fuzzy number.

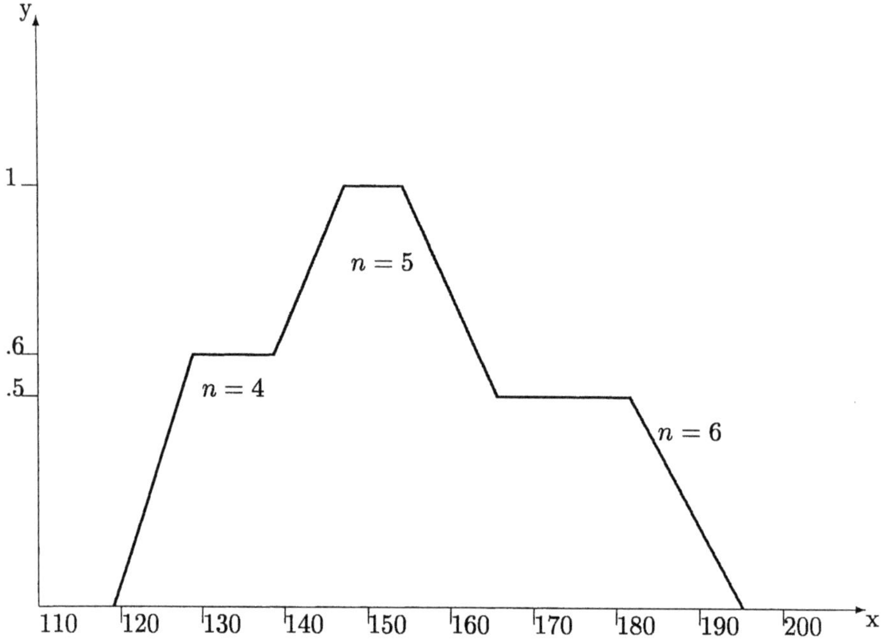

Figure 4.1: The Value $\overline{S}_e = \overline{A}(1+\overline{r})^{\overline{n}}$ in Example 4.1.1

4.2 Present Value

We wish to find the present value of a future amount A, n periods in the future, if r is the interest rate per period. Let S be the present value of A. There are two ways to compute S. The first method is that the present value of A equals S_1, if you can invest S_1 today, at rate r per period, so that in n periods it accumulates to A. That is, S_1 solves

$$S_1(1+r)^n = A \ . \tag{4.13}$$

The second method is to solve equation (4.13) for S_1 giving

$$S_2 = A(1+r)^{-n} \ . \tag{4.14}$$

In non–fuzzy mathematics $S_1 = S_2$, however the two methods may produce different results for fuzzy mathematics.

We now substitute $\overline{A}$ for A and $\overline{r}$ for r and solve $\overline{S}_1$ and $\overline{S}_2$. Quite often in finance future cash amounts A are estimated. One usually employs educated guesses, based on expected values or other statistical techniques, to obtain a value for A. Fuzzy mathematics allows an alternative to having to use an exact value for A. $\overline{A}$ will be a trapezoidal fuzzy number $(a_1/a_2, a_3/a_4)$ meaning that the future amount is approximately between \$ a_2 and \$ a_3.

We first solve for $\overline{S}_1$ and we may obtain three solutions $\overline{S}_{1c}$, $\overline{S}_{1e}$ and $\overline{S}_{1I}$.

The classical solution solves

$$\overline{S}_{1c}(1+\overline{r})^n = \overline{A} \ . \tag{4.15}$$

Taking α–cuts we obtain

$$[s_{1c1}(\alpha), s_{1c2}(\alpha)][w_1(\alpha), w_2(\alpha)] = [a_1(\alpha), a_2(\alpha)] \tag{4.16}$$

where

$$w_i(\alpha) = (1 + r_i(\alpha))^n \ , \tag{4.17}$$

$i = 1, 2$. Now $\overline{A} \geq 0$, so $\overline{S}_{1c} \geq 0$. We get

$$s_{1c1}(\alpha) = \frac{a_1(\alpha)}{(1+r_1(\alpha))^n} \ , \tag{4.18}$$

$$s_{1c2}(\alpha) = \frac{a_2(\alpha)}{(1+r_2(\alpha))^n} \ . \tag{4.19}$$

The "interval" $[s_{1c1}(\alpha), s_{1c2}(\alpha)]$ may, or may not, define a fuzzy number $\overline{S}_{1c}$

Continuing with $\overline{S}_1$ the solution $\overline{S}_{1e}$ fuzzifies $a/(1+r)^n$. Alfa–cuts of $\overline{S}_{1e}$ are

$$s_{1e1}(\alpha) = \min\left\{ \frac{a}{(1+r)^n} \mid a \in \overline{A}[\alpha], r \in \overline{r}[\alpha] \right\} , \tag{4.20}$$

$$s_{1e2}(\alpha) = \max\left\{ \frac{a}{(1+r)^n} \mid a \in \overline{A}[\alpha], r \in \overline{r}[\alpha] \right\} . \tag{4.21}$$

Then

$$\overline{S}_{1e}[\alpha] = \left[\frac{a_1(\alpha)}{(1+r_2(\alpha))^n}, \frac{a_2(\alpha)}{(1+r_1(\alpha))^n} \right] \ . \tag{4.22}$$

Lastly, $\overline{S}_{1I}[\alpha]$ comes from

$$\frac{\overline{A}[\alpha]}{(1+\overline{r}[\alpha])^n} \ , \tag{4.23}$$

after substituting the intervals for $\overline{A}[\alpha]$ and $\overline{r}[\alpha]$ and simplifying using interval arithmetic. We see that $\overline{S}_{1e} = \overline{S}_{1I}$.

Turning to $\overline{S}_2$ we have only $\overline{S}_{2e}$ and $\overline{S}_{2I}$. $\overline{S}_{2e} = \overline{A}(1+\overline{r})^{-n}$ using the extension principle. Hence, $\overline{S}_{2e} = \overline{S}_{1e}$. Also, $\overline{S}_{2I}$ is just equation (4.23). Therefore, $\overline{S}_{1e} = \overline{S}_{1I} = \overline{S}_{2e} = \overline{S}_{2I}$. $\overline{S}_{1c}$, if it exists, is the only solution that always satisfies equation (4.13), for $\overline{A}$ and $\overline{r}$ substituted for A and r, evaluated by α–cuts and interval arithmetic.

Since $\overline{S}_{1c}$ satisfies the present value equation our solution strategy is: (1) choose $\overline{S}_{1c}$ when it exists; and (2) use $\overline{S}_{1e} = \overline{S}_{2e}$ when $\overline{S}_{1c}$ does not exist.

Example 4.2.1

Let $\overline{A} = (8000/10000, 12000/14000)$ and $\overline{r} = (0.05/0.06, 0.06/0.07)$. If $n = 10$, then $\overline{S}_{1c}$ exists since

$$s_{1c1}(\alpha) = \frac{(8000 + 2000\alpha)}{(1.05 + 0.01\alpha)^{10}} \tag{4.24}$$

$$s_{1c2}(\alpha) = \frac{(14000 - 2000\alpha)}{(1.07 - 0.01\alpha)^{10}} \tag{4.25}$$

$0 \leq \alpha \leq 1$, defines the α–cuts of a trapezoidal shaped fuzzy number.

Now we allow n to be a positive discrete fuzzy set $\overline{n}$, since the number of interest periods into the future is uncertain. We have two equations to consider for present value of $\overline{A}$

$$\overline{S}_1(1+\overline{r})^{\overline{n}} = \overline{A}\ , \tag{4.26}$$

and

$$\overline{S}_2 = \frac{\overline{A}}{(1+\overline{r})^{\overline{n}}}\ , \tag{4.27}$$

from equations (4.13) and (4.14). We will not use equation (4.26) for two reasons: (1) $\overline{S}_{1c}$ may not exist; and (2) more importantly $(1+\overline{r})^{\overline{n}}$ may not be a fuzzy number (Example 4.1.1) so that α–cuts can be a union of intervals and then we cannot use standard interval arithmetic. When α–cuts of $(1+\overline{r})^{\overline{n}}$ are not a single interval, we cannot solve equation (4.26) for the α–cuts of $\overline{S}_{1c}$. So, we only have $\overline{S}_{2e}$ to consider. We know $\overline{S}_{1e} = \overline{S}_{2e}$ and we do not calculate $\overline{S}_{1I}$ and $\overline{S}_{2I}$ (no intervals for α–cuts of $\overline{n}$). From the extension principle

$$\overline{S}_{2e}(x) = \max\left\{\pi(a,r,n) \mid a(1+r)^{-n} = x\right\}, \tag{4.28}$$

for π given by equation (4.12). If we write $\overline{S}_{n_i} = \overline{A}(1+\overline{r})^{-n_i}$, then

Theorem 4.2 $\overline{S}_{2e}(x) = \max_{1 \leq i \leq K} \{\min\{\overline{S}_{n_i}(x), \lambda_i\}\}$.

Proof 4.2 *Same as Theorem 4.1.*

$\overline{S}_{2e}$ need not be a fuzzy number.

4.3 Annuities

We consider only ordinary (regular) annuities where: (1) the payment period equals the interest period; and (2) the equal periodic payments A are at the end of each period for n periods. We first look at the future value of an annuity and then the present value.

4.3.1 Future Value

The future value S of the regular annuity of n equal payments A at rate r per period is

$$S = A(1+r)^{n-1} + A(1+r)^{n-2} + \cdots + A(1+r) + A \ , \tag{4.29}$$

or

$$S = Aq(n,r) \ , \tag{4.30}$$

where

$$q(n,r) = \frac{(1+r)^n - 1}{r} \ . \tag{4.31}$$

Fuzzifying equation (4.29) we wish to find $\overline{S}$ where

$$\overline{S} = \overline{A}(1+\overline{r})^{n-1} + \overline{A}(1+\overline{r})^{n-2} + \cdots + \overline{A}(1+\overline{r}) + \overline{A} \ , \tag{4.32}$$

There are two solutions $\overline{S}_e$ and $\overline{S}_I$. $\overline{S}_e$ is based on the extension principle and its α–cuts are

$$s_{e1}(\alpha) = \min\left\{aq(n,r) \mid a \in \overline{A}[\alpha], r \in \overline{r}[\alpha] \ \right\}, \tag{4.33}$$

$$s_{e2}(\alpha) = \max\left\{aq(n,r) \mid a \in \overline{A}[\alpha], r \in \overline{r}[\alpha] \ \right\}. \tag{4.34}$$

Since S is an increasing function of a and r we obtain

$$\overline{S}_e[\alpha] = [a_1(\alpha)q(n, r_1(\alpha)), a_2(\alpha)q(n, r_2(\alpha))] \ , \tag{4.35}$$

$0 \leq \alpha \leq 1$. $\overline{A} \geq 0$ and $\overline{r}$ is in $[0,1]$.

Now $\overline{S}_I$ will not equal $\overline{S}_e$. $\overline{S}_I[\alpha]$ is $A[\alpha]q(n, \overline{r}[\alpha])$ and we obtain an α–cut of $\overline{S}_I$ as

$$\left[a_1(\alpha)\left(\frac{(1+r_1(\alpha))^n - 1}{r_2(\alpha)}\right), a_2(\alpha)\left(\frac{(1+r_2(\alpha))^n - 1}{r_1(\alpha)}\right)\right] \ . \tag{4.36}$$

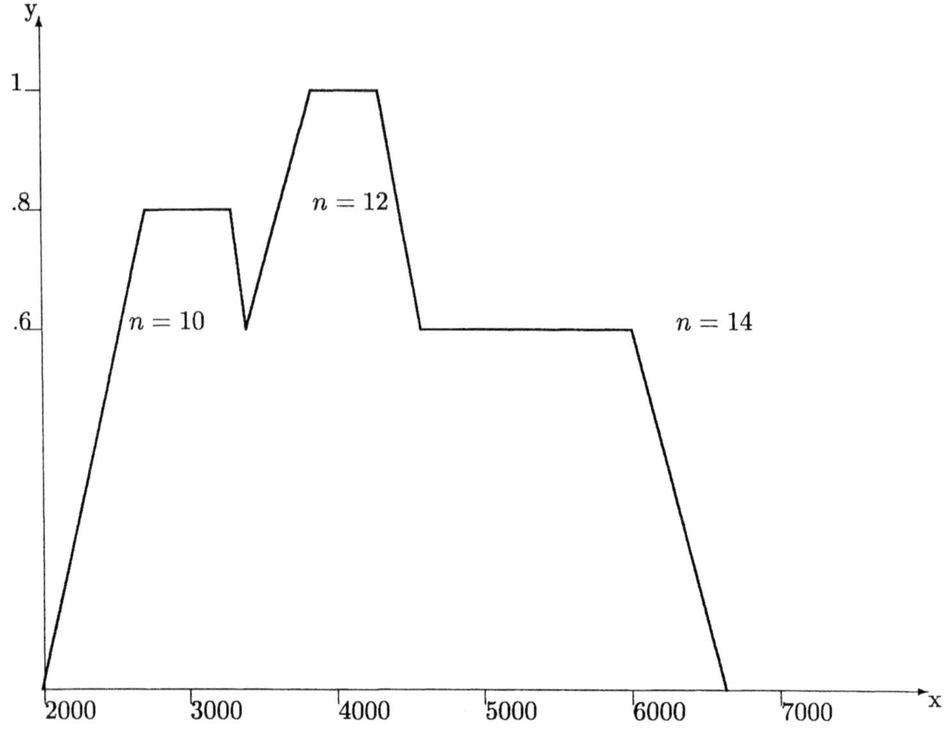

Figure 4.2: Future Value of a Fuzzy Annuity in Example 4.3.1.1

Clearly, $\overline{S}_e \leq \overline{S}_I$. We will use $\overline{S}_e$ as the solution.

Let $n = \overline{n}$, the termination date is fuzzy. Let $\overline{S}_e$ be $\overline{S}_{n_i}$ when we use $n = n_i$ in equation (4.32), $1 \leq i \leq K$. Then we obtain the result, as in Theorem 4.1 and 4.2, of

$$\overline{S}_e(x) = \max_{1 \leq i \leq K} \left\{ \min\{\overline{S}_{n_i}(x), \lambda_i\} \right\} . \tag{4.37}$$

Example 4.3.1.1

Let $\overline{A} = (150/190, 200/220)$, $\overline{r} = (0.06/0.09, 0.10/0.11)$ and $n = n_1 = 10$ for $\lambda_1 = 0.8$, $n = n_2 = 12$ for $\lambda_2 = 1.0$, $n = n_3 = 14$ for $\lambda_3 = 0.6$. The future value of this fuzzy annuity $\overline{S}_e$ is shown in Figure 4.2. It is not a fuzzy number.

4.3.2 Present Value

The present value S of a regular annuity is

$$S = A(1+r)^{-1} + A(1+r)^{-2} + \cdots + A(1+r)^{-n} , \tag{4.38}$$

or

$$S = A\beta(n,r) \ , \tag{4.39}$$

for

$$\beta(n,r) = \frac{1-(1+r)^{-n}}{r} \ . \tag{4.40}$$

The present value of a fuzzy annuity is

$$\overline{S} = \overline{A}(1+\overline{r})^{-1} + \overline{A}(1+\overline{r})^{-2} + \cdots + \overline{A}(1+\overline{r})^{-n} \ . \tag{4.41}$$

We simply fuzzify equation (4.39). Alpha–cuts of $\overline{S}_e$ are

$$s_{e1}(\alpha) = \min\left\{a\beta(n,r) \mid a \in \overline{A}[\alpha], r \in \overline{r}[\alpha]\right\} \ , \tag{4.42}$$

$$s_{e2}(\alpha) = \max\left\{a\beta(n,r) \mid a \in \overline{A}[\alpha], r \in \overline{r}[\alpha]\right\} \ . \tag{4.43}$$

Now S is an increasing function of A and a decreasing function of r so

$$\overline{S}_e[\alpha] = [a_1(\alpha)\beta(n,r_2(\alpha)), a_2(\alpha)\beta(n,r_1(\alpha))] \ , \tag{4.44}$$

$0 \leq \alpha \leq 1$. We again get $\overline{S}_e \subset \overline{S}_I$ and $\overline{S}_e \neq \overline{S}_I$. It is a good, and short, exercise to evaluate

$$\overline{A}[\alpha]\beta(n,\overline{r}(\alpha)) \ , \tag{4.45}$$

which is $\overline{S}_I[\alpha]$ and see $\overline{S}_e \subsetneq \overline{S}_I$. We suggest $\overline{S}_e$ as the present value.

With $n = \overline{n}$ fuzzy we obtain similar result as before (Theorem 4.1, 4.2 and equation (4.37)).

4.4 Portfolio Analysis

Two commonly used methods of comparing (mutually exclusive) investment alternatives are net present value (NPV) and the internal rate of return (IRR). Let $\mathcal{A} = A_0, A_1, \ldots, A_n$ be a given , or estimated, net cash flow of a proposed investment project over n periods of time. If an $A_i < 0$ then $-A_i$ is the net investment (total investment minus total return) made in the project at the end of the i–th period. If $A_i > 0$, then A_i is the net return (total return minus total investment) from the project at the end of the i–th period. We assume that $A_0 < 0$ because this is an investment project that always starts off with an initial investment. We also assume the time periods equal the interest period. We will first discuss the NPV method. See [10], [15] – [17], [21], [22] and [24] for other approaches to fuzzy portfolio analysis.

4.4.1 NPV Method

In the NPV method (present worth method, discounted present value method) one finds the present value of all future net returns discounted at the appropriate cost of capital to firm (r_0), minus any initial cash outlays. Therefore, NPVof the cash flow $\mathcal{A} = A_0, A_1, \ldots, A_n$ is

$$\mathrm{NPV}(\mathcal{A}, n) = \sum_{i=0}^{n} A_i(1+r_0)^{-i} . \tag{4.46}$$

Suppose we have a number of investment proposals under consideration with cash flows $\mathcal{A}$, $\mathcal{B}, \ldots$. One ranks these proposals with NPV from highest NPV to lowest NPV and selects investments in that order until the investment capital of the firm is exhausted. At first glance this procedure appears sound but do not forget that all future cash flows have to be estimated and that interest rate r_0 may also be estimated.

Therefore, we now consider a fuzzy cash flow $\overline{\mathcal{A}} = \overline{A}_0, \ldots, \overline{A}_n$ with $\overline{r}_0$ a fuzzy interest rate representing the cost of capital for the firm. $\overline{A}_0$ is a negative fuzzy number and the other $\overline{A}_i$ are either negative, or positive, fuzzy numbers. The fuzzy net present value of $\overline{\mathcal{A}}$ is

$$\mathrm{NPV}(\overline{\mathcal{A}}, n) = \overline{A}_0 + \sum_{i=1}^{n} \overline{A}_i(1+\overline{r}_0)^{-i} . \tag{4.47}$$

We first evaluate this expression using the extension principle giving $NPV_e(\overline{\mathcal{A}}, n)$ whose α–cuts are

$$npv_{e1}(\alpha) = \min\left\{a_0 + \sum_{i=1}^{n} a_i(1+r_0)^{-i} \;\middle|\; a_i \in \overline{A}_i[\alpha], 0 \le i \le n, r_0 \in \overline{r}_0[\alpha]\right\} , \tag{4.48}$$

$$npv_{e2}(\alpha) = \max\left\{a_0 + \sum_{i=1}^{n} a_i(1+r_0)^{-i} \;\middle|\; a_i \in \overline{A}_i[\alpha], 0 \le i \le n, r_0 \in \overline{r}_0[\alpha]\right\} . \tag{4.49}$$

We see that the expression $a_0 + \sum_{i=1}^{n} a_i(1+r_0)^{-i}$ is increasing in a_i, $0 \le i \le n$. If $\overline{A}_i[\alpha] = [a_{i1}(\alpha), a_{i2}(\alpha)]$, $0 \le i \le n$, then we substitute $a_{i1}(\alpha)$ ($a_{i2}(\alpha)$) for a_i to get $npv_{e1}(\alpha)$ ($npv_{e2}(\alpha)$). However, it is difficult to see what value of r_0 to use for $npv_{ei}(\alpha)$, $i = 1, 2$. So we need to design some numerical procedure in order to compute the α–cuts of $\mathrm{NPV}_e(\overline{\mathcal{A}}, n)$.

Next we employ α–cuts and interval arithmetic to evaluate equation (4.47) giving $\mathrm{NPV}_I(\overline{\mathcal{A}}, n)$. We need to remember two things about interval arithmetic in computing $\overline{A}_i(1+\overline{r}_0)^{-i}$. They are $[a_1, b_1] \cdot [a_2, b_2] = [a_1a_2, b_1b_2]$ for positive intervals but $[a_1, b_1][a_2, b_2] = [a_1b_2, b_1a_2]$ if $[a_1, b_1]$ is negative and $[a_2, b_2]$ is positive. Let $\overline{r}_0[\alpha] = [r_1(\alpha), r_2(\alpha)]$. So $\overline{A}_i[\alpha](1+\overline{r}_0[\alpha])^{-i}$ becomes

$$[a_{i1}(\alpha), a_{i2}(\alpha)][(1+r_2(\alpha))^{-i}, (1+r_1(\alpha))^{-i}] , \tag{4.50}$$

where $[a_{i1}(\alpha), a_{i2}(\alpha)]$ may be positive or negative.

Project $\overline{\mathcal{A}}$	Project $\overline{\mathcal{B}}$
$\overline{A}_0 = (-1100/-1000, -1000/-900)$	$\overline{B}_0 = (-1100/-1000, -1000/-900)$
$\overline{A}_1 = (450/500, 500/550)$	$\overline{B}_1 = (50/100, 100/150)$
$\overline{A}_2 = (350/390, 410/450)$	$\overline{B}_2 = (150/190, 210/250)$
$\overline{A}_3 = (250/280, 320/350)$	$\overline{B}_3 = (250/280, 320/350)$
$\overline{A}_4 = (150/170, 230/250)$	$\overline{B}_4 = (350/370, 430/450)$
$\overline{A}_5 = (50/60, 140/150)$	$\overline{B}_5 = (450/470, 530/550)$
	$\overline{B}_6 = (550/570, 630/650)$

Table 4.1: Fuzzy Net Cash Flows for Examples 4.4.1.1 and 4.4.1.2

Example 4.4.1.1

Two fuzzy net cash flows are given in Table 4.1. We will compute $\text{NPV}_e(\overline{\mathcal{A}}, 5)$ and $\text{NPV}_I(\overline{\mathcal{A}}, 5)$ for the $\overline{\mathcal{A}}$ in Table 4.1. Notice that the uncertainty about the values of the a_i grows as we go further into the future. For $\overline{r}_0$ we will use $(0.08/0.10, 0.10/0.12)$.

$\text{NPV}_I(\overline{\mathcal{A}}, 5)$ is shown in Figure 4.3. $\text{NPV}_I(\overline{\mathcal{A}}, 5)$ is easy to compute since $\overline{A}_i > 0$, $1 \le i \le 5$. The α–cuts of $\text{NPV}_I(\overline{\mathcal{A}}, 5)$ are

$$npv_{I1}(\alpha) = a_{01}(\alpha) + \sum_{i=1}^{n} a_{i1}(1 + r_2(\alpha))^{-i} , \qquad (4.51)$$

$$npv_{I2}(\alpha) = a_{02}(\alpha) + \sum_{i=1}^{n} a_{i2}(1 + r_1(\alpha))^{-i} . \qquad (4.52)$$

Now each $a_i \in \overline{A}_i[\alpha]$ is positive, $1 \le i \le n$, $\alpha \in [0,1]$, for project $\overline{\mathcal{A}}$ in Table 4.1. So the expression $a_0 + \sum_{i=1}^{n} a_i(1 + r_0)^{-i}$ is a decreasing function of r_0. Hence, equation (4.48) ((4.49)) becomes equation (4.51) ((4.52)) and $\text{NPV}_e(\overline{\mathcal{A}}, 5) = \text{NPV}_I(\overline{\mathcal{A}}, 5)$. The same happens for project $\overline{\mathcal{B}}$ in Table 4.1. For this reason $\text{NPV}_I(\overline{\mathcal{A}}, 5)$ is labeled $\text{NPV}_e(\overline{\mathcal{A}}, 5)$ in Figure 4.3.

In Figure 4.3 $\text{NPV}_e(\overline{\mathcal{A}}, 5)$ appears as a trapezoidal fuzzy number, but in fact it is a trapezoidal shaped fuzzy number which deviates very slightly from a trapezoidal fuzzy number.

The termination date of the project may be uncertain so that $n = \overline{n}$ becomes fuzzy. Recall $\overline{n}$ takes on values $n_1 < n_2 < \cdots < n_K$ with possibilities $\lambda_1, \ldots, \lambda_K$. We assume we have a fuzzy cash flow $\overline{\mathcal{A}} = \overline{A}_0, \ldots, \overline{A}_N$ where $N = \overline{n}$. We only have $\text{NPV}_e(\overline{\mathcal{A}}, \overline{n})$ now. The net present value is found using the extension principle. As before (Theorem 4.1, 4.2 and equation (4.37)) we obtain for $\overline{A}_i > 0$, $1 \le i \le n$ [5].

$$\text{NPV}_e(\overline{\mathcal{A}}, \overline{n}) = \max_{1 \le i \le K} \{\min\{\text{NPV}_e(\overline{\mathcal{A}}, n_i), \lambda_i\}\} . \qquad (4.53)$$

Now suppose we have a number of investment alternatives under consideration with fuzzy cash flows $\overline{\mathcal{A}}$, $\overline{\mathcal{B}}$, $\overline{\mathcal{C}}, \ldots$, with fixed termination dates $n_a, n_b, n_c, \ldots$, respectively. We will use the extension principle of determining their net present values. So we compute the fuzzy numbers $\mathrm{NPV}_e(\overline{\mathcal{A}}, n_a)$, $\mathrm{NPV}_e(\overline{\mathcal{B}}, n_b)$, $\mathrm{NPV}_e(\overline{\mathcal{C}}, n_c), \ldots$. We need to do two things: (1) discard a project $\overline{D}$ unless $\mathrm{NPV}_e(\overline{D}, n_d) > \overline{0}$, where $\overline{0}$ is an appropriate fuzzy zero; and (2) rank all the alternatives, not discarded, from "best" to "worst". An appropriate fuzzy zero depends on the amounts of money in the cash flows. For example, if the amounts are in the millions, $\overline{0} = (-100000/0.0, 0.0/100000)$ could be a choice for fuzzy zero.

Ranking fuzzy numbers has always been an important topic in fuzzy set theory. If $\overline{M}$ and $\overline{N}$ are two fuzzy numbers there have been numerous proposals for defining $\overline{M} < \overline{N}$, $\overline{M} \leq \overline{N}$. See [12] for an excellent review of this literature up to 1992. There continues to be many new proposals published every year. A recent review of ranking techniques is in [13]. We will employ our method discussed in [3] and [4]. Define

$$v(\overline{M} \geq \overline{N}) = \max_{x \geq y} \{\min\{\overline{M}(x), \overline{N}(y)\}\} \ . \tag{4.54}$$

We write $\overline{M} > \overline{N}$ if $v(\overline{M} \geq \overline{N}) = 1$ and $v(\overline{N} \geq \overline{M}) < \theta$ where θ is some fixed value in $(0, 1]$. We have usually used $\theta = 0.8$. If $\overline{M} > \overline{N}$ is not true and $\overline{N} > \overline{M}$ is not true we write $\overline{M} \approx \overline{N}$. $\overline{M} \geq \overline{N}$ means $\overline{M} > \overline{N}$ or $\overline{M} \approx \overline{N}$. To rank a collection $\overline{M}_i$, $1 \leq i \leq m$, of fuzzy numbers we construct the non–empty disjoint subsets H_i, $1 \leq i \leq L$, of $M = \{\overline{M}_i | 1 \leq i \leq m\}$ so that the union of the H_i, $1 \leq i \leq L$, is M. The highest ranked fuzzy numbers belong to H_1, the next highest belong to H_2, etc. Now H_1 contains all the undominated $\overline{M}_i$ in M. That is, $\overline{M}_i \in H_1$ if $\overline{M}_j > \overline{M}_i$ is not true, $1 \leq j \leq m$. H_2 is all the undominated $\overline{M}_i$ in $M - H_1$, etc. Properties of $>$, $\geq$, $\approx$, and the H_i may be found in [3] and [4]. We then use this method to rank the undiscarded net present values.

Example 4.4.1.2

We compare the two fuzzy cash flows in Table 4.1. We use the same value of $\overline{r}_0$ as in Example 4.4.1.1. The graph of $\mathrm{NPV}_e(\overline{\mathcal{A}}, 5)$ and $\mathrm{NPV}_e(\overline{\mathcal{B}}, 6)$ are in Figure 4.3. Recall that $\mathrm{NPV}_I(\overline{\mathcal{A}}, 5) = \mathrm{NPV}_e(\overline{\mathcal{A}}, 5)$ and $\mathrm{NPV}_I(\overline{\mathcal{B}}, 6) = \mathrm{NPV}_e(\overline{\mathcal{B}}, 6)$ for Table 4.1. A reasonable $\overline{0}$ would be $(-200/0, 0/200)$ and we see, using $\theta = 0.8$, that both net present values exceed $\overline{0}$. Also from Figure 4.3 using the horizontal line at $\alpha = 0.8$, we conclude $\mathrm{NPV}_e(\overline{\mathcal{B}}, 6) \approx \mathrm{NPV}_e(\overline{\mathcal{A}}, 5)$. In fact we calculated that : (1) the height of the intersection of $\overline{0}$ and $\mathrm{NPV}_e(\overline{\mathcal{A}}, 5)$ on the interval $[-117, 200]$ was $0.69 < \theta = 0.8$; and (2) the height on the intersection of $\mathrm{NPV}_e(\overline{\mathcal{A}}, 5)$ and $\mathrm{NPV}_e(\overline{\mathcal{B}}, 6)$ on the interval $[-1.4, 559]$ was $0.92 > \theta = 0.8$.

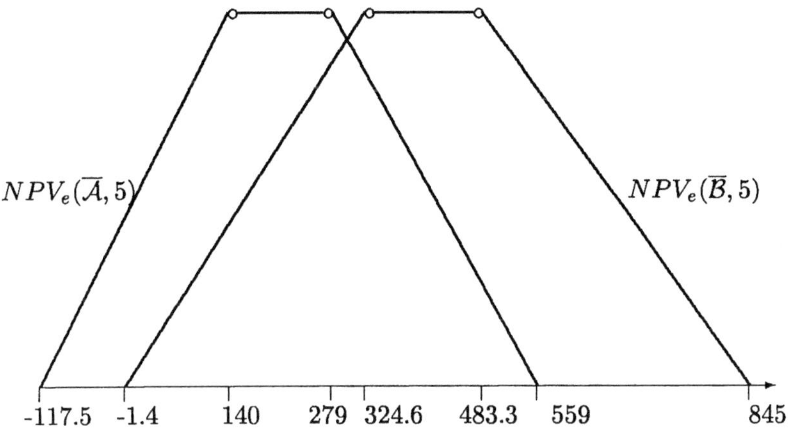

Figure 4.3: Fuzzy Net Present Values in Examples 4.4.1.1 and 4.4.1.2

If the termination dates are uncertain then we have $\mathrm{NPV}_e(\overline{A}, \overline{n}_a)$, $\mathrm{NPV}_e(\overline{B}, \overline{n}_b), \ldots$. When the fuzzy cash flow numbers are positive, $i \geq 1$, we may use equation (4.53) to get the net present value. Otherwise, the extension principle must be used. The net present value need not be a fuzzy number but we may still use the ranking method discussed above to rank the projects from best H_1, to worst H_L.

4.4.2 IRR Method

We next turn to the IRR method of ranking investment alternatives. Given a future cash flow $\mathcal{A} = A_0, \ldots, A_n$ the internal rate of return $\mathrm{IRR}(\mathcal{A}, n)$ is any solution $r > -1$ to

$$\sum_{i=1}^{n} A_i(1+r)^{-i} = -A_0 \ . \qquad (4.55)$$

We have assumed that $A_0 < 0$, being the initial cash outlay, so $-A_0$ is positive. If the cash flow has one sign change, then equation (4.55) has a unique solution for $r > -1$. If the cash flow has more than one sign change, then equation (4.55) may have no solution or it could have multiple solutions. When a cash flow has multiple internal rates of return, the IRR method is considered not applicable to ranking investment alternatives. Assuming all projects have unique IRR, then those projects with IRR $> r_0$ are ranked from highest to lowest according to their IRR value. The firm then accepts these projects in this order until its investment capital is depleted. If it is found that conflicts exist between the NPVand IRR procedures, then one should consult the financial literature to see what the experts recommend. See [9], and the references in this paper, for more details on the IRR technique.

Now assume the cash flow $\overline{A} = \overline{A}_0, \overline{A}_1, \ldots, \overline{A}_n$ is fuzzy. $\overline{A}_0$ is a negative fuzzy number and the other $\overline{A}_i$ may be positive or negative fuzzy numbers. The fuzzy internal rate return IRR$(\overline{A}, n)$ is a fuzzy interest rate $\overline{r}$ that makes the present value of all future cash amounts equal to the initial cash outlay. Therefore, the fuzzy number $\overline{r}$ satisfies

$$\sum_{i=1}^{n} \overline{A}_i(1+\overline{r})^{-i} = -\overline{A}_0 \; . \tag{4.56}$$

We need to solve this equation, using α–cuts and interval arithmetic , for $\overline{r}_c = \mathrm{IRR}_c(\overline{A}, n)$ (if it exists). However, too often $\overline{r}_c$ will not exist.

Example 4.4.2.1

Let $-\overline{A}_0 = (100/110, 110/120)$, $\overline{A}_1 = \overline{A}_2 = (190/200, 300/310)$. Then $(-\overline{A}_0)[\alpha] = [100 + 10\alpha, 120 - 10\alpha]$, $\overline{A}_1[\alpha] = \overline{A}_2[\alpha] = [190 + 10\alpha, 310 - 10\alpha]$. If $\overline{r}_c[\alpha] = [r_{c1}(\alpha), r_{c2}(\alpha)]$, then we get

$$(190+10\alpha)((1+r_{c2}(\alpha))^{-1} + (1+r_{c2}(\alpha))^{-2}) = 100+10\alpha \; , \tag{4.57}$$
$$(310-10\alpha)((1+r_{c1}(\alpha))^{-1} + (1+r_{c1}(\alpha))^{-2}) = 120-10\alpha \; . \tag{4.58}$$

Let $\alpha = 1$ and solve for $r_{c2}(1)$ and $r_{c1}(1)$. The equations are

$$(1+r_{c2}(1))^{-1} + (1+r_{c2}(1))^{-2} = \frac{11}{20} \; , \tag{4.59}$$
$$(1+r_{c1}(1))^{-1} + (1+r_{c1}(1))^{-2} = \frac{11}{30} \; . \tag{4.60}$$

Solution is $r_{c2}(1) = 1.535 < 2.505 = r_{c1}(1)$. But for a fuzzy number $\overline{r}_c$ we must have $r_{c1}(1) \leq r_{c2}(1)$. Hence, $\overline{r}_c$ does not exist.

Next we try to find $\overline{r}_e = \mathrm{IRR}_e(\overline{A}, n)$. But we would need a formula for r to fuzzify. Solve equation (4.55) for r giving

$$r = f(A_0, A_1, \ldots, A_n) \; . \tag{4.61}$$

However, there is no general solution for r. We need to take a different approach to get $\overline{r}_e$. Define

$$\Omega(\alpha) = \left\{ r \mid \sum_{i=1}^{n} a_i(1+r)^{-i} = -a_0, a_i \in \overline{A}_i[\alpha], 0 \leq i \leq n \right\} . \tag{4.62}$$

Under certain circumstances $\Omega(\alpha)$ defines the α–cuts of a fuzzy number $\overline{r}_e$ ([7], Theorem 3). What is needed is for the internal rate of return equation (4.55) to produce a unique solution. We know if $a_i > 0$, $1 \leq i \leq n$, then there is a unique solution. So, if $\overline{A}_i > 0$, $1 \leq i \leq n$, and $\overline{A}_0 < 0$, then

$$\sum_{i=1}^{n} a_i(1+r)^{-i} = -a_0 , \tag{4.63}$$

has a unique solution for r for all $a_i \in \overline{A}_i[\alpha]$, $0 \leq i \leq n$, $0 \leq \alpha \leq 1$. In this case $\Omega(\alpha)$ defines the α–cuts of a fuzzy internal rate of return number $\mathrm{IRR}_e(\overline{\mathcal{A}}, n)$. We also know that $\mathrm{IRR}_c(\overline{\mathcal{A}}, n) \leq \mathrm{IRR}_e(\overline{\mathcal{A}}, n)$ ([7],Theorem 4) when $\overline{r}_c$ exists.

Example 4.4.2.2

Let us use the fuzzy cash flow $\overline{\mathcal{A}}$ in Table 4.1. We try to find $\overline{r}_c = \mathrm{IRR}_c(\overline{\mathcal{A}}, 5)$. But, as in Example 4.4.2.1, we discover $r_{c2}(1) < r_{c1}(1)$ so $\overline{r}_c$ does not exist.

$\overline{\mathcal{A}}$ satisfies the assumptions for $\Omega(\alpha)$ to give the α–cuts of $\overline{r}_e = \mathrm{IRR}_e(\overline{\mathcal{A}}, n)$. Let the α–cuts of $\mathrm{IRR}_e(\overline{\mathcal{A}}, 5)$ be $[irr_{e1}(\alpha), irr_{e2}(\alpha)]$. Since $\overline{A}_i > 0$, $1 \leq i \leq n$, we compute these α–cuts as follows: (1) $irr_{e1}(\alpha)$ is the unique solution for r in equation (4.63) using $a_{i1}(\alpha)$ for a_i , $1 \leq i \leq n$ and $a_{02}(\alpha)$ for a_0; and (2) $irr_{e2}(\alpha)$ is the unique solution for r in equation (4.63) using $a_{i2}(\alpha)$ for a_i, $1 \leq i \leq n$ and $a_{01}(\alpha)$ for a_0.

Next we consider $\overline{\mathcal{B}}$ in Table 4.1. Again $\mathrm{IRR}_c(\overline{\mathcal{B}}, 6)$ does not exist. We see $\overline{B}_i > 0$, $1 \leq i \leq 6$, and $\overline{B}_0 < 0$ so $\mathrm{IRR}_e(\overline{\mathcal{B}}, 6)$ exists with α–cuts given by $\Omega(\alpha)$, substituting $\overline{B}_i$ for $\overline{A}_i$. We also found α–cuts of $\mathrm{IRR}_e(\overline{\mathcal{B}}, 6)$, the same method as was used for project $\overline{\mathcal{A}}$ discussed above, and both fuzzy internal rates of return are in Figure 4.4. We see that $\mathrm{IRR}_e(\overline{\mathcal{A}}, 5) \approx \mathrm{IRR}_e(\overline{\mathcal{B}}, 6)$.

If we have a number of fuzzy cash flows $\overline{\mathcal{A}}$, $\overline{\mathcal{B}}$, $\overline{\mathcal{C}}, \ldots$ we compute $\mathrm{IRR}_e(\overline{\mathcal{A}}, n_a)$, $\mathrm{IRR}_e(\overline{\mathcal{B}}, n_b)$, $\mathrm{IRR}_e(\overline{\mathcal{C}}, n_c), \ldots$ and then rank the proposals using these fuzzy numbers. We probably would not be using IRR_c since we expect it not to exist for at least one of the fuzzy cash flows. However, we need to discard a proposal $\overline{\mathcal{D}}$ if $\mathrm{IRR}_e(\overline{\mathcal{D}}, n_d) > \overline{r}_0$ is not true, where $\overline{r}_0$ is the cost of capital to the firm. In Example 4.4.2.2 let $\overline{r}_0 = (0.08/0.10, 0.10/0.12)$ from Example 4.4.1.1. Then we see that $\mathrm{IRR}_e(\overline{\mathcal{B}}, 6) > \overline{r}_0$ from Figure 4.4 and

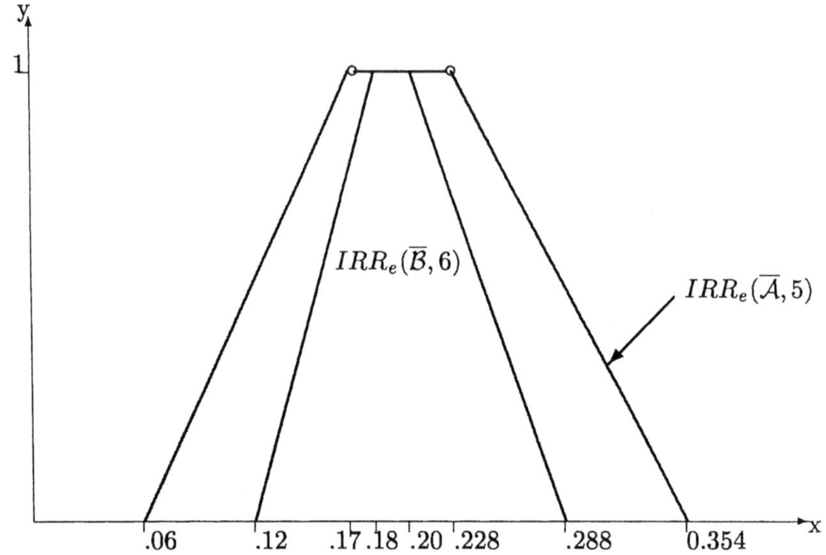

Figure 4.4: Fuzzy Internal Rates of Return in Example 4.4.2.2

we calculate that the height on the intersection of $\overline{r}_0$ and $\mathrm{IRR}_e(\overline{\mathcal{A}}, 5)$ on the interval $[0.0, 0.12]$ is 0.45 so that $\mathrm{IRR}_e(\overline{\mathcal{A}}, 5) > \overline{r}_0$ also.

We do not compute $\mathrm{IRR}_I(\overline{\mathcal{A}}, n_a)$. What if the termination date is uncertain so that $n = \overline{n}$ is fuzzy? The extension principle can be used to get the membership function of $\overline{r}_e = \mathrm{IRR}_e(\overline{\mathcal{A}}, \overline{n})$ as follows

$$\overline{r}_e(x) = \max\left\{ \pi(a_0, \ldots, a_n, n) \mid \sum_{i=1}^{n} a_i(1+x)^{-i} = a_0 \right\}, \tag{4.64}$$

for

$$\pi(a_0, \ldots, a_n, n) = \min\{\, \overline{A}_0(a_0), \ldots, \overline{A}_n(a_n), \overline{n}(n) \,\}\ . \tag{4.65}$$

If $\overline{A}_i > 0$, $1 \le i \le N$, $N = \overline{n}$, and $\overline{A}_0 < 0$, the expression for $\overline{r}_e(x)$ can be simplified as in Theorems 4.1.1 and 4.2.1 and equation (4.37).

4.5 Summary and Conclusions

In the first three sections we looked at future value, present value, future value of a regular annuity and the present value of a regular annuity when the cash amounts, interest rates and number of compounding can all be fuzzy. Then in the last section we fuzzified the net present value method (NPV) and the internal rate of return method (IRR), which are two procedures used to compare investment alternatives. This seems a natural application of fuzzy sets because both methods (NPV and IRR) involve estimated future cash flows which can easily be modeled as fuzzy numbers. The examples of fuzzy NPV and IRR used Table 4.1 where the future fuzzy cash amounts were all positive ($i \geq 1$) which made the computations relatively easy. If some of the future fuzzy cash amounts are negative, then the computations become much more difficult, and we would probably employ our evolutionary algorithm to obtain approximate solutions.

Bibliography

[1] C. von Altrock: Fuzzy Logic and NeuroFuzzy Applications in Business and Finance, Prentice Hall, Upper Saddle River, N.J., 1997.

[2] M. Bojadziev and G. Bojadziev: Fuzzy Logic for Business, Finance and Management, World Scientific Press, Singapore, 1997.

[3] J.J. Buckley: Ranking Alternatives using Fuzzy Numbers, Fuzzy Sets and Systems, 15 (1985), pp. 21–31.

[4] J.J. Buckley: Fuzzy Hierarchical Analysis, Fuzzy Sets and Systems, 17 (1985), pp. 233–247.

[5] J.J. Buckley: Fuzzy Mathematics of Finance, Fuzzy Sets and Systems, 21 (1987), pp. 257–273.

[6] J.J. Buckley: Portfolio Analysis Using Possibility Distributions, in: E. Sanchez and L. Zadeh (Eds.): Approximate Reasoning in Intelligent Systems, Decision and Control, Pergamon Press, Oxford, UK, 1987, pp. 69–76.

[7] J.J. Buckley: Solving Fuzzy Equations in Economics and Finance, Fuzzy Sets and Systems, 48 (1992), pp. 289–296.

[8] J.J. Buckley: Fuzzy Mathematics of Finance, in: D. Dubois, H. Prade and R. Yager (Eds.), Fuzzy Sets for Intelligent Systems, Morgan Kaufman, San Francisco, 1993, pp. 840–848.

[9] J.J. Buckley and R.A. Fleck: Determining the Number of Internal Rats of Return, Computer Journal, 21 (1978), pp. 373–377.

[10] C. Carlsson and R. Fuller: Capital Budgeting Problems with Fuzzy Cash Flows, Mathware and Soft Computing, 6 (1999), pp. 81–89.

[11] M. Li Calzi: Towards a General Setting for the Fuzzy Mathematics of Finance, Fuzzy Sets and Systems, 35 (1990), pp. 265–280.

[12] S.J. Chen and C.-L. Hwang: Fuzzy Multiple Attribute Decision Making: Methods and Applications, Springer-Verlag, N.Y., 1992.

[13] D. Dubois and H. Prade (eds.): Fundamentals of Fuzzy Sets, Kluwer Academic Publishers, Boston, 2000, Chapter 10 "Fuzzy Interval Analysis" by D.Dubois, E.Kerre, R.Mesiar and H.Prade, pp. 483–582.

[14] M. Fedrizzi, M. Fedrizzi and W. Ostasiewicz: Towards Fuzzy Modelling in Economics, Fuzzy Sets and Systems, 54 (1993), pp. 259–268.

[15] C. Fourali: Fuzzy Logic and the Quality of Assessment of Portfolios, Fuzzy Sets and Systems, 68 (1994), pp. 123–140.

[16] C. Kahraman and E. Tolga: Capital Budgeting Techniques Under Fuzzy Information, Proc. Tenth Int. Conf. Tools with Artifical Intelligence, Nov. 10–12, 1998, Taipei, Taiwan, pp. 242–249.

[17] D. Kuchta: Fuzzy Capital Budgeting, Fuzzy Sets and Systems, 111 (2000), pp. 367–385.

[18] R. Kuo, L. Lee and C. Lee: Intelligent Stock Trading Decision Support System through the Integration of Artificial Neural Networks and Fuzzy Delphi Models, J. Computational Intelligence in Finance, 6.2 (1998), pp. 24–34.

[19] Y.-J. Lai and C.-L. Hwang: Possibilistic Linear Programming for Managing Interest Rate Risk, Fuzzy Sets and Systems, 54 (1993), pp. 135–146.

[20] M. Mohammadian, M. Kingham and B. Bignall: Hierarchical and Feed-Forward Fuzzy Logic Systems for Interest Rate Prediction, J. Computational Intelligence in Finance, 6.3 (1998), pp. 5–11.

[21] R. Östermark: A Fuzzy Control Model for Dynamic Portfolio Management, Fuzzy Sets and Systems, 78 (1996), pp. 243–254.

[22] M.A. Parra, A.B. Terol and M.V.R. Ura: A Fuzzy Goal Programming Approach to Portfolio Selection, European J. Operational Research, 133 (2001), pp. 287–297.

[23] R.A. Ribeiro, H.-J. Zimmermann, R.R. Yager and J. Kacprzyk (eds.): Soft Computing in Financial Engineering, Physica-Verlag, Heidelberg, 1999.

[24] H. Tanaka, P. Guo and I.B. Turksen: Portfolio Selection Based on Fuzzy Probabilities and Possibilitiy Disributions, Fuzzy Sets and Systems, 111 (2000), pp. 387–389.

[25] M. Tarrazo: Practical Applications of Approximate Equations in Finance and Economics, Quorum Books, Westport, Conn., 2001.

Chapter 5

Fuzzy Non–Linear Regression

This chapter is concerned with fuzzy non–linear regression. In fuzzy linear regression one tries to find the "best" values of triangular (trapezoidal) shaped fuzzy numbers $\overline{A}$ and $\overline{B}$ so that the fuzzy linear function $\overline{Y} = \overline{A}\ \overline{X} + \overline{B}$ "explains" the fuzzy data $(\overline{X}_i, \overline{Z}_i)$, $1 \leq i \leq p$. In fuzzy non–linear regression we are looking for a fuzzy polynomial, or a fuzzy exponential, or a fuzzy logarithmic, ..., function that "explains" the data.

There have been numerous articles on fuzzy linear regression. For example, see ([3]–[13],[16], and [18] which is multivariate) and the references in these publications. In contrast to fuzzy linear regression, there have been only a few articles on fuzzy non–linear regression. See ([3], [10], [14],[19] and [20]), and the references in these papers, for a survey of fuzzy non–linear regression. What these researchers ([19], [20]) in fuzzy non–linear regression were concerned about was data of the form $(x_i, \overline{Z}_i)$, $1 \leq i \leq p$, for the x_i crisp (non–fuzzy), generated by some unknown non–linear function and they used a neural net to model the unknown function. In our approach our data is completely fuzzy and we are looking for an explicit function (polynomial, exponential, logarithmic, ...) to model the unknown non–linear function. In [14] the authors used the regression model $y = (a + b\exp(-x))^{-1}$ for fuzzy number data $(\overline{X}_i, \overline{Z}_i)$, $1 \leq i \leq p$. The object was to find a and b to minimize a special error function so that $a + b\exp(\overline{X}_i) \approx 1/\overline{Z}_i$, for all i.

We will use an evolutionary algorithm (EA) in fuzzy non–linear regression. A few authors have employed genetic algorithms in fuzzy regression. In [15] the authors use a genetic algorithm to classify the input data for fuzzy linear regression. Also, in [17] the authors use a genetic algorithm to design a fuzzy rule based system to approximate the values of some unknown, crisp, function. Paper [19] also employs a genetic algorithm. We plan to use an EA to choose the "best" fuzzy function, from a library of fuzzy functions, to explain the fuzzy data.

In the first section of this chapter we discuss univariate (one independent variable) fuzzy non–linear regression followed by multivariate fuzzy non–linear regression in the second section. We will use mostly non–negative fuzzy numbers in this chapter. We restrict ourselves to non–negative fuzzy numbers when we want algebraic expressions to have the same value using the extension principle or using α–cuts and interval arithmetic. For example, $\overline{Y} = \overline{A}\,\overline{X}^2 + \overline{B}\,\overline{X} + \overline{C}$ produces the same $\overline{Y}$ using the extension principle, or α–cuts and interval arithmetic, when $\overline{A}$, $\overline{X}$, $\overline{B}$ and $\overline{C}$ are all non–negative.

What is normally done in regular (crisp) linear regression is to derive formulas for the estimates of the parameters in the regression model so that you can construct confidence intervals, do hypothesis testing, etc. Some of this can also be done in fuzzy linear regression. However, in our fuzzy non–linear regression we have no formulas for the parameters so we are unable to do any "statistical" analysis of the regression models. We use a directed search algorithm to find the "best" member in the library to model the data.

Section 5.1 is based on [1] and Section 5.2 is based on [2]. We use mostly triangular (shaped) fuzzy numbers in 5.1 but trapezoidal (shaped) fuzzy numbers are used in 5.2.

5.1 Univariate Non–Linear Fuzzy Regression

Let $\mathcal{T}$denote all triangular fuzzy numbers and $\mathcal{F}_0$ is all triangular shaped fuzzy numbers. Also, let $\mathcal{F}$ be all fuzzy numbers. We shall not be working a lot with general fuzzy numbers and all we need to know about $\overline{N}$ in $\mathcal{F}$ is that its membership function $y = \overline{N}(x)$ need not be continuous but $\overline{N}[\alpha]$ is always a closed, bounded, interval. A function $F : \mathcal{T} \rightarrow \mathcal{F}$ will be written $\overline{Y} = F(\overline{X}; \overline{K}_1, \ldots, \overline{K}_n)$ where $\overline{X}$ is the variable in $\mathcal{T}$ and the $\overline{K}_i$, $1 \leq i \leq n$, are constants also in $\mathcal{T}$. For example, the fuzzy quadratic $\overline{Y} = \overline{K}_2\,\overline{X}^2 + \overline{K}_1\,\overline{X} + \overline{K}_0$ is such a function. $\overline{Y}$ will be in $\mathcal{F}$ and possibly also in $\mathcal{F}_0$. Ω will be some fixed collection of $F : \mathcal{T} \rightarrow \mathcal{F}$ called our library. We will discuss the composition of Ω later in this and the next section.

The data is $(\overline{X}_i, \overline{Z}_i)$, $1 \leq i \leq p$, with $\overline{X}_i \in \mathcal{T}$ and $\overline{Z}_i \in \mathcal{F}$. The restricted fuzzy non–linear regression problem is to find the "best" $F \in \Omega$ that models (explains) the data. To define "best" we need to define a distance measure on $\mathcal{F}$. If $\overline{N}$, $\overline{M}$ are in $\mathcal{F}$, the distance between $\overline{M}$ and $\overline{N}$ will be taken to be

$$D(\overline{M}, \overline{N}) = \sup_{0 \leq \alpha \leq 1} \{\max\{|m_1(\alpha) - n_1(\alpha)|, |m_2(\alpha) - n_2(\alpha)|\}\} \;, \tag{5.1}$$

where $\overline{M}[\alpha] = [m_1(\alpha), m_2(\alpha)]$, $\overline{N}[\alpha] = [n_1(\alpha), n_2(\alpha)]$. We measure the "best" through the error function

$$E(F) = \frac{1}{p} \sum_{i=1}^{p} D^2(\overline{Y}_i, \overline{Z}_i) \;, \tag{5.2}$$

for $F \in \Omega$, $\overline{Y}_i = F(\overline{X}_i; \overline{K}_1, \ldots, \overline{K}_n)$, $1 \leq i \leq p$. The restricted (to Ω) fuzzy non-linear regression problem is to find $F^* \in \Omega$ so that

$$\inf_{F \in \Omega} (E(F)) = E(F^*) . \tag{5.3}$$

Now we describe the basic library. First, set Ω_1 to be all $\overline{K}_2 \, \overline{X} + \overline{K}_1$ for $\overline{K}_1, \overline{K}_2 \in \mathcal{T}$. Ω_1 contains all linear fuzzy functions. Ω_2 will be all

$$\overline{K}_n \, \overline{X}^n + \overline{K}_{n-1} \, \overline{X}^{n-1} + \cdots + \overline{K}_1 \, \overline{X} + \overline{K}_0 , \tag{5.4}$$

for $\overline{K}_i \in \mathcal{T}$, $0 \leq i \leq n$, and $n \leq \Delta$, Δ some positive integer. Ω_2 contains all fuzzy polynomials of degree less than, or equal to, Δ. Ω_3 is all

$$\overline{K}_3 \exp(\overline{K}_2 \, \overline{X}) + \overline{K}_1 , \tag{5.5}$$

for $\overline{K}_i \in \mathcal{T}$, $1 \leq i \leq 3$. Fuzzy exponentials make up Ω_3. Lastly, Ω_4 is all

$$\overline{K}_3 \ln(\overline{X} + \overline{K}_2) + \overline{K}_1 , \tag{5.6}$$

for $\overline{K}_i \in \mathcal{T}$, $1 \leq i \leq 3$, with $\overline{X} + \overline{K}_2 \geq 1$. If $\overline{X} + \overline{K}_2 = (a/b/c)$, we require $a \geq 1$. Then

$$\Omega = \bigcup_{i=1}^{4} \Omega_i . \tag{5.7}$$

5.1.1 Testing the EA

To test the EA (evolutionary algorithm) we picked a $G(\overline{X}; \overline{K}_1, \ldots, \overline{K}_n) \in \Omega$ to generate the data and then asked the EA to find G. We did this two ways: (1) first the data was $(\overline{X}_i, \overline{Z}_i)$, $\overline{Z}_i = G(\overline{X}_i; \overline{K}_1, \ldots, \overline{K}_n)$, $1 \leq i \leq p$; and (2) secondly, the data $(\overline{X}_i, \overline{Z}_i)$, $\overline{Z}_i = G(\overline{X}_i; \overline{K}_1, \ldots, \overline{K}_n) + \varepsilon_i$, $1 \leq i \leq p$, had "noise" in it with ε_i distributed $N(0, \sigma^2)$. We will now only discuss the first case.

For the no noise case there were four tests: (1) $\overline{Z} = \overline{K}_1 \, \overline{X} + \overline{K}_2$; (2) $\overline{Z} = \overline{K}_2 \, \overline{X}^2 + \overline{K}_1 \, \overline{X} + \overline{K}_0$; (3) $\overline{Z} = \overline{K}_3 \exp(\overline{K}_2 \, \overline{X}) + \overline{K}_1$; and (4) $\overline{Z} = \overline{K}_3 \ln(\overline{X} + \overline{K}_2) + \overline{K}_1$ for given values of the $\overline{K}_i$. Consider case (2) above where the "unknown" function was $G(\overline{X}; \overline{K}_0, \overline{K}_1, \overline{K}_2) = \overline{K}_2 \, \overline{X}^2 + \overline{K}_1 \, \overline{X} + \overline{K}_0$. The EA found $F(\overline{X}; \overline{C}, \overline{B}, \overline{A}) = \overline{A} \, \overline{X}^2 + \overline{B} \, \overline{X} + \overline{C}$ where $\overline{A} \approx \overline{K}_2$, $\overline{B} \approx \overline{K}_1$ and $\overline{C} \approx \overline{K}_0$. In all tests the EA essentially found the unknown G.

The EA used to discover the correct function is in two parts. In the data let $\overline{X}_i = (x_{i1}/x_{i2}/x_{i3})$ in $\mathcal{T}$ and $\overline{Z}_i$ is unimodal meaning there is a unique value of z, say z_i, so that $\overline{Z}_i(z_i) = 1$, $1 \leq i \leq p$. The first EA, EA_1, attempts to fit a crisp linear, polynomial of degree $\leq \Delta$, exponential and logarithmic function to the crisp data (x_{i2}, z_i), $1 \leq i \leq p$. The model with the smallest error was chosen for the next phase. For example, if $\overline{Z} = \overline{K}_2 \, \overline{X}^2 + \overline{K}_1 \, \overline{X} + \overline{K}_0$, then EA_1 would pick $f(x) = ax^2 + bx + c$ for some a, b, c. The second EA,

EA_2, would look for $\overline{A}$, $\overline{B}$, and $\overline{C}$ so that $F(\overline{X};\overline{C},\overline{B},\overline{A}) = \overline{A}\,\overline{X}^2 + \overline{B}\,\overline{X} + \overline{C}$ would fit the data, or $E(F)$ would be acceptably small. But, in order to make EA_2 run faster, we first cut down on the search space. Based on the results of EA_1 intervals J_i where computed for the unknown parameters $\overline{K}_i$. That is, for $\overline{Z} = \overline{K}_2\,\overline{X}^2 + \overline{K}_1\,\overline{X} + \overline{K}_0$, EA_2 only looked in J_1 for $\overline{A}$, J_2 for $\overline{B}$ and J_3 for $\overline{C}$. In all cases (with or without noise) the EA essentially found the unknown G in Ω.

5.1.2 Application

Let us describe how we would like to test our EA on a G not in Ω. Even though we do not have any numerical results yet for this testing, we will still present our outline for this research.

First we will change Ω to be just fuzzy polynomials of degree $\leq \Delta$. The reason for this change is that fuzzy polynomials are universal approximators. Let us now explain what is meant by universal approximator.

Let $f(x; k_1, \ldots, k_n)$ map $[a, b]$ into $\mathbb{R}$ where variable x is in $[a, b]$ and the constants k_i are in intervals I_i, $1 \leq i \leq n$. Let $\Gamma = [a, b] \times \prod_{i=1}^{n} I_i$. f is assumed to be continuous on Γ into $\mathbb{R}$. For example

$$f(x; k_1, \ldots, k_5) = \frac{k_1 \sin(k_2 x + k_3)}{k_4 x^2 + k_5}, \tag{5.8}$$

for $x \in [0, 2\pi]$, $k_1, k_2, k_3 \in [-10, 10]$, $k_4, k_5 \in [1, 10]$. Let $\mathcal{C}$ be all such continuous f from Γ into $\mathbb{R}$.

Extend each f in $\mathcal{C}$ to $F(\overline{X}; \overline{K}_1, \ldots, \overline{K}_n)$ using the extension principle, for $\overline{X}$ in $[a, b]$ and $\overline{K}_i \in I_i$, $1 \leq i \leq n$. Also, $\overline{X}$, $\overline{K}_1, \ldots, \overline{K}_n \in \mathcal{T}$. Let $\overline{\mathcal{C}}$ be all these extensions. Also, let $\overline{Z} = F(\overline{X}; \overline{K}_1, \ldots, \overline{K}_n)$.

Define $\mathcal{P}$ to be all polynomials $p(x; a_0, \ldots, a_m) = a_m x^m + \cdots + a_1 x + a_0$. The extension principle extends p to

$$\overline{Y} = P(\overline{X}; \overline{A}_1, \ldots, \overline{A}_m) = \overline{A}_m\,\overline{X}^m + \cdots + \overline{A}_1\,\overline{X} + \overline{A}_0\ , \tag{5.9}$$

for $\overline{X} \in \mathcal{T}$, $\overline{A}_i \in \mathcal{F}_0$ for all i, $\overline{X} \in [a, b]$. Let $\overline{\mathcal{P}}$ be all such extensions. Notice, for a certain technical reason [1], the $\overline{A}_i$ are in $\mathcal{F}_0$. The following result was shown in [1]. Choose $F \in \overline{\mathcal{C}}$ and then fix all the values of the $\overline{K}_i$ in I_i, $1 \leq i \leq n$, so that they now become constants. Let $\varepsilon > 0$. There is a $P \in \overline{\mathcal{P}}$ so that

$$D(\overline{Z}, \overline{Y}) < \varepsilon\ , \tag{5.10}$$

for all $\overline{X}$ in $[a, b]$. Equation (5.10) is the statement that fuzzy polynomials, with coefficients in $\mathcal{F}_0$, are universal approximators for $F(\overline{X}; \overline{K}_1, \ldots, \overline{K}_n)$, with $\overline{X} \in \mathcal{T}$ and in $[a, b]$, and $\overline{K}_i$ constants in I_i and $\mathcal{T}$, $1 \leq i \leq n$.

What this means is that if the data was generated by $F \in \overline{\mathcal{C}}$, $\overline{K}_i$ constant for all i, there is a fuzzy polynomial P so that $E(P) < \varepsilon^2$, which is small since usually $0 < \varepsilon < 1$. For this reason we choose Ω to be all fuzzy polynomials of degree $\leq \Delta$.

Let us now explain why we will be using only non-negative fuzzy numbers in $\mathcal{T}$ or $\mathcal{F}_0$. The above results on universal approximators is for the extension principle. The extension principle is used to evaluate $F(\overline{X};\overline{K}_1,\ldots,\overline{K}_n)$ and $P(\overline{X};\overline{A}_1,\ldots,\overline{A}_n)$. However, in our EA α-cuts and interval arithmetic are used. But for non-negative fuzzy numbers both methods produce the same results for $\overline{Z}$ and $\overline{Y}$. Therefore, we may use this universal approximator result in our fuzzy non-linear regression.

Let us see how our EA might approximate fuzzy data generated by

$$g(x;c,b,a) = a\cos x + b\sin x + cx \ , \tag{5.11}$$

which is oscillation $(a\cos x + b\sin x)$ plus a "drift" cx. Assume $x \in [0, 2\pi]$ and a, b, $c \in [0, 10]$. In fact, choose $a = 4$, $b = 3$ and $c = 1$ so we get

$$g(x) = 4\cos x + 3\sin x + x \ , \tag{5.12}$$

$0 \le x \le 2\pi$. Use the extension principle and extend g to

$$G(\overline{X};\overline{C},\overline{B},\overline{A}) = \overline{A}\cos(\overline{X}) + \overline{B}\sin(\overline{X}) + \overline{C}\,\overline{X} \ , \tag{5.13}$$

for $\overline{A} = (3/4/5)$, $\overline{B} = (2/3/4)$, $\overline{C} = (0.5/1.0/1.5)$ and $\overline{X} = (x_1/x_2/x_3)$ in $[0, 2\pi]$.

The data set will be $(\overline{X}_i, \overline{Z}_i)$, $0 \le i \le 20$, where $\overline{X}_i = ((i-1)\delta/i\delta/(i+1)\delta)$, $0 \le i \le 20$, $\delta = \pi/10$. To handle the end points of the interval $[0, 2\pi]$ we have $\overline{X}_0$ centered at zero and $\overline{X}_{20}$ centered at 2π. The $\overline{Z}_i$ are computed as

$$\begin{aligned} z_{i1}(\alpha) &= \min\big\{a\cos x + b\sin x + c \ \big| \\ a \in \overline{A}[\alpha], b &\in \overline{B}[\alpha], c \in \overline{C}[\alpha], x \in \overline{X}_i[\alpha]\big\} \ , \end{aligned} \tag{5.14}$$

$$\begin{aligned} z_{i2}(\alpha) &= \max\big\{a\cos x + b\sin x + c \ \big| \\ a \in \overline{A}[\alpha], b &\in \overline{B}[\alpha], c \in \overline{C}[\alpha], x \in \overline{X}_i[\alpha]\big\} \ , \end{aligned} \tag{5.15}$$

$0 \le \alpha \le 1$, where $\overline{Z}_i[\alpha] = [z_{i1}(\alpha), z_{i2}(\alpha)]$. Of course, we cannot compute $\overline{Z}_i[\alpha]$ for all α in $[0, 1]$ so we will only obtain these α-cuts for $\alpha = 0, 0.1, \ldots, 0.9, 1.0$.

We wish to find $\overline{Y} = P(\overline{X};\overline{A}_0,\ldots,\overline{A}_m) = \overline{A}_m\overline{X}^m + \cdots + \overline{A}_1\overline{X} + \overline{A}_0$, $\overline{A}_i \in \mathcal{F}_0$ and $\overline{X} \in \mathcal{T}$, so that

$$D(\overline{Z}_i, \overline{Y}_i) < \varepsilon = 0.1 \ , \tag{5.16}$$

$1 \le i \le 20$, where $\overline{Y}_i = P(\overline{X}_i;\overline{A}_0,\ldots,\overline{A}_m)$. We have picked $\varepsilon = 0.1$. We first find crisp $p(x;a_0,\ldots,a_m) = a_m x^m + \cdots + a_1 x + a_0$, of minimal degree m, so that

$$\big|g(x;c,b,a) - p(x;a_0,\ldots,a_m)\big| < \varepsilon = 0.1 \ , \tag{5.17}$$

for all x in $[0, 2\pi]$. This is always possible because crisp polynomials are universal approximators for continuous functions. For example suppose m

turns out to be 10. Then our EA searches for $\overline{A}_0, \ldots, \overline{A}_{10}$ in $\mathcal{F}_0$ so that equation (5.16) is true. To cut down on the search space we first need to estimate intervals J_i for $\overline{A}_i$, $1 \leq i \leq 10$. We also first look for the $\overline{A}_i$ in $\mathcal{T}$.

Since we cannot evaluate expressions for variables in infinite sets, we need to discretize equations (5.14) – (5.17). In equations (5.14) – (5.16) we use $\alpha = 0.0, 0.1 \ldots, 0.9, 1.0$. In equation (5.17) we use discrete x values in the set $[0, 2\pi]$ evenly spread from zero to 2π.

The above plan will need to be carried out for selected G not in Ω.

5.2 Multivariate Non-Linear Fuzzy Regression

Let $\mathcal{T}$ be all trapezoidal fuzzy numbers, $\mathcal{F}_0$ is all trapezoidal shaped fuzzy numbers and $\mathcal{F}$ denotes all fuzzy numbers. In this section we will only be working with two independent variables, so we will be interested in functions like $z = f(x_1, x_2; k_1, \ldots, k_n)$ for variables $x_1 \in [a,b]$ and x_2 in $[c,d]$ and constants k_i in intervals I_i, $1 \leq i \leq n$. Assume f is a continuous function mapping $[a,b] \times [c,d] \times \prod_{i=1}^{n} I_i$ into $\mathbb{R}$. Using the extension principle we extend f into $\overline{Z} = F(\overline{X}_1, \overline{X}_2; \overline{K}_1, \ldots, \overline{K}_n)$ where $\overline{X}_1$, $\overline{X}_2$, $\overline{K}_1, \ldots, \overline{K}_n$ are all in $\mathcal{T}$, $\overline{X}_1$ in $[a,b]$, $\overline{X}_2$ in $[c,d]$ and the $\overline{K}_i$ are in I_i, $1 \leq i \leq n$. Let $\overline{\mathcal{C}}$ be all such fuzzy functions F.

As in the previous section we can show that fuzzy polynomials in two variables are universal approximators for $\overline{\mathcal{C}}$. Let $p(x_1, x_2; a_0, \ldots, a_m)$ be a polynomial in x_1 and x_2 with coefficients $a_0, \ldots, a_m$. Extend p into P so that $\overline{Y} = P(\overline{X}_1, \overline{X}_2; \overline{A}_0, \ldots, \overline{A}_m)$ where $\overline{X}_1$ and $\overline{X}_2$ are in $\mathcal{T}$, the $\overline{A}_i \in \mathcal{F}_0$ for all i, and $\overline{X}_1 \in [a,b]$, $\overline{X}_2 \in [c,d]$. Let $\overline{\mathcal{P}}$ be the set of all such polynomials P.

The following result was proven in [2]. Choose $F \in \overline{\mathcal{C}}$ and then fix all the values of the $\overline{K}_i$ so that they now become constants. Given $\varepsilon > 0$, there is a P in $\overline{\mathcal{P}}$ so that

$$D(\overline{Z}, \overline{Y}) < \varepsilon \ , \tag{5.18}$$

for all $\overline{X}_1$ in $[a,b]$ and $\overline{X}_2$ in $[c,d]$. This means that $\overline{\mathcal{P}}$ is a universal approximator for $F(\overline{X}_1, \overline{X}_2; \overline{K}_1, \ldots, \overline{K}_n)$ in $\overline{\mathcal{C}}$ where the variables are $\overline{X}_1$, $\overline{X}_2$ and the constants are $\overline{K}_1, \ldots, \overline{K}_n$.

Let the data be $((\overline{X}_{1i}, \overline{X}_{2i}), \overline{Z}_i)$, $1 \leq i \leq p$, for $\overline{X}_{1i}$ and $\overline{X}_{2i}$ in $\mathcal{T}$, $\overline{Z}_i$ in $\mathcal{F}$. The library Ω will be all fuzzy polynomials $P(\overline{X}_1, \overline{X}_2; \overline{A}_0, \ldots, \overline{A}_m)$ where the degree in $\overline{X}_1 \leq \Delta$ and the degree in $\overline{X}_2 \leq \Delta$, Δ some positive integer. The degree on $\overline{X}_i$ is the highest power of $\overline{X}_i$ in $P(\overline{X}_1, \overline{X}_2; \overline{A}_0, \ldots, \overline{A}_m)$, $i = 1, 2$. Let the data be generated by F in $\overline{\mathcal{C}}$. That is $\overline{Z}_i = \overline{F}(\overline{X}_{1i}, \overline{X}_{2i}; \overline{K}_1, \ldots, \overline{K}_n)$, for the constants $\overline{K}_i$, $1 \leq i \leq n$. Then we can find a P in $\overline{\mathcal{P}}$, Δ sufficiently large, so that

$$D(\overline{Z}_i, \overline{Y}_i) < \varepsilon \ , \tag{5.19}$$

$1 \leq i \leq p$, where $\overline{Y}_i = P(\overline{X}_{1i}, \overline{X}_{2i}; \overline{A}_0, \ldots, \overline{A}_m)$, $1 \leq i \leq p$. Then

$$E(P) < \varepsilon^2 \ , \tag{5.20}$$

where $E(\cdot)$ is the error function of equation (5.2). But first we tested our evolutionary algorithm (EA) on F in Ω.

5.2.1 Testing

We again used only non–negative fuzzy numbers, as in Section 5.1, so that evaluations using α–cuts and interval arithmetic (used in the EA) produce the same results as evaluations using the extension principle.

The library consists of fuzzy polynomials with degree ≤ 4. That is, $\overline{Y} = P(\overline{X}_1, \overline{X}_2; \overline{A}_0, \ldots, \overline{A}_m)$ is in Ω if degree$(\overline{X}_1) \leq 4$ and degree$(\overline{X}_2) \leq 4$. We ran three tests: (1) F in Ω was linear $(\overline{K}_1\overline{X}_1 + \overline{K}_2\overline{X}_2 + \overline{K}_3)$; (2) F was quadratic (degree$(\overline{X}_1) = 2 =$ degree$(\overline{X}_2)$); and (3) we chose a cubic. In each case F generated the data without "noise" and with "noise" (see Subsection 5.1.1).

In the no "noise" case the data was $((\overline{X}_{1i}, \overline{X}_{2i}), \overline{Z}_i)$, for $\overline{Z}_i = F(\overline{X}_{1i}, \overline{X}_{2i}; \overline{K}_1, \ldots, \overline{K}_n)$, $1 \leq i \leq p$. The objective was for the EA to find F.

The EA has two parts: (1) EA_1 was to discover the structure of F (linear, quadratic,...) and (2) EA_2 was to approximate the constants $\overline{K}_i$, $1 \leq i \leq n$. In the data let $\overline{X}_{ji} = (x_{ji1}/x_{ji2}, x_{ji3}/x_{ji4})$, $j = 1, 2$ and $1 \leq i \leq p$. Since F is a fuzzy polynomial with coefficients in $\mathcal{F}_0$ and variables in $\mathcal{T}$ we have $\overline{Z}_i \approx (z_{i1}/z_{i2}, z_{i3}/z_{i4})$, $1 \leq i \leq p$. We consider two sets of crisp data $D_1 = ((x_{1i2}, x_{2i2}), z_{i2})$ and $D_2 = ((x_{1i3}, x_{2i3}), z_{i3})$, $1 \leq i \leq p$. EA_1 first tries to fit a linear function to D_1, let the error be E_{11}. The "error" is the usual mean sum of squares. E_{12} is the error for fitting a linear function to D_2. Next EA_1 fits a quadratic to D_1 with error E_{21} and for D_2 error E_{22}. EA_1 also generates E_{31} and E_{32} for cubics and E_{41}, E_{42} for fourth degree polynomials in x_1 and x_2. Let $E_i^* = \max\{E_{i1}, E_{i2}\}$, $1 \leq i \leq 4$, and we pick the model with minimum E_i^*. Since $F \in \Omega$, one E_i^* will be small, the rest will be large, assuming our EA_1 can find small error solutions.

Suppose the minimum E_i^* was E_2^*. EA_2 looks for values of the $\overline{A}_i$ in

$$\overline{A}_5\,\overline{X}_1^{\;2} + \overline{A}_4\,\overline{X}_1\,\overline{X}_2 + \overline{A}_3\,\overline{X}_2^{\;2} + \overline{A}_2\,\overline{X}_1 + \overline{A}_1\,\overline{X}_2 + \overline{A}_0\ , \tag{5.21}$$

to minimize the error function in equation (5.2). We find $\overline{A}_i \approx \overline{K}_i$, $0 \leq i \leq 5$. In all tests the EA essentially found the F in Ω. To reduce the search space for EA_2 we first determined intervals $\overline{J}_i$ for each $\overline{A}_i$ and the search was then restricted to $\overline{A}_i$ in $\overline{J}_i$, for all i.

5.2.2 Application

Let us again describe how we would like to test our EA on a fuzzy function not in the library, or not a fuzzy polynomial. Even though we do not have any numerical results for this testing, we will still present our outline for this research.

We will allow

$$f(x_1, x_2; k_1, \ldots, k_4) = k_1 \cos x_1 + k_2 \sin x_2 + k_3 x_1 + k_4 x_2 \ , \tag{5.22}$$

to generate the data, $x_1, x_2 \in [0, 2\pi]$. To fix the values of the k_i we choose

$$f(x_1, x_2) = 4 \cos x_1 + 3 \sin x_2 + 2x_1 + x_2 \ , \tag{5.23}$$

for x_1, x_2 in $[0, 2\pi]$. Extend f to F

$$F(\overline{X}_1, \overline{X}_2; \overline{K}_1, \ldots, \overline{K}_4) = \overline{K}_1 \cos \overline{X}_1 + \overline{K}_2 \sin \overline{X}_2 + \overline{K}_3\, \overline{X}_1 + \overline{K}_4\, \overline{X}_2 \ , \tag{5.24}$$

for $\overline{K}_1 = (3/4/5)$, $\overline{K}_2 = (2/3/4)$, $\overline{K}_3 = (1/2/3)$, $\overline{K}_4 = (0.5/1.0/1.5)$, and $\overline{X}_i = (x_{i1}/x_{i2}/x_{i3})$, $i = 1, 2$, and $\overline{X}_1$, $\overline{X}_2$ in $[0, 2\pi]$.

The data $(\overline{X}_{1i}, \overline{X}_{2i})$ will cover the square $[0, 2\pi] \times [0, 2\pi]$. Let $\overline{N}_i = ((i-1)\delta/i\delta/(i+1)\delta)$, $0 \leq i \leq 5$, for $\delta = 2\pi/5$. Then $(\overline{X}_{1i}, \overline{X}_{2i})$ is $(\overline{N}_0, \overline{N}_0)$, $(\overline{N}_1, \overline{N}_0), \ldots, (\overline{N}_5, \overline{N}_0)$, $(\overline{N}_0, \overline{N}_1), \ldots, (\overline{N}_5, \overline{N}_1), \ldots, (\overline{N}_5, \overline{N}_5)$ with 36 members. $\overline{Z}_i$ is computed as in equations (5.14) and (5.15). We actually find $\overline{Z}_i[\alpha]$ only for $\alpha = 0.0, 0.1, \ldots, 0.9, 1.0$.

We first wish to find a polynomial $p(x_1, x_2; a_0, \ldots, a_m)$ so that the difference between f and p over $[0, 2\pi] \times [0, 2\pi]$ is less than $\varepsilon = 0.1$. We will use the following "Grid". Grid $= \{(0,0), (0, \delta), \ldots, (0, 5\delta), (\delta, 0), \ldots, (\delta, 5\delta), \ldots, (5\delta, 5\delta)\}$. In practice we find p so that the difference between f and p is less than ε on the Grid.

Suppose p turns out to have degree 5 in x_1 and x_2. Let $a_0, \ldots, a_m$ be the coefficients in p. Using the results above we determine intervals J_i for the $\overline{A}_i$, $0 \leq i \leq m$. Then we use our EA to solve for the best $\overline{A}_i$ in J_i, $0 \leq i \leq m$, so that $P(\overline{X}_1, \overline{X}_2; \overline{A}_0, \ldots, \overline{A}_m)$ approximates $F(\overline{X}_1, \overline{X}_2; \overline{K}_1, \cdots, \overline{K}_4)$ on the data set. We first try for $\overline{A}_i \in \mathcal{T}$ but we may have to let $\overline{A}_i$ belong to $\mathcal{F}_0$.

The above plan will need to be carried out for selected functions not in the library.

5.3 Conclusions and Results

In this chapter we presented a general strategy for fuzzy regression. We place certain fuzzy functions in our library Ω. Given some fuzzy data our algorithm searches the library Ω for a fuzzy function that best explains the fuzzy data. Experimental results show that the method works quite well. However, the experiments run so far all involved choosing a fuzzy function in the library to generate the fuzzy data. Future experiments need to use a fuzzy function not from the library.

Other areas to look at are : (1) remove the restriction that all the fuzzy numbers must be non-negative; and (2) better methods to reduce the search space to speed up the algorithm.

Bibliography

[1] J.J. Buckley and Th. Feuring: Linear and Non–Linear Fuzzy Regression: Evolutionary Algorithm Solutions, Fuzzy Sets and Systems, 112 (2000), pp. 381–394.

[2] J.J. Buckley, Th. Feuring and Y. Hayashi: Multivariate Fuzzy Regression: An Evolutionary Algorithm Approach, Int. J. Uncertainty, Fuzziness and Knowledge Based Systems, 7 (1999), pp. 83–98.

[3] Y.-H.O. Chang: Hybrid Fuzzy Least-Squares Regression Analysis and Its Reliability Measures, Fuzzy Sets and Systems, 119 (2001), pp. 225–246.

[4] Y.-H.O. Chang and B.M. Ayyub: Fuzzy Regression Methods - a Comparative Assessment, Fuzzy Sets and Systems, 119 (2001), pp. 187–203.

[5] P. Diamond: Fuzzy Least Squares, Information Sciences, 46 (1988), pp. 141–157.

[6] P. Diamond and H.Tanaka: Fuzzy Regression Analysis, in: R.Slowinski (ed.), Fuzzy Sets in Decision Analysis, Operations Research and Statistics, Kluwer Academic Publishers,,Boston,,1998, pp. 349–390.

[7] D.H. Hong, J.-K. Song and H. Young Do: Fuzzy Least-Squares Linear Regression Analysis Using Shape Preserving Operations, Information Sciences, 138 (2001), pp.185–193.

[8] D.H. Hong, S. Lee and H.Y. Do: Fuzzy Linear Regression Analysis for Fuzzy Input-Output Data Using Shape-Preserving Operations, Fuzzy Sets and Systems, 122 (2001), pp. 513–526.

[9] M. Inuiguchi, M. Sakawa and S. Ushiro: Mean–Absolute–Deviation–Based Fuzzy Linear Regression Analysis by Level Sets Automatic Deduction from Data, Proc. FUZZ-IEEE'97, Barcelona, Spain, July 1-5, 1997, Vol. 1, pp. 829-834.

[10] H. Ishibuchi and M. Nii: Fuzzy Regression using Asymmetric Fuzzy Coefficients and Fuzzified Neural Networks, Fuzzy Sets and Systems, 119 (2001), pp. 273–290.

[11] J. Kacprzyk and M. Fedrizzi: Fuzzy Regression Analysis, Omnitech Press, Warsaw and Physica–Verlag, Heidelberg, 1992.

[12] H.T. Lee and S.H. Chen: Fuzzy Regression Model with Fuzzy Input and Output Data for Manpower Forcasting, Fuzzy Sets and Systems, 119 (2001), pp. 205–213.

[13] M. Ming, M. Friedman, A. Kandel: General Fuzzy Least Squares, Fuzzy Sets and Systems, 88 (1997), pp. 107–118.

[14] X. Ruoning: S–Curve Regression Model in Fuzzy Environment, Fuzzy Sets and Systems, 90 (1997), pp. 317–326.

[15] Y.-J. Seo, Y.-M. Park, S.-G. Hwang, K.-P. Park: Fuzzy Regression Using Genetic Algorithms, Proc. Fifth IFSA, Seoul, July 4-9, 1993, pp. 513-516.

[16] H. Tanaka and H. Lee: Fuzzy Linear Regression Combining Central Tendency and Possibilistic Properties, Proc. FUZZ-IEEE'97, Barcelona, Spain, 1997, Vol. 1, pp. 63-68.

[17] L. Wang, L. Zhang, H. Itoh, H. Seki: A Fuzzy Regression Method Based on Genetic Algorithms, Proc. 3rd Int. Conf. Fuzzy Logic, Neural Nets and Soft Computing, Iizuka, Aug. 1–7, Japan, 1994, pp. 471–472.

[18] R. Xu and C. Li: Multidimensional Least-Squares Fitting with a Fuzzy Model, Fuzzy Sets and Systems, 119 (2001), pp. 215–223.

[19] Y. Yabuuchi, J. Watada and T. Tatsumi: Fuzzy Regression Analysis of Data with Error, Japanese J. of Fuzzy Theory and Systems, 6 (1994), pp. 673–685.

[20] X. Zhang, S. Omachi and H. Aso: Fuzzy Regression Analysis Using RFLN and its Application, Proc. FUZZ-IEEE'97, Barcelona, July 1–5, Spain, 1997, Vol. 1, pp. 51–56.

Chapter 6

Operations Research

The applications of fuzzy sets and fuzzy logic to operations research has become a vast area of research. The journal of Operations Research [46] lists 51 major classifications for papers submitted for publication. Some of these major classifications, like "Programming", has numerous secondary classifications. Obviously, we can cover only a few applications in this chapter. In the next section we discuss fuzzy linear programming followed by PERT, inventory control, queuing theory and finally network analysis. We refer the reader to the books ([17],[19],[51],[55],[62]) and paper [26] for other applications, and to the proceedings of IFSA/NAFIPS [27] for recent presentations in the area. It is interesting to note the recent growth in the applications to scheduling as seen by [42], [52] and the conference [18].

6.1 Fuzzy Linear Programming

Fuzzy linear programming has long been an area of application of fuzzy sets. Consider the classical linear program

$$\begin{aligned} &\max Z = c_1 x_1 + \cdots + c_n x_n \\ \text{subject to:} \quad & \\ &a_{i1}x_1 + \cdots + a_{in}x_n \leq b_i \ , \quad 1 \leq i \leq m \\ &x_i \geq 0 \ , \text{for all } i \ . \end{aligned} \tag{6.1}$$

We need to have values for all the parameters c_i, a_{ij} and b_i to completely specify the optimization problem. Many of these must be estimated and are therefore uncertain. It is then natural to model these uncertain parameters using fuzzy numbers. The problem then becomes a fuzzy linear programming problem.

We are going to allow all the parameters to be fuzzy and we obtain what we have called the fully fuzzified linear programming problem [5]. Many researchers ([15],[21],[34],[39],[48], [49]) have looked at parts of this problem: (1) the a_{ij} and b_i can be fuzzy; or (but not both) (2) the c_i can be fuzzy. The fully fuzzified linear program is

$$
\begin{aligned}
max\overline{Z} = \overline{C}_1\overline{X}_1 + \cdots + \overline{C}_n\overline{X}_n& \\
\text{subject to:}& \\
\overline{A}_{i1}\overline{X}_1 + \cdots + \overline{A}_{in}\overline{X}_n \leq \overline{B}_i,\ 1 \leq i \leq m,& \\
\overline{X}_i \geq 0, \text{for all i.}&
\end{aligned}
\tag{6.2}
$$

where the $\overline{C}_i$, $\overline{A}_{ij}$ and $\overline{B}_i$ can all be triangular fuzzy numbers. Not every single parameter need be fuzzy; but we shall assume the some of the $\overline{C}_i$ are fuzzy and some of the $\overline{A}_{ij}$ and $\overline{B}_i$ are fuzzy. Since the parameters are fuzzy, the variables $\overline{X}_i$ will be triangular shaped fuzzy numbers.

This section is now concerned with solving the optimization problem in equation (6.2). But first we must do two things: (1) explain what we mean by max $\overline{Z}$ since $\overline{Z}$ will also be a triangular shaped fuzzy number; and (2) decide on how we will evaluate the inequality ($\leq$) between fuzzy numbers $\overline{E}_i \leq \overline{B}_i$, where $\overline{E}_i = \overline{A}_{i1}\ \overline{X}_1 + \cdots + \overline{A}_{in}\ \overline{X}_n$.

6.1.1 Maximize $\overline{Z}$

$\overline{Z}$ will be a triangular shaped fuzzy number like the one shown in Figure 6.1. We cannot maximize a fuzzy number. The problem is similar to that encountered in finance where $\overline{Z}$ would be a random variable and we want to "maximize" $\overline{Z}$. In finance one does not maximize random variable $\overline{Z}$ but instead maximizes the expected value of $\overline{Z}$, minimizes the variance of $\overline{Z}$, or maximizes the skewness of $\overline{Z}$ to the right of its expected value. We propose doing an analogous procedure of (refer to Figure 6.1): (1) maximize z_2; (2) minimize A_1, the area under the graph from z_1 to z_2; and (3) maximize A_2, the area from z_2 to z_3. The reason for this decision is: (1) we wish to make $\overline{Z}$ as large as possible so we make its central value (core) a maximum; (2) we want the possibility of exceeding z_2 to be as large as possible so we maximize A_2; and (3) we wish the possibility of obtaining less than z_2 to be as small as possible so we minimize A_1. So for max $\overline{Z}$ we have a multiobjective optimization problem

$$(\sup z_2, \sup A_2, \inf A_1)\ , \tag{6.3}$$

for feasible $\overline{X}_1$, $\overline{X}_2, \ldots, \overline{X}_n$, where sup = supremum and inf = infimum. We must use sup and inf because, at this point, we do not know if $\max z_2$, $\max A_2$ or $\min A_1$ exist.

In a multiobjective optimization problem a solution is a feasible vector $\overline{X} = (\overline{X}_1, \ldots, \overline{X}_n)$ which is undominated with respect to the objectives in equation (6.3). Before we can define the concept of "undominated" we first need to introduce some new concepts, notations, etc.

Let $\mathcal{F}$ be the set of feasible vectors $\overline{X}$. That is, $\mathcal{F}$ consists of all $\overline{X} = (\overline{X}_1, \ldots, \overline{X}_n)$ so that $\overline{X}_i \geq 0$ for all i and $\overline{A}_{i1}\ \overline{X}_1 + \cdots + \overline{A}_{in}\ \overline{X}_n \leq \overline{B}_i$,

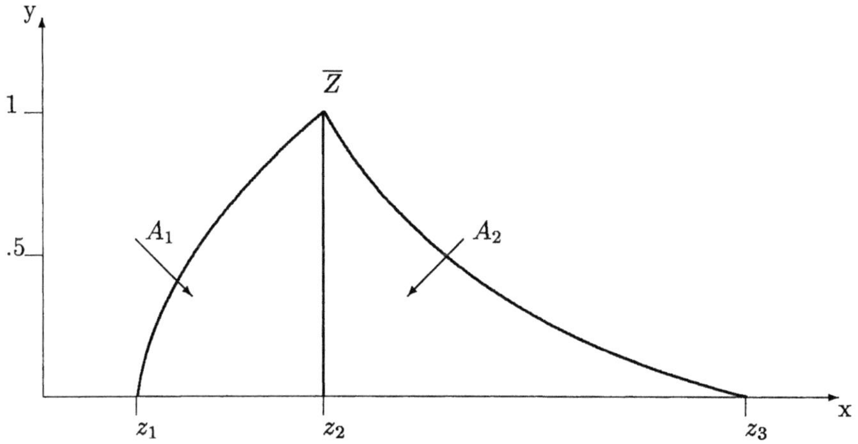

Figure 6.1: Graphical Description of z_2, A_2 and A_1

$1 \le i \le m$. We will discuss methods of evaluating $\le$ between fuzzy numbers in the next section, so for now let us assume that we have an interpretation of $\le$ so that $\mathcal{F}$ is defined. $\mathcal{F}$ could be empty so we now assume that $\overline{C}_i \geq 0$, $\overline{A}_{ij} \geq 0$, and $\overline{B}_i \geq 0$, all i and j, so that $\overline{X}_i = 0$ (crisp, real number zero) all i, is feasible. Having all the parameter fuzzy sets non–negative assures us that at least $\overline{X} = 0$ (zero vector) is feasible and $\mathcal{F}$ is non–empty.

We must make one important assumption about the "$\leq$" we will use between fuzzy numbers. For $\overline{X}$ in $\mathcal{F}$ we assume that $\overline{Z} = \overline{C}_1\,\overline{X}_1 + \cdots \overline{C}_n\,\overline{X}_n$ is bounded. What this means is that there is a $M > 0$ so that $\overline{Z}$ is a fuzzy subset of the interval $[0, M]$. The $\le$ that we will use guarantees that $\overline{Z}$ is bounded. Otherwise, the areas (A_1, A_2) may be unbounded.

Now let $\sup z_2 = b_1$, $\sup A_2 = b_2$ and $\sup A_1 = b_3$ for $\overline{X}$ in $\mathcal{F}$. All of these supremums exist and are finite because z_2, A_2 and A_1 are all bounded above by M. Since $\overline{X} = 0$ is feasible, a possible value for $\overline{Z}$ is zero and hence $\min z_2 = \min A_2 = \min A_1 = 0$ for feasible $\overline{X}$.

It will be convenient, for the rest of this section on fuzzy linear programming, to change our objective of $\inf A_1$ to a supremum. Let $A'_1 = b_3 - A_1$ so that $\inf A_1$ is equivalent to $\sup A'_1$. The multiobjective problem becomes

$$(\sup z_2, \sup A_2, \sup A'_1)\ , \tag{6.4}$$

for $\overline{X}$ in $\mathcal{F}$.

The point (b_1, b_2, b_3) is called the ideal point of the problem since it is usually unattainable. If there is a feasible $\overline{X}$ so that $z_2 = b_1$, $A_2 = b_2$ and $A'_1 = b_3$, then we are finished, the solution is $\overline{X}$. So let us assume that no feasible $\overline{X}$ can make $z_2 = b_1$, $A_2 = b_2$, $A'_1 = b_3$ simultaneously.

Now we may specify the undominated set in $\mathcal{F}$. If $\overline{X}_a$ and $\overline{X}_b$ are two feasible vectors, let $\overline{Z}_a$ ($\overline{Z}_b$) be the value of the objective function using $\overline{X}_a$ ($\overline{X}_b$). Next, let $(z_{2a}, A_{2a}, A'_{1a})$ $((z_{2b}, A_{2b}, A'_{1b}))$ be the value of (z_2, A_2, A'_1)

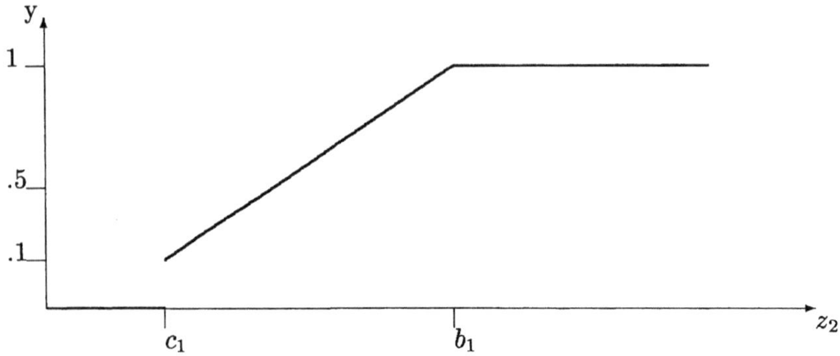

Figure 6.2: Fuzzy Goal $\overline{G}_1$ for z_2

computed from $\overline{Z}_a$ ($\overline{Z}_b$). We say $\overline{X}_a$ dominates $\overline{X}_b$ if $z_{2a} \geq z_{2b}$, $A_{2a} \geq A_{2b}$, $A'_{1a} \geq A'_{1b}$ and at least one of these inequalities is a strict inequality. For example, if $z_{2a} = z_{2b}$, $A_{2a} > A_{2b}$, $A'_{1a} = A'_{1b}$, then $\overline{X}_a$ dominates $\overline{X}_b$. Let UD be all the undominated $\overline{X}$ in $\mathcal{F}$. We have been unable, in general, to show that UD is always non–empty [5]. However, for UD non–empty a solution to the multiobjective optimization problem will be any $\overline{X}$ in UD. In [5] we defined "solution sequences" when UD was empty, to approximate a solution. However, we shall not go into any of these details in this chapter. We feel that in the usual application areas, as in the examples presented below in Section 6.1.4, UD is non–empty. So, assuming UD not empty we need an algorithm to explore the solutions in the undominated set. By defining a special fuzzy goal for each objective we will have a method of generating solutions in UD.

For each objective we set up a fuzzy goal $\overline{G}_i$. We set up the $\overline{G}_1$ for z_2 as follows: if $0 \leq c_1 < b_1$, then

$$\overline{G}_1(z_2) = \begin{cases} 0 & : \quad z_2 < c_1 \;, \\ 0.1 + 0.9\left(\frac{z_2 - c_1}{b_1 - c_1}\right) & : \quad c_1 \leq z_2 \leq b_1 \;, \\ 1 & : \quad z_2 \geq b_1 \;, \end{cases} \tag{6.5}$$

and if $c_1 = b_1$

$$\overline{G}_1(z_2) = \begin{cases} 0 & : \quad z_2 < b_2 \;, \\ 1 & : \quad z_2 \geq b_1 \;. \end{cases} \tag{6.6}$$

The graph of $\overline{G}_1$, for $0 \leq c_1 < b_1$, is shown in Figure 6.2.

The structure of $\overline{G}_2$, for A_2, is similar to $\overline{G}_1$: if $0 \leq c_2 < b_2$, then

$$\overline{G}_2(A_2) = \begin{cases} 0 & : \quad A_2 < c_2 \;, \\ 0.1 + 0.9\left(\frac{A_2 - c_2}{b_2 - c_2}\right) & : \quad c_2 \leq A_2 \leq b_2 \;, \\ 1 & : \quad A_2 \geq b_2 \;, \end{cases} \tag{6.7}$$

and if $c_2 = b_2$

$$\overline{G}_2(A_2) = \begin{cases} 0 & : \quad A_2 < b_2 \;, \\ 1 & : \quad A_2 \geq b_2 \;. \end{cases} \tag{6.8}$$

$\overline{G}_3$ for A'_1 is the same as $\overline{G}_1$ and $\overline{G}_2$. If $0 \leq c_3 < b_3$, then

$$\overline{G}_3(A'_1) = \begin{cases} 0 & : \quad A'_1 < c_3 \ , \\ 0.1 + 0.9\left(\frac{A'_1 - c_3}{b_3 - c_3}\right) & : \quad c_3 \leq A'_1 \leq b_3 \ , \\ 1 & : \quad A'_1 \geq b_3 \ , \end{cases} \tag{6.9}$$

and if $c_3 = b_3$

$$\overline{G}_3(A'_1) = \begin{cases} 0 & : \quad A'_1 < b_3 \ , \\ 1 & : \quad A'_1 \geq b_3 \ . \end{cases} \tag{6.10}$$

Now we set up the following single objective optimization problem

$$\begin{aligned} &\sup \left(\overline{G}_1(z_2) \cdot \overline{G}_2(A_2) \cdot \overline{G}_3(A'_1)\right) , \\ &\text{subject to: } \overline{X} \text{ feasible} . \end{aligned} \tag{6.11}$$

It was shown in [5] that as each c_i varied between zero and b_i, $1 \leq i \leq 3$, solutions to equation (6.11) will explore the whole undominated set. However, to make this scheme operational we have two problems to solve: (1) find values for the b_i; and (2) use appropriate values for the c_i. We estimate the b_i with values b'_i, $b'_i > b_i$, $1 \leq i \leq 3$. The above results are still valid using b'_i in place of b_i in the fuzzy goals $\overline{G}_i$, $1 \leq i \leq 3$. If we choose the c_i values too large, now $0 \leq c_i \leq b'_i$, we can get $\overline{G}_1(z_2) = \overline{G}_2(A_2) = \overline{G}_3(A'_1) = 0$ all feasible $\overline{X}$. For example, if $c_i = b'_i$ all i with $b'_i > b_i$ all i, then surely the objective function in equation (6.11) is always zero and the solution set is all of $\mathcal{F}$. So, if we get all the fuzzy goals identically zero on $\mathcal{F}$ we reduce their c_i values until all the goals can be positive for some $\overline{X}$ in $\mathcal{F}$. Then we can use the optimization problem in (6.11) to find undominated solutions.

However, we do not know of an algorithm that will give solutions to (6.11). So, we will employ a directed search algorithm procedure called an evolutionary algorithm to generate approximate solutions. Our general purpose evolutionary algorithm is described in Chapter 13 at the end of the book and how we are adapted it to this problem is discussed in Section 6.1.3. What we want from our evolutionary algorithm, which is an iterative process, is a sequence of feasible vectors $\overline{X}^{(k)}$ converging to an undominated solution, or a solution to (6.11), for given values of c_i, $1 \leq i \leq 3$.

6.1.2 Fuzzy Inequality

In this section we will discuss some methods of evaluating $\overline{M} \leq \overline{N}$, where $\overline{M}$ and $\overline{N}$ are fuzzy numbers, that we may use in our evolutionary algorithm. $\overline{M}$ will be $\sum_{j=1}^{n} \overline{A}_{ij}\overline{X}_j = \overline{E}_i$, the left hand side of a fuzzy constraint, and $\overline{N}$ will be $\overline{B}_i$ the right hand side of the fuzzy constraint. The methods used here will be different from those used in Sections 4.4 and 6.5.

This section is not intended to be a survey of all methods presented in the literature of evaluating $\overline{M} \leq \overline{N}$. See [15] for an excellent review of this literature up to 1992. But there continues to be more and more proposals published on how to evaluate $\overline{M} \leq \overline{N}$. One can find many of these publications in the journal Fuzzy Sets and Systems (see reference [21]). Our criteria for choosing a method of evaluating $\overline{M} \leq \overline{N}$ is: (1) $\overline{Z}$ must be bounded; (2) it is fairly easily incorporated into our evolutionary algorithm; and (3) it seems "reasonable" to the authors. We want procedures that will produce a "true" or "false" for $\overline{M} \leq \overline{N}$ given two fuzzy numbers. In this way we can test a given $\overline{X}$, $\overline{X}_i \geq 0$, as being feasible or not feasible.

Obviously, using different methods of evaluating $\overline{M} \leq \overline{N}$, each one satisfying our criteria mentioned above, can give different solutions (undominated, feasible $\overline{X}$) to the multiobjective optimization problem. So, one might consider sensitivity analysis on "$\leq$", but we shall not do this with the two examples in Section 6.1.4.

Twenty-three different ranking methods were presented in [15]. Almost all of these procedures had to be rejected because they did not guarantee that $\overline{Z}$ would be bounded. The problem was that all of these methods assumed that in evaluating $\overline{M} \leq \overline{N}$, $\overline{M}$ and $\overline{N}$ are triangular fuzzy numbers. But in evaluating $\overline{E}_i \leq \overline{B}_i$ the fuzzy number $\overline{E}_i$ is not a triangular fuzzy number. $\overline{E}_i$ is a triangular shaped fuzzy number like the one shown in Figure 6.3.

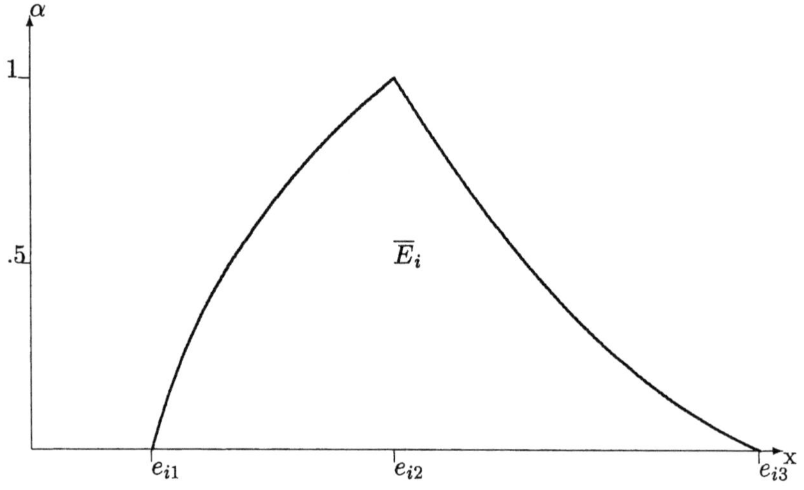

Figure 6.3: The Shape of $\overline{E}_i = \sum_{j=1}^{n} \overline{A}_{ij}\, \overline{X}_j$

Because the graph of $\overline{E}_i$ is concave up on the interval $[e_{i2}, e_{i3}]$ almost all of the methods of evaluating fuzzy inequality in [15] would say that $\overline{E}_i \leq \overline{B}_i$ is true when, holding everything else fixed, e_{i3} grows larger and larger. This means that $\overline{E}_i$ can be unbounded and still have $\overline{E}_i \leq \overline{B}_i$ true. It follows that the $\overline{X}_i$ can be unbounded which implies that $\overline{Z}$ is unbounded.

First Method

We first need to present the fuzzy max (written $\overline{\max}$) of two fuzzy numbers. If $\overline{O} = \overline{\max}(\overline{M}, \overline{N})$, then

$$\overline{O}(z) = \sup \left\{ \min(\overline{M}(x), \overline{N}(y)) | \max(x, y) = z \right\} . \tag{6.12}$$

The authors in [33] give a detailed study of the properties of $\overline{\max}$ and $\overline{\min}$ (fuzzy min).

Next we define the Hamming distance between $\overline{M}$ and $\overline{N}$. The Hamming distance, $d(\overline{M}, \overline{N})$, is defined as

$$d(\overline{M}, \overline{N}) = \int_{-\infty}^{\infty} |\overline{M}(x) - \overline{N}(x)| dx . \tag{6.13}$$

Clearly, d is a metric (distance measure) on the space of continuous fuzzy numbers (those whose membership function is continuous).

Then we say $\overline{M} \leq \overline{N}$ is true whenever

$$d(\overline{N}, \overline{\max}(\overline{M}, \overline{N})) \leq d(\overline{M}, \overline{\max}(\overline{M}, \overline{N})) . \tag{6.14}$$

This is simply a fuzzification of $x \leq y$ if and only if $\max(x, y) = y$ for real x, y. A numerical example showing $\overline{M} \leq \overline{N}$ by this method is in ([33], p. 407 - 408). We will call this procedure for evaluating fuzzy inequalities Kerre's method.

The authors in [5] now show that using this procedure of evaluating $\overline{M} \leq \overline{N}$: (1) $\overline{Z}$ is bounded; and (2) how to estimate the b_i' so that $b_i' > b_i$, $1 \leq i \leq 3$.

Second Method

The second method of ranking fuzzy numbers we focused on is presented by Chen in [15]. A score is computed for each fuzzy number which is needed for ranking. The fuzzy set with the highest score is the largest fuzzy number. In order to rank triangular shaped fuzzy numbers $\overline{N} \cong (n_1/n_2/n_3)$ and $\overline{M} \cong (m_1/m_2/m_3)$ Chen defined a fuzzy max and a fuzzy min where the supports of fuzzy max and min is $[x_{\min}, x_{\max}]$ where

$$x_{\min} = \min(n_1, m_1) , \tag{6.15}$$
$$x_{\max} = \max(n_3, m_3) . \tag{6.16}$$

Fuzzy min and fuzzy max are triangular fuzzy numbers with membership degree one at the left and the right limit of the support, respectively (see Figure 6.4). The membership functions are

$$\mu_{\min}(x) = \begin{cases} \frac{x - x_{\max}}{x_{\min} - x_{\max}} & : \ x_{\min} \leq x \leq x_{\max} , \\ 0 & : \ \text{otherwise} \end{cases} \tag{6.17}$$

$$\mu_{\max}(x) = \begin{cases} \frac{x - x_{\min}}{x_{\max} - x_{\min}} & : \ x_{\min} \leq x \leq x_{\max} , \\ 0 & : \ \text{otherwise} . \end{cases} \tag{6.18}$$

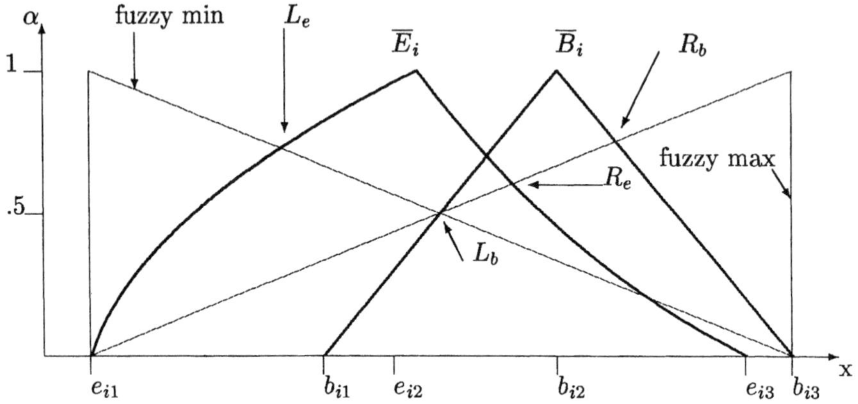

Figure 6.4: Ranking Fuzzy Numbers Based on Chen's Method

We continue to use $\overline{M} = \overline{E}_i$ and $\overline{N} = \overline{B}_i$. The intersection points between fuzzy max and $\overline{M}$ and $\overline{N}$ as well as the intersection points between fuzzy min and $\overline{M}$ and $\overline{N}$ are needed for computing the final scores. We compute

$$\mu_R(\overline{M}) = \sup_x(min(\mu_{\max}(x), \overline{M}(x))), \tag{6.19}$$

$$\mu_L(\overline{M}) = \sup_x(min(\mu_{\min}(x), \overline{M}(x))), \tag{6.20}$$

$$\tag{6.21}$$

where $\mu_R(\overline{M})$ indicates the max of the intersection point between fuzzy max and $\overline{M}$ and $\mu_L(\overline{M})$ stands for the left score which is given by the max intersection point with fuzzy min. The larger $\mu_R(\overline{M})$ is, the higher $\overline{M}$ should be ranked. On the other hand a high value of $\mu_L(\overline{M})$ and $\overline{M}$ is close to the fuzzy min, and therefore should be ranked lower. By combining both scores we get the final rating

$$\mu_T(\overline{M}) = \frac{1}{2}\left(\mu_R(\overline{M}) + (1 - \mu_L(\overline{M})\right) . \tag{6.22}$$

Similarly, we get $\mu_T(\overline{N})$. We then say that $\overline{M} \le \overline{N}$ is true if $\mu_T(\overline{M}) \le \mu_T(\overline{N})$.

In Figure 6.4 we used the notation $L_e = \mu_L(\overline{E}_i)$, $R_e = \mu_R(\overline{E}_i)$, $L_b = \mu_L(\overline{B}_i)$ and $R_b = \mu_R(\overline{B}_i)$. Using this method of evaluating $\overline{M} \le \overline{N}$ the authors in [5] now show $\overline{Z}$ is bounded and how to estimate $b_i' > b_i$, $1 \le i \le 3$.

6.1.3 Evolutionary Algorithm

Chapter 13, at the end of the book, contains an explanation of our general purpose evolutionary algorithm. What we wish to do here is show how it was adapted to find approximate undominated solutions to the multiobjective optimization problem using the two methods of evaluating $\overline{M} \le \overline{N}$ discussed in the previous Section 6.1.2

Each element in the population is described by a set of triangular fuzzy numbers $\overline{X}_i$, $1 \leq i \leq n$ which are in the feasible set. Let $\overline{X}_i = (x_{i1}/x_{i2}/x_{i3})$, $1 \leq i \leq n$. In order to compute the sum and product of fuzzy numbers, α–cuts for each fuzzy number are calculated. The α - levels are used to produce the corresponding sum or product. Since the product of triangular fuzzy numbers may not be triangular, but instead triangular shaped fuzzy numbers, we need to store all these α–cuts so that we can use these triangular shaped fuzzy numbers in further calculations. An additional value is added to each population member which represents its mutation rate. Therefore, the population members are vectors $\pi \in R^{3n+1}$. An element in the population looks like

$$\pi = (x_{11}, x_{12}, x_{13}, x_{21}, \cdots, x_{n1}, x_{n2}, x_{n3}, \sigma), \tag{6.23}$$

where σ stands for the mutation rate for this population member. We usually start with $\sigma \approx 0.3$, and then experiment with other initial values to get better fitness values.

In the applications in the next section we used a population size of 2000. First, all elements of the population are randomly chosen according to the constraints. In order to get a suitable starting population, we randomly choose the central values of $\overline{X}_i$, x_{i2}, close to the solutions of the corresponding crisp linear programming problem. An individual is taken into the starting population only if it satisfies the constraints.

After initialization the evolution starts with the selection process. The fitness function is given in equation (6.11). The closer the value of the fitness function is to one the greater is the fitness of this population member. In the applications in the next section we always selected 300 of the fittest individuals to generate the next population.

Crossover and mutation, used to obtain the next population, are basically as described in Chapter 13. However, crossover was adapted to only points between two encoded fuzzy numbers so that after the crossover operation we do have n triangular fuzzy numbers $\overline{X}_i$ and all we need to do is check to see if it is feasible. The mutation process was also adapted so that we always get $x_{i1} < x_{i2} < x_{i3}$, $1 \leq i \leq n$.

6.1.4 Applications

Here we present two applications of our evolutionary algorithm solution to the fully fuzzified linear program in equation (6.2). In both cases we use the problem in equation (6.11) to get undominated solutions using both methods of evaluating $\overline{M} \leq \overline{N}$ presented in Section 6.1.2. The first application is the standard product mix problem and the second is the standard diet problem. The second application is a minimization problem so we will have to adjust the fuzzy goals in Section 6.1.1.

Department

		D_1	D_2	D_3
	P_1	6	12	2
Product	P_2	8	8	4
	P_3	3	6	1

Table 6.1: Approximate Times Product P_i is in Department D_j

Product Mix Problem

A company produces three products P_1, P_2 and P_3 each of which must be processed through three departments D_1, D_2 and D_3. The approximate time, in hours, each P_i spends in each D_j is given in Table 6.1.

Each department has only so much time available each week. These times can vary slightly from week to week so the following numbers are estimates of the maximum time available per week, in hours, for each department: (1) for D_1 288 hours; (2) 312 hours for D_2; and (3) D_3 has 124 hours. Finally, the selling price for each product can vary a little due to small discounts to certain customers but we have the following average selling prices: (1) \$6 per unit for P_1; (2) \$8 per unit for P_2, and (3) for P_3 \$6/unit. The company wants to determine the number of units to produce for each product per week to maximize its revenue.

Since all the numbers given are uncertain, we will model the problem as a fully fuzzified linear program. We substitute a triangular fuzzy number for each value given where the peak of the fuzzy number is at the number given. So, we have the following fully fuzzified linear program

$$\max \overline{Z} = (5.8/6/6.2)\overline{X}_1 + (7.5/8/8.5)\overline{X}_2 + (5.6/6/6.4)\overline{X}_3 \tag{6.24}$$

subject to:

$$(5.6/6/6.4)\overline{X}_1 + (7.5/8/8.5)\overline{X}_2 + (2.8/3/3.2)\overline{X}_3 \leq (283/288/293)\ ,$$
$$(11.4/12/12.6)\overline{X}_1 + (7.6/8/8.4)\overline{X}_2 + (5.7/6/6.3)\overline{X}_3 \leq (306/312/318)\ ,$$
$$(1.8/2/2.2)\overline{X}_1 + (3.8/4/4.2)\overline{X}_2 + (0.9/1/1.1)\overline{X}_3 \leq (121/124/127)\ ,$$
$$\overline{X}_1, \overline{X}_2, \overline{X}_3 \geq 0\ ,$$

where the $\overline{X}_i$ are triangular fuzzy numbers for the amount to produce for P_i per week. We change this to

$$\sup[\overline{G}_1(z_2) \cdot \overline{G}_2(A_2) \cdot \overline{G}_3(A_1')], \tag{6.25}$$

subject to the same constraints.

Next we find estimates of b_i to construct the fuzzy goals. This depends on which inequality relation $\leq$ we will use.

Using b_i' we next select values for c_i. We decide on three separate selections for the c_i in order to minimally explore the undominated set.

	Chen's Inequality	Kerre's Inequality
b'	2936	2637
$c_1/c_2/c_3$	100 /100 / 400	
Best Fitness	0.156	0.028
z_2	$\overline{G}_1(1409.93) = 0.516$	$\overline{G}_1(392.33) = 0.204$
A_2	$\overline{G}_2(750.06) = 0.306$	$\overline{G}_2(235.57) = 0.148$
A_1	$\overline{G}_3(39.31) = 0.986$	$\overline{G}_3(200.15) = 0.919$
$c_1/c_2/c_3$	100 /200 / 400	
Best Fitness	0.142	0.019
z_2	$\overline{G}_1(1307.54) = 0.483$	$\overline{G}_1(163.05) = 0.122$
A_2	$\overline{G}_2(800.39) = 0.297$	$\overline{G}_2(352.82) = 0.156$
A_1	$\overline{G}_3(36.59) = 0.987$	$\overline{G}_3(83.21) = 0.967$
$c_1/c_2/c_3$	150 /100 / 300	
Best Fitness	0.154	0.026
z_2	$\overline{G}_1(1445.78) = 0.519$	$\overline{G}_1(392.32) = 0.188$
A_2	$\overline{G}_2(732.45) = 0.301$	$\overline{G}_2(235.58) = 0.148$
A_1	$\overline{G}_3(40.26) = 0.986$	$\overline{G}_3(200.12) = 0.923$

Table 6.2: Results for the Product Mix Problem Using Different Values for c_i (i=1,2,3)

We used the solution to the crisp linear program to randomly generate the initial values of the $\overline{X}_i$ in the evolutionary algorithm. The crisp linear program is the one obtained by using the peak values (where the membership function is one) of all the $\overline{C}_i$, $\overline{A}_{ij}$ and $\overline{B}_i$. The crisp solution is $x_1 = 0$ for (P_1), $x_2 = 27$ (for P_2) and $x_3 = 16$ (for P_3) with $\max z = 312$. The results of the evolutionary algorithm, for the different selections of $\leq$ and the different values of the c_i are presented in Table 6.2.

In the b' row the single value means $b'_1 = b'_2 = b'_3$ equals the value given.

The results in Table 6.2 were surprising in that Chen's method of evaluating $\leq$ produced much better results for z_2, A_2, and A_1. It looks like the feasible set is much larger using Chen's method.

Figures 6.5 and 6.6 show the "optimal" $\overline{X}_i$, for both methods, for one value of $c = (c_1, c_2, c_3)$. In Figure 6.5, $\overline{X}_1$ was not shown because it was very close to zero.

Figure 6.7 shows the value of the objective function $\overline{Z}$, corresponding to the values of the $\overline{X}_i$ given in Figure 6.5. The pictures (Figures 6.7 and 6.8) imply that $\overline{Z}$ is a triangular fuzzy number, but in fact it is a triangular shaped fuzzy number very close to being a triangle. Figure 6.8 shows the value of the objective function $\overline{Z}$, for the values of the $\overline{X}_i$ shown in Figure 6.6.

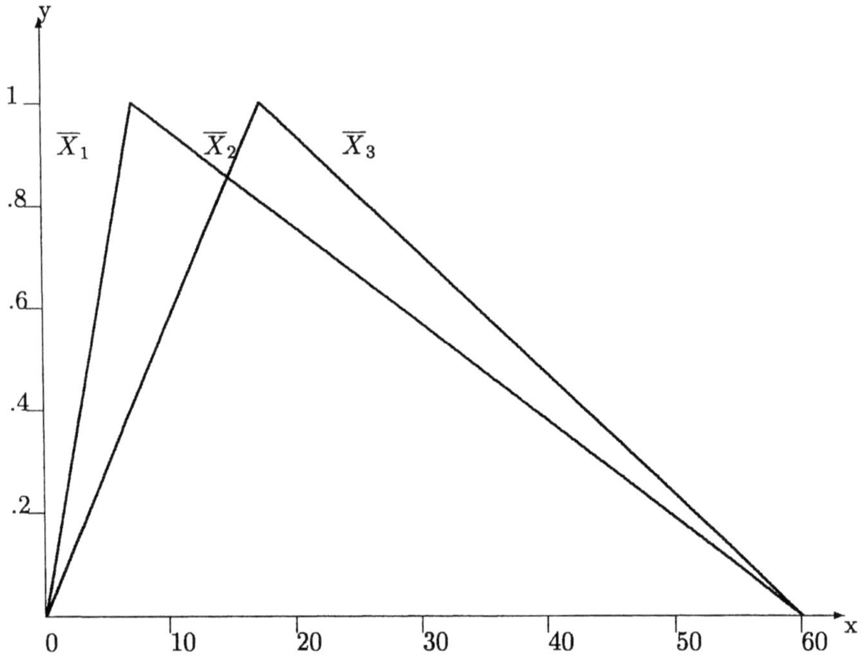

Figure 6.5: $\overline{X}_1, \overline{X}_2, \overline{X}_3$ Obtained with Kerre's Inequality ($c_1 = 100, c_2 = 200, c_3 = 400$), Product Mix Problem

Diet Problem

A farmer has three products P_1, P_2 and P_3 which he plans to mix together to feed his pigs. He knows the pigs require a certain amount of food F_1 and F_2 per day. Table 6.3 presents estimates of the units of F_1 and F_2 available, per gram of P_1, P_2 and P_3.

Also, each pig should have approximately at least 54 units of F_1 and approximately at least 60 units of F_2, per day. The costs of P_1, P_2 and P_3 vary slightly from day to day but the average costs are: (1) 0.08\$ per gram of P_1; (2) P_2 is 0.09\$ per gram; and (3) 0.10\$ /gram for P_3.

The farmer wants to know how many grams of P_1, P_2 and P_3 he should

		Food F_1	Food F_2
Product	P_1	2.5	5
Product	P_2	4.5	3
Product	P_3	5	10

Table 6.3: Approximate Units of Food F_j in Product P_i

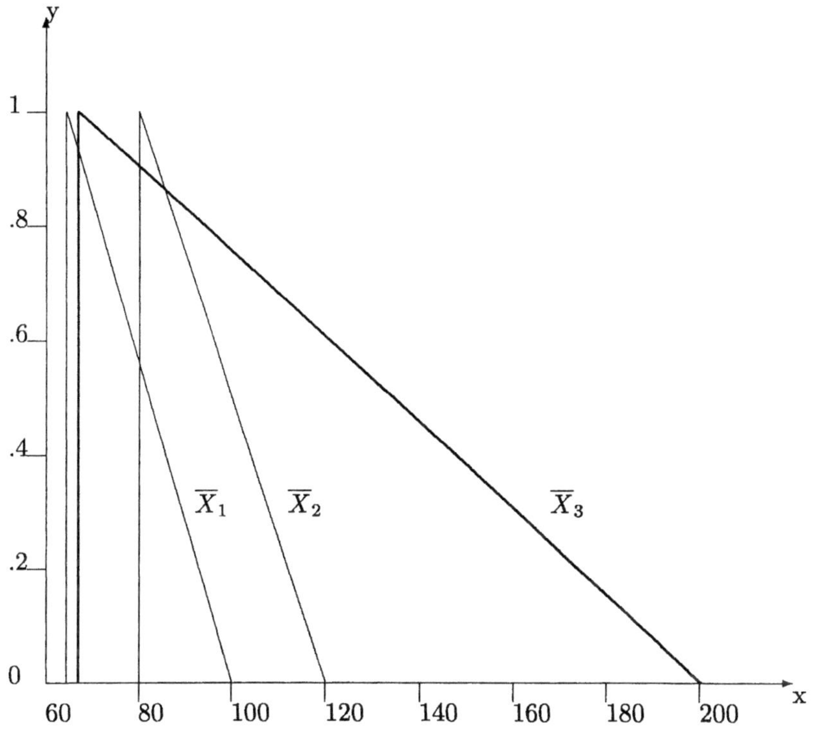

Figure 6.6: $\overline{X}_1, \overline{X}_2, \overline{X}_3$ Obtained with Chen's Inequality ($c_1 = 150, c_2 = 100, c_3 = 300$), Product Mix Problem

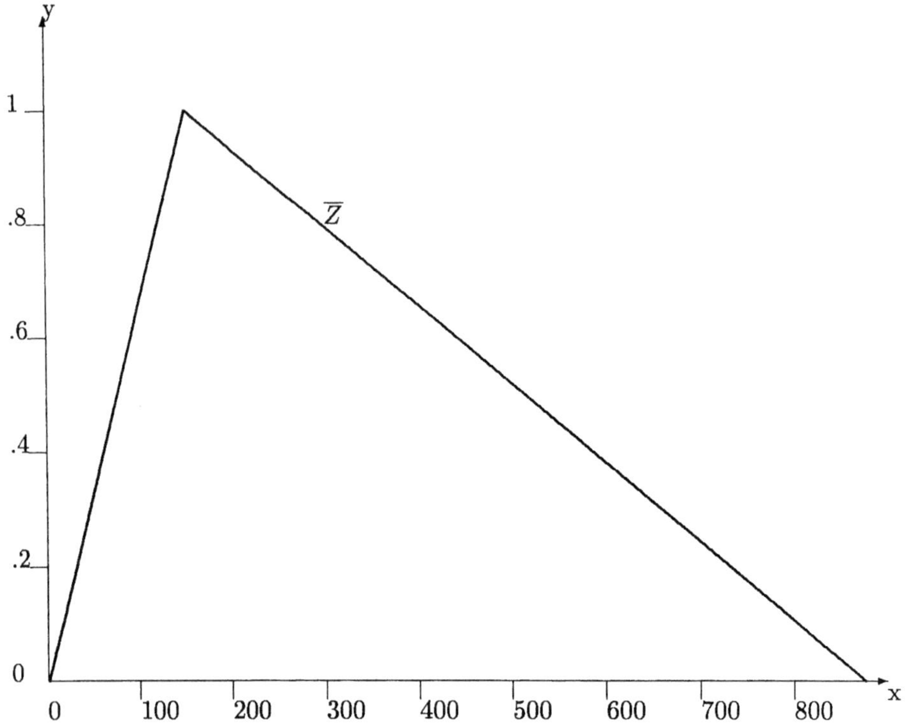

Figure 6.7: $\overline{Z}$ Obtained with Kerre's Inequality ($c_1 = 100, c_2 = 200, c_3 = 400$), Product Mix Problem

mix together each day, so his pigs will get the approximate minimums, to minimize his costs. Since all the numbers are uncertain we substitute triangular fuzzy numbers, whose peak is at the given value, for all the parameters to give a fully fuzzified linear program. The problem becomes

$$\begin{aligned} \min \overline{Z} &= (7/8/9)\overline{X}_1 + (8/9/10)\overline{X}_2 + (9/10/11)\overline{X}_3 \qquad (6.26)\\ &\text{subject to:}\\ &\quad (2/2.5/3)\overline{X}_1 + (4/4.5/5)\overline{X}_2 + (4.5/5/5.5)\overline{X}_3 \geq (50/54/58)\ ,\\ &\quad (4.5/5/5.5)\overline{X}_1 + (2.5/3/3.5)\overline{X}_2 + (9/10/11)\overline{X}_3 \geq (56/60/64)\ ,\\ &\quad \overline{X}_1, \overline{X}_2, \overline{X}_3 \geq 0\ , \end{aligned}$$

where $\overline{X}_i$ is the amount of P_i, in grams, to use per day, for each pig.

Changing to a multiobjective fuzzy linear program we obtain

$$(\inf z_2, \inf A_2, \inf A_1')\ . \qquad (6.27)$$

We define fuzzy goals for the minimization problem as shown in Figures 6.9 through 6.11. Using these goals our problem now becomes

$$\sup(\overline{G}_1(z_2) \cdot \overline{G}_2(A_2) \cdot \overline{G}_3(A_1'))\ , \qquad (6.28)$$

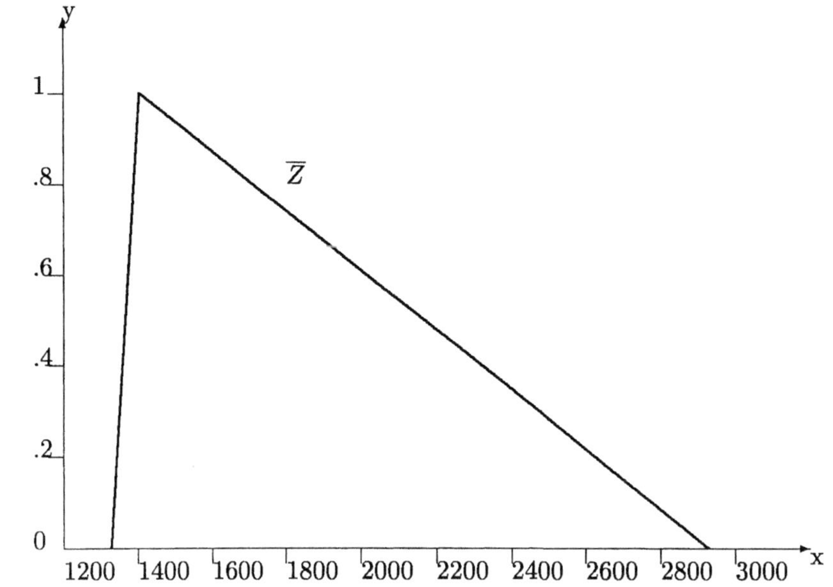

Figure 6.8: $\overline{Z}$ Obtained with Chen's Inequality ($c_1 = 150, c_2 = 100, c_3 = 300$), Product Mix Problem

for $\overline{X}$ feasible. Because of the way the fuzzy goals are setup this is a maximization (supremum) problem in equation (6.28) and not a minimization problem.

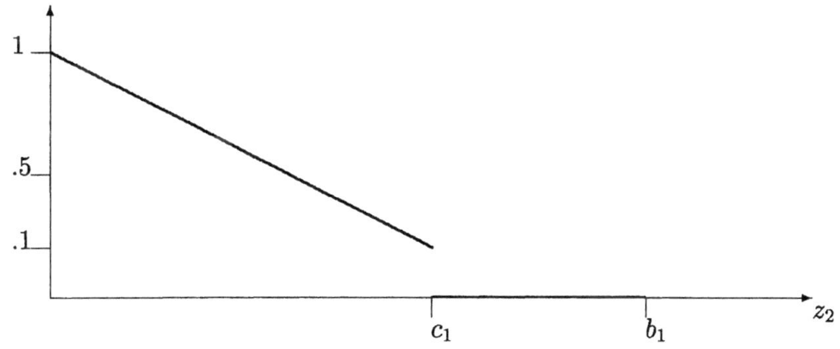

Figure 6.9: Fuzzy Goal $\overline{G}_1$ for the Diet Problem

There are a number of other changes, that we will now briefly discuss, that are needed for a minimization problem. The definition of dominance is changed in an obvious manner for minimization. We now evaluate $\overline{M} \geq \overline{N}$ for the constraints instead of $\overline{M} \leq \overline{N}$ discussed in Section 6.1.2. The feasible set is still non–empty if we assume $\overline{C}_i \geq 0$, $\overline{A}_{ij} \geq 0$, $\overline{B}_i \geq 0$ for all i, j. The ideal point becomes $(0, 0, 0)$ which we assume is unattainable. We still have

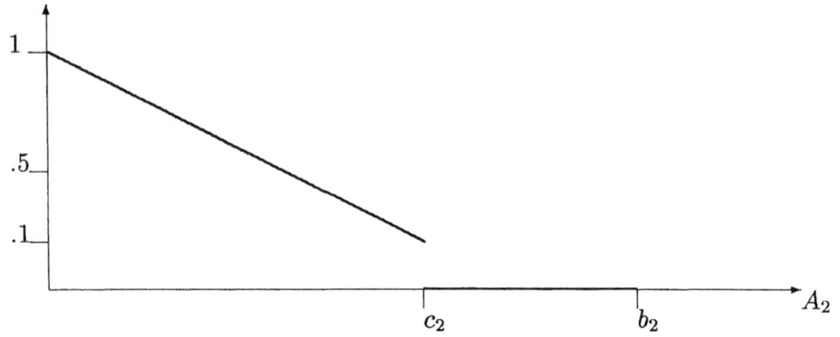

Figure 6.10: Fuzzy Goal $\overline{G}_2$ for the Diet Problem

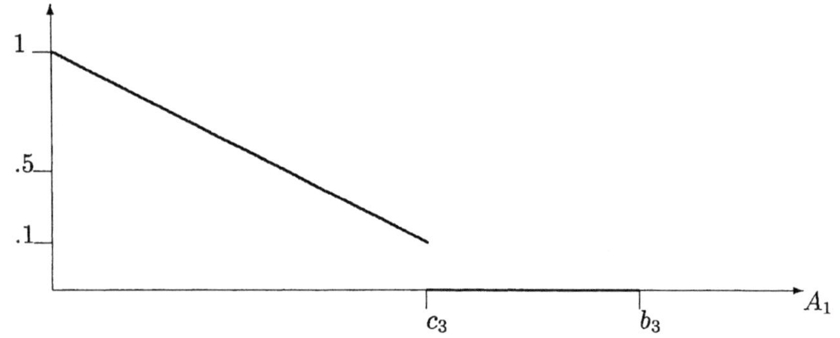

Figure 6.11: Fuzzy Goal $\overline{G}_3$ for the Diet Problem

to estimate the b_i with b_i' and choose the c_i to explore the undominated set.

We again used the solution to the crisp linear program to randomly generate the initial values of the $\overline{X}_i$ in the evolutionary algorithm. The crisp solution is $x_1 = 0$ (for P_1), $x_2 = 8$ (for P_2) and $x_3 = 3.6$ (for P_3) with minimum $z = 108$ (in cents).

The results are shown in Table 6.4. Again, the b' now means $b_1' = b_2' = b_3'$ equals the value given.

It appears for this min problem, Kerre's method gives slightly better results. We wanted to use the same c_i values for both methods, but in the first row of results in Table 6.4 a value of $c_3 = 900$ in Chen's method produced all fuzzy goals zero, so we had to increase the value of c_3 until all fuzzy goals $\overline{G}_i$, $1 \leq i \leq 3$, where not zero. This explains the different c_3 values for the two methods.

Figures 6.12 and 6.13 display the "optimal" $\overline{X}_i$, for both methods, for one value of c. It is interesting to note that $\overline{X}_1$ and $\overline{X}_3$ are reversed between Figures 6.12 and 6.13.

Figures 6.14 and 6.15 present the value of the objective function $\overline{Z}$ for the values of the $\overline{X}_i$ shown in Figures 6.12 and 6.13, respectively. Again, $\overline{Z}$

	Chen's Inequality	Kerre's Inequality
b'	1111	951
$c_1/c_2/c_3$	400 / 400 / 1050	400 / 400 / 900
Best Fitness	0.069	0.083
z_2	$\overline{G}_1(209.41) = 0.529$	$\overline{G}_1(173.45) = 0.610$
A_2	$\overline{G}_2(\ 32.34) = 0.927$	$\overline{G}_2(\ 9.93) = 0.978$
A_1	$\overline{G}_3(108.40) = 0.141$	$\overline{G}_3(\ 89.87) = 0.139$
$c_1/c_2/c_3$	300 / 450 / 1050	300 / 450 / 900
Best Fitness	0.038	0.070
z_2	$\overline{G}_1(150.34) = 0.549$	$\overline{G}_1(119.74) = 0.641$
A_2	$\overline{G}_2(197.73) = 0.604$	$\overline{G}_2(\ 6.91) = 0.986$
A_1	$\overline{G}_3(\ 77.82) = 0.117$	$\overline{G}_3(\ 62.05) = 0.111$
$c_1/c_2/c_3$	450 / 300 / 1050	450 / 300 / 900
Best Fitness	0.077	0.088
z_2	$\overline{G}_1(246.40) = 0.507$	$\overline{G}_1(192.61) = 0.615$
A_2	$\overline{G}_2(\ 13.52) = 0.960$	$\overline{G}_2(\ 10.95) = 0.967$
A_1	$\overline{G}_3(127.51) = 0.157$	$\overline{G}_3(\ 99.80) = 0.149$

Table 6.4: Results for the Diet Problem Using Different Values for c_i (i=1,2,3)

is a triangular shaped fuzzy number very close to being an exact triangle.

6.1.5 Summary and Conclusions

In this section we discussed solutions to the fully fuzzified linear programming problem where all the parameters and variables are triangular fuzzy numbers. We changed the problem of $\max(\overline{Z})$, the fuzzy number value of the objective function, into a single crisp objective problem (equation 6.11)) using fuzzy goals. Solutions are undominated solutions to the multiobjective problem (equation (6.4)). We also discussed methods of evaluating fuzzy inequalities and picked two procedures, that insured that $\overline{Z}$ was bounded, to be used in the two applications. The computations were sufficiently complex that we employed an evolutionary algorithm to compute approximate solutions in the applications.

The problem was made easier by assuming all the fuzzy numbers were non-negative triangular fuzzy numbers. So, future research should be concerned with allowing for negative fuzzy numbers, allowing for trapezoidal (shaped) fuzzy numbers, and other methods of evaluating fuzzy inequalities. The results described here have been generalized to fully fuzzified multiobjective linear programming in [10].

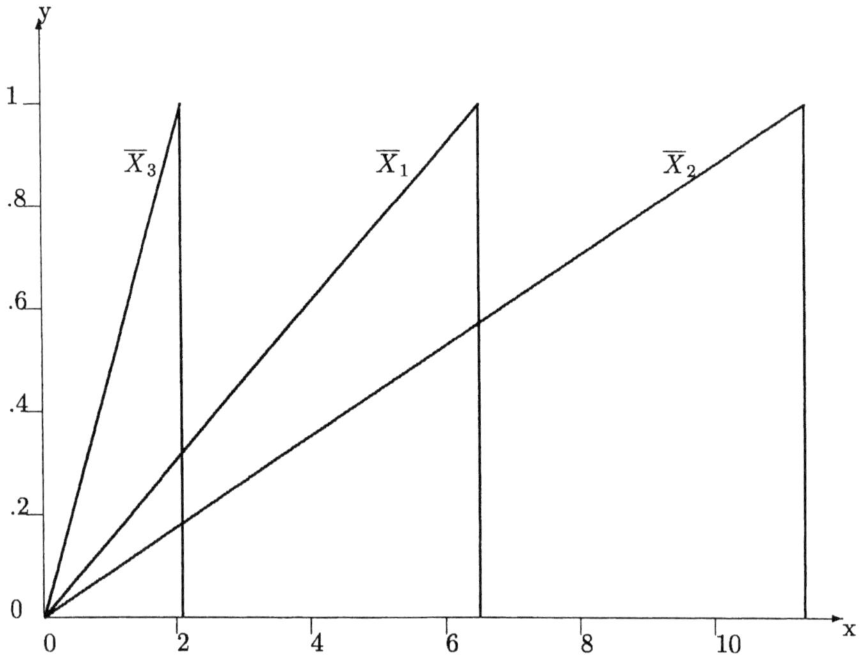

Figure 6.12: $\overline{X}_1, \overline{X}_2, \overline{X}_3$ Obtained with Kerre's Inequality ($c_1 = 400, c_2 = 400, c_3 = 900$), Diet Problem

6.2 Fuzzy PERT

PERT stands for "Project Evaluation and Review Technique". A project defines a combination of interrelated activities (jobs) that must be completed in a certain order before the entire project can be completed. The project that we will concentrate on in this section is shown in Figure 6.16.

The jobs are interrelated in a sequence such that some jobs cannot start until others are finished. In Figure 6.16 we see that jobs J_{35} and J_{34} coming from node number 3, cannot start until all jobs leading into node 3, which are J_{13} and J_{23}, have been completed. In general, jobs require time and resources for completion. We will not discuss the allocation of resources to jobs in this section so the concept of "crashing" ([16],[54]) will not be used. So, attached to each job J_{ij} is only its time to completion $t_{ij} \geq 0$. Let T_{ij} denote a variable whose values t_{ij} are possible job times for job J_{ij}.

In general, job times are uncertain and in PERT T_{ij} is a random variable, while in fuzzy PERT, T_{ij} is a fuzzy variable. Let T_i be the time all jobs entering node number i are finished and is therefore the earliest time all jobs leaving this node may start. We wish to find T_8 giving the time of project duration. We see from Figure 6.12 that $T_8 = \max\{T_{68}, T_{78}\}$, $T_7 = \max\{T_{47}, T_{57}\}$, etc. In stochastic PERT, T_8 will be a random variable, so one

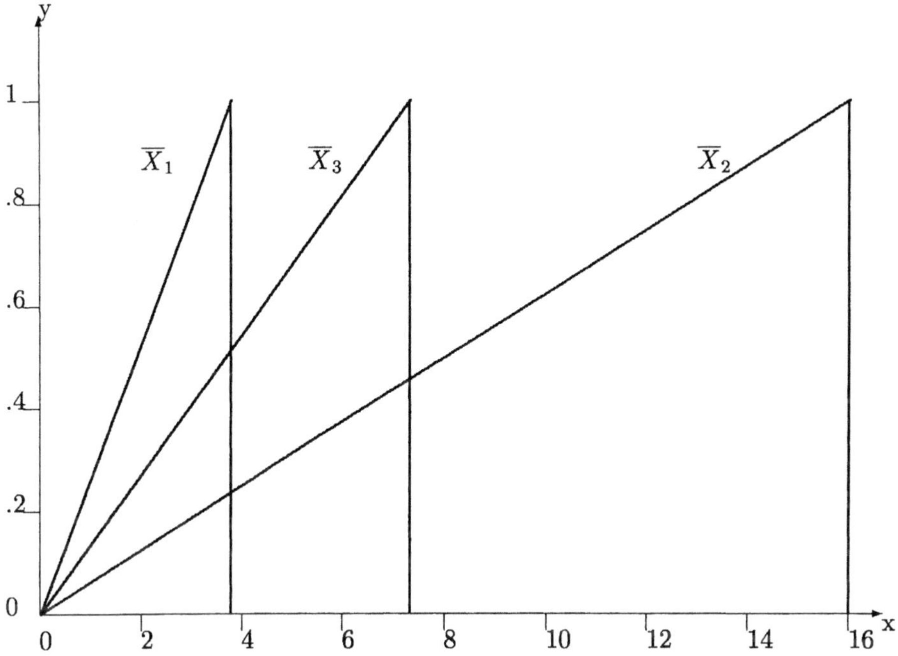

Figure 6.13: $\overline{X}_1, \overline{X}_2, \overline{X}_3$ Obtained with Chen's Inequality ($c_1 = 450, c_2 = 300, c_3 = 1050$), Diet Problem

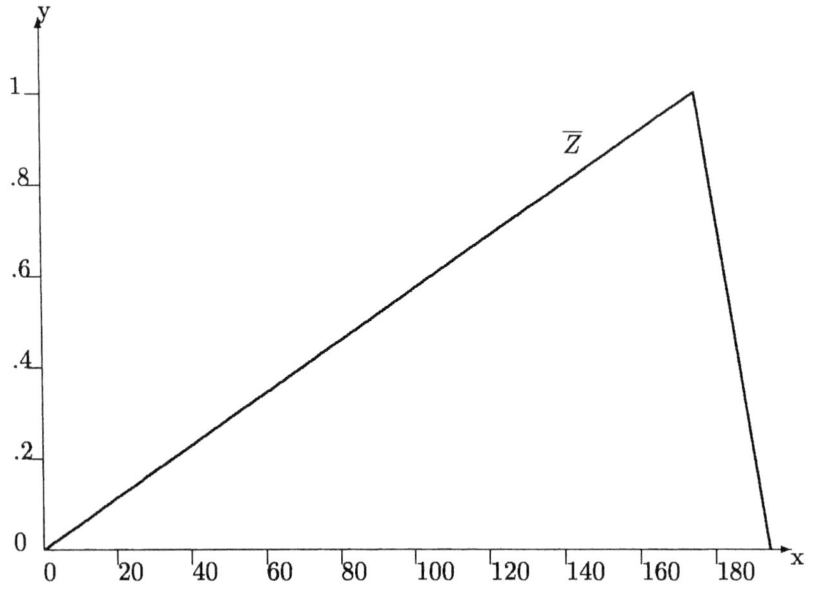

Figure 6.14: $\overline{Z}$ Obtained with Kerre's Inequality ($c_1 = 400, c_2 = 400, c_3 = 900$), Diet Problem

would want to find its probability density function so that we can compute Prob$[T_8 \leq \lambda]$ for various possible due date times λ. We may have a constraint like Prob$[T_8 > \lambda] \leq 0.05$. However, in probabilistic PERT it is very difficult, in general, to find the probability density of project duration and one usually uses simulation to construct a histogram approximating the unknown density. In fuzzy PERT T_8 is a fuzzy variable and we wish to find its possibility distribution so that we can find Poss$[T_8 \leq \lambda]$ for values of λ. We now show that there is an algorithm to find the possibility distribution of T_8. The only assumption needed is for the job times to be non–interactive, which is analogous to independence for random variables. For fuzzy PERT we will consider two cases: (1) job times are trapezoidal fuzzy numbers; and (2) job times are discrete fuzzy sets.

There are a number of other items usually computed in PERT such as floats, earliest start time for a job, latest start time for a job, etc. We will not be concerned with any of these computations and the reader may consult ([16],[54]) for probabilistic PERT, and ([20], and the references in this paper ,and [12],[13], [43], [58], [59] for fuzzy PERT), for further details. This section is based on [2].

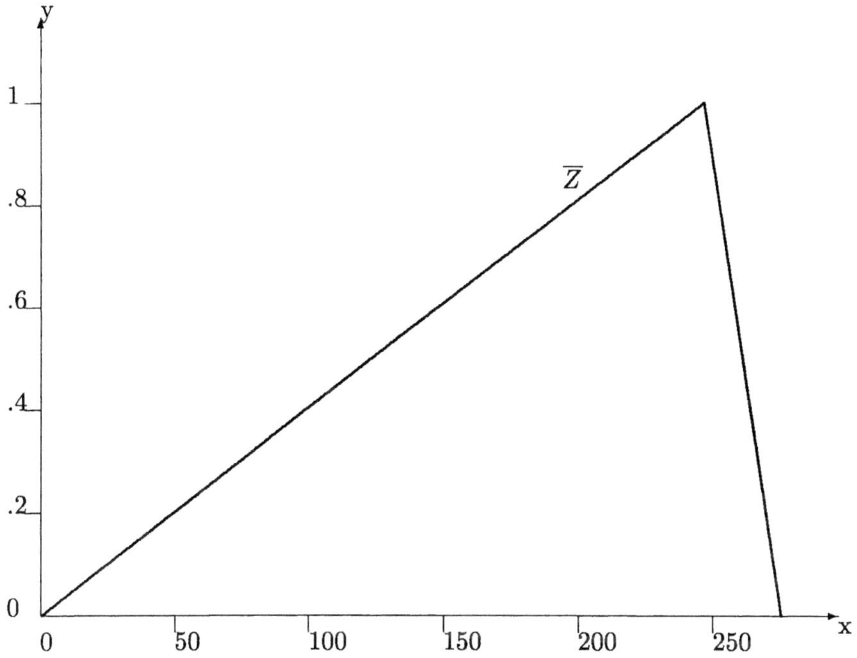

Figure 6.15: $\overline{Z}$ Obtained with Chen's Inequality ($c_1 = 450, c_2 = 300, c_3 = 1050$), Diet Problem

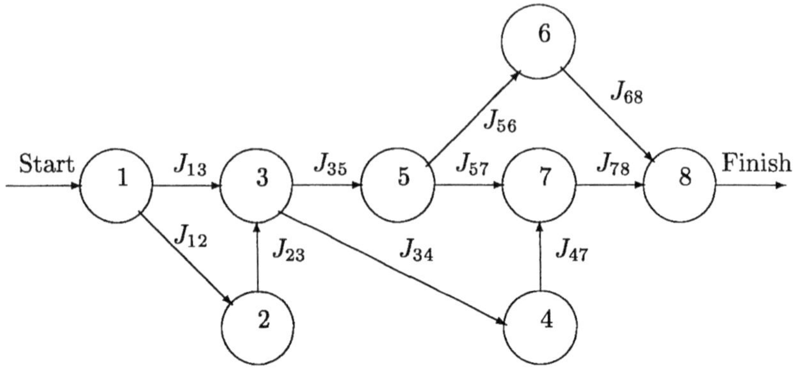

Figure 6.16: Project Network

6.2.1 Job Times Fuzzy Numbers

Let the possibility distribution for T_{ij} be the trapezoidal fuzzy number $\overline{M}_{ij} = (a_{ij}/m_{ij}^1, m_{ij}^2/b_{ij})$. We say that the possibility that $T_{ij} = x$ is equal to $\overline{M}_{ij}(x)$ and this is written as $\text{Poss}[T_{ij} = x] = \overline{M}_{ij}(x)$. In the definition of $\overline{M}_{ij}$: (1) a_{ij} is the optimistic job time for J_{ij}, or the shortest time possible if everything goes well; (2) m_{ij}^1, m_{ij}^2 are the lower, and upper, bounds respectively, of the most likely, or expected, job time; and (3) b_{ij} is the pessimistic job time for J_{ij}, or the longest time possible when everything goes wrong. We assume that $0 \leq a_{ij} < m_{ij}^1 \leq m_{ij}^2 < b_{ij}$ for all jobs. So job times may be triangular when $m_{ij}^1 = m_{ij}^2$. These fuzzy numbers $\overline{M}_{ij}$ can be obtained from expert opinion and if so, one might want to only solicit the numbers a_{ij}, m_{ij}^1, m_{ij}^2 and b_{ij} from experts.

To determine the possibility distribution of project duration we first form the joint possibility distribution for all jobs. Let $< t_{ij} >$ be a vector of possible job times for all the jobs in the project. The joint distribution $\mu(< t_{ij} >)$ is defined, for non–interactive fuzzy variables T_{ij}, as

$$\mu(< t_{ij} >) = \min\left\{\overline{M}_{ij}(t_{ij}) \mid \text{for all jobs}\right\} , \tag{6.29}$$

if $a_{ij} < t_{ij} < b_{ij}$ for all jobs. If a t_{ij} is not in the interval (a_{ij}, b_{ij}), then $\mu(< t_{ij} >) = 0$. Next let

$$\Gamma_x = \{< t_{ij} > \mid T_8 = x\} , \tag{6.30}$$

which is the set of all vectors of job times that produce project duration equal to x. Then

$$\text{Poss}[T_8 = x] = \sup\{\mu(< t_{ij} >) \mid < t_{ij} > \in \Gamma_x\} , \tag{6.31}$$

because $\mu(< t_{ij} >)$ is a possibility distribution. In general, if X is a universal set and $\overline{A}$ is a possibility distribution on X ($\overline{A}$ is a fuzzy subset of X) and

Job	a_{ij}	m_{ij}^1	m_{ij}^2	b_{ij}
J_{12}	4	5	5	6
J_{13}	7	8	8	10
J_{23}	2	4	4	5
J_{35}	3	4	5	6
J_{34}	3	5	5	6
J_{47}	3	5	5	6
J_{56}	2	5	6	7
J_{57}	4	5	5	6
J_{78}	4	5	5	8
J_{68}	3	5	5	6

Table 6.5: Possibility Distributions for Job Times in Example 6.2.1

E is a (crisp) subset of X, then Poss$[E] = \sup\{\overline{A}(x) \mid x \in E\}$. This is how we obtained equation (6.31). The possibility distribution for T_8 is a fuzzy set $\overline{M}_8$ so that Poss$[T_8 = x] = \overline{M}_8(x)$. Equation (6.31) is used to construct $\overline{M}_8$.

We now present a computational method of computing $\overline{M}_8$. Let

$$L_{ij}(y) = (m_{ij}^1 - a_{ij})y + a_{ij} \ , \tag{6.32}$$

$$R_{ij}(y) = (m_{ij}^2 - b_{ij})y + b_{ij} \ , \tag{6.33}$$

for $0 \le y \le 1$. Now let us assume we have a software package, which we will call $\mathcal{L}$, which computes project duration T_8 given $< t_{ij} >$. The project in Figure 6.16 is fixed, all we input into $\mathcal{L}$ is $< t_{ij} >$. Let $\mathcal{L}(< t_{ij} >) = T_8$. For each fixed y in $[0,1]$ let $\mathcal{L}(< L_{ij}(y) >) = t_8^1(y)$ and $\mathcal{L}(< R_{ij}(y) >) = t_8^2(y)$. The $t_8^1(y)$ and $t_8^2(y)$ are project duration times. Then it was shown in [2] that

$$\overline{M}_8(x) = \begin{cases} y & : \text{ if } x = t_8^1(y),\ 0 \le y < 1 \ , \\ 1 & : \text{ if } t_8^1(1) \le x \le t_8^2(1) \ , \\ y & : \text{ if } x = t_8^2(y), 0 \le y < 1, \\ 0 & : \text{ otherwise } . \end{cases} \tag{6.34}$$

Example 6.2.1.1

The project network is in Figure 6.16 with fuzzy job times presented in Table 6.5.

We ran $\mathcal{L}$ on $L_{ij}(y)$ and $R_{ij}(y)$ for $y = 0, 0.1, \ldots, 0.9, 1.0$. Using these results and equation (6.34) we constructed $\overline{M}_8$ shown in Figure 6.17.

Notice that the slope of $\overline{M}_8$ in Figure 6.17 changes at $t_8 = 20$ and at $t_8 = 27.5$ both corresponding to $\alpha = 0.5$. In both cases the slope changes due to different critical paths being created in the network. A critical path is a path through the network from start to finish so that the delay of the completion of any job on this path will cause a delay in the completion of the project. It is easy to compute $\text{Poss}[T_8 \le \lambda]$ because: (1) $\text{Poss}[T_8 \le \lambda] = \overline{M}_8(\lambda)$ if $\lambda \le 24$; and (2) $\text{Poss}[T_8 \le \lambda] = 1$ if $\lambda \ge 24$. We define $\text{Poss}[T_8 \le \lambda] = \max\{\overline{M}_8(x) | x \le \lambda\}$.

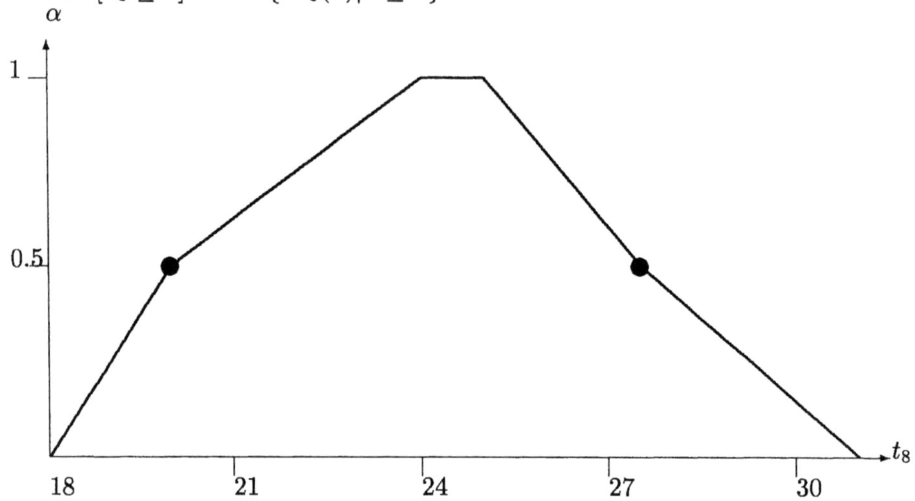

Figure 6.17: Possibility Distribution for Project Duration in Example 6.2.1.1

6.2.2 Job Times Discrete Fuzzy Sets

Now job times t_{ij} will be non–negative integers. The possibility distribution for a job time might look like

$$\left\{ \frac{0.2}{2}, \frac{0.5}{3}, \frac{1.0}{4}, \frac{0.6}{5}, \frac{0.1}{6} \right\} . \tag{6.35}$$

The computational algorithm to find the possibility distribution of project duration is more complicated than the one for trapezoidal number job times and we will omit it. The interested reader needs to consult [2] for the complete details.

6.2.3 Summary

This section develops possibilistic PERT as an alternative to probabilistic PERT. Job times are fuzzy variables defined by possibility distributions instead of random variable specified by probability distributions. We showed that it is not too difficult to determine the possibility distribution of project duration but other possibility distributions (float, earliest start time, etc.) appear to be considerably more difficult to calculate [20].

It is interesting to compare the techniques of possibility theory and probability theory. In possibility theory we use non–interactive, minimum, and

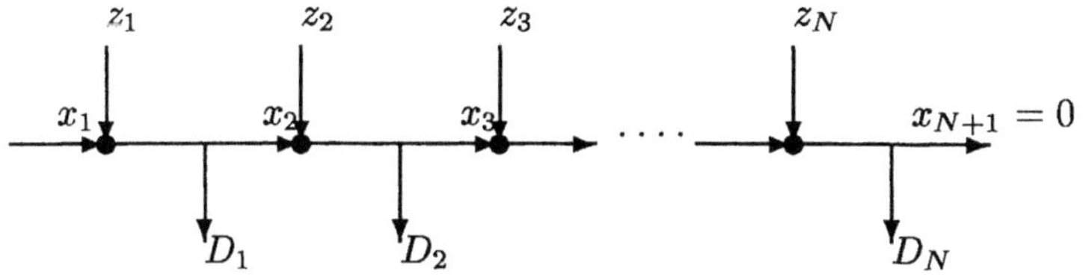

Figure 6.18: Inventory Problem

supremum in place of the probabilistic concepts of independence, product, and integration. It turns out in PERT certain calculations become easier in the possibilistic model than in the probabilistic model. Possibly this can be generalized in other network models.

6.3 Fuzzy Inventory Control

Most of the applications of fuzzy sets to inventory control have been in the area of the EOQ, or the economic order quantity problem. See ([14], [22] [24],[31],[32],[35], [38],[40],[53], [57], [60], [61]), and the references in these papers, for a survey of this literature on fuzzy inventory control. In this section we will apply fuzzy sets to model uncertainty in the single item, N-period, inventory control problem. The crisp problem will be outlined below. We first assume that demand is known (not fuzzy) but the ordering cost and holding cost can be fuzzy. This is in Section 6.3.1. In the next Section 6.3.2 demand will be fuzzy, shortages are allowed and the penalty cost for shortages is also fuzzy. Finally, we add to the model in Section 6.3.2 backorders, maximize profit instead of minimizing cost since we now sell the product, and discount all future monies to their present value in Section 6.3.3. All of this section is based on ([7],[8]).

Let us now describe the inventory problem we shall be studying. The flow is shown in Figure 6.18. The incoming inventory x_1 will always be a given real number. The variables are the $z_1, z_2, \ldots, z_N$, which are the amounts we are to order each period. The z_i, $1 \leq i \leq N$, will always be non–negative integers. If we allow the z_i to be fuzzy we will have to defuzzify them in the end, so we will have them non–negative integers. The D_i represents the demand in the i-th period and the x_i, $2 \leq i \leq N$, stands for the outgoing inventory, which will be the starting inventory for the next period, in the

i-th period. So we must have

$$x_{i+1} = x_i + z_i - D_i \ , \tag{6.36}$$

$1 \leq i \leq N$.

There are only N periods and at the end of the planning horizon we assume that the final inventory will be zero. That is, we want $x_{N+1} = 0$. When D_i is fuzzy, the x_i will also be fuzzy from equation (6.36), so we will adjust this crisp constraint of zero final inventory in Section 6.3.2.

Another basic assumption throughout Section 6.3 is that we have zero delivery lag with instant replenishment at the start of each period.

We wish to minimize the total inventory cost over the N periods. This cost is made up of four components: (1) purchase cost; (2) ordering cost; (3) holding cost; and (4) shortage cost. We now discuss the first three in detail. The fourth one will be discussed later in Section 6.3.2.

We assume that we are to buy the item and we do not produce it ourselves. There may, or may not, be price breaks. A simple model, which we shall use is

$$c_i(z_i) = \begin{cases} 10z_i & : \ 0 \leq z_i \leq L_i \\ 10L_i + 5(z_i - L_i) & : \ z_i > L_i \ . \end{cases} \tag{6.37}$$

This means in the i-th period we pay \$ 10 per unit for the first L_i units and then we pay \$ 5/unit for each additional unit. We will assume these are all known numbers and they will not be fuzzy.

The ordering cost K_i in the i-th period is the cost of placing the order, checking up on the order, and putting the items into inventory when they arrive. This number is always difficult to estimate so we will model it using a fuzzy number. Then the total cost of obtaining z_i units at the start of the i-th period is

$$\overline{C}_i(z_i) = \begin{cases} 0 & : \ z_i = 0 \\ \overline{K}_i + c_i(z_i) & : \ z_i > 0 \ , \end{cases} \tag{6.38}$$

for fuzzy $\overline{K}_i$.

The holding cost is assumed to be proportional to the ending inventory $x_{i+1} = x_i + z_i - D_i$. The model may be readily extended to cover any holding cost function $H_i(x_{i+1})$ by replacing x_{i+1} by $H_i(x_{i+1})$. For example, holding cost may be modeled as proportional to $(x_i + x_{i+1})/2$. In our first model (Section 6.3.1) no shortages are allowed and D_i is crisp so all the x_i will be non-negative real numbers. In our other models shortages are allowed and $\overline{D}_i$ can be fuzzy, so certain adjustments in the development must be made. Let h_i be the holding cost per unit for the i-th period. This number, depending on interest on invested capital, depreciation, etc., is very difficult to determine exactly so it will be fuzzy. For non-negative x_i and crisp D_i the holding cost for the i-th period is

$$\overline{h}_i(x_i + z_i - D_i) \ , \tag{6.39}$$

for fuzzy $\overline{h}_i$, since $x_{i+1} = x_i + z_i - D_i$.

When no shortages are allowed let

$$\overline{TC}_i = \overline{C}_i(z_i) + \overline{h}_i(x_i + z_i - D_i) \ , \tag{6.40}$$

and

$$\overline{Z} = \sum_{i=1}^{N} \overline{TC}_i \ . \tag{6.41}$$

We wish to find z_i, $1 \leq i \leq N$, to minimize $\overline{Z}$. The constraint is $x_{N+1} = 0$.

Now $\overline{Z}$ is a fuzzy set so we can not minimize it. What we shall do is first change $\min(\overline{Z})$ into a multiobjective optimization problem, and then into a single objective problem as in Section 6.1. Let the support of $\overline{Z}$ be $[u_1, u_3]$ and let the center of the core of $\overline{Z}$ be at u_2. Define

$$A_1 = \int_{u_1}^{u_2} \overline{Z}(x)dx \ , \tag{6.42}$$

and

$$A_2 = \int_{u_2}^{u_3} \overline{Z}(x)dx \ , \tag{6.43}$$

where $\overline{Z}(x)$ denotes the membership function of $\overline{Z}$ evaluated at $x \in \mathbb{R}$. In place of $\min(\overline{Z})$ we have the multiobjective problem

$$(\max A_1, \min u_2, \min A_2) \ , \tag{6.44}$$

because: (1) we wish to maximize the possibility of getting values less than u_2 ($\max A_1$); and (2) we wish to minimize the possibility of getting values more than u_2 ($\min A_2$). We now change the multiobjective problem into a single objective problem as follows

$$\min \{\lambda_1[M - A_1] + \lambda_2 u_2 + \lambda_3 A_2\} \ , \tag{6.45}$$

where $\lambda_i > 0$, $\lambda_1 + \lambda_2 + \lambda_3 = 1$, and M is a large positive number so that $\max A_1$ is equivalent to $\min[M - A_1]$.

We used a different method of changing the multiobjective optimization problem into a single objective problem in Section 6.1.1 (see equation (6.11)). The procedure above, equation (6.45), is another way to obtain a single objective. However, it is well–known, see [23], that one may not be able to explore the whole undominated set using equation (6.45). You will get different undominated solutions by choosing different values of the $\lambda_i > 0$, $\lambda_1 + \lambda_2 + \lambda_3 = 1$, but there are examples where you cannot obtain all undominated solutions by varying the λ_i values. Nevertheless, we will use this method for fuzzy inventory control.

The model discussed in Section 6.3.1, together with a numerical example, is to find the z_i to minimize the expression in equation (6.45) subject to $x_{N+1} = 0$ given: (1) x_1, $c_i(z_i)$, N, D_i, λ_i and M all crisp; and (2) $\overline{h}_i$ and $\overline{K}_i$ fuzzy.

The changes made in Section 6.3.2 is to allow shortages and have $\overline{D}_i$ fuzzy. There will be no backlogging (back orders) so shortages result in lost sales and loss of customer's goodwill. The penalty cost, due to shortages, is usually very difficult to estimate so it will be modeled by a fuzzy number $\overline{p}_i$ for the period i. We still minimize the expression in equation (6.45), with $\overline{D}_i$, $\overline{K}_i$, $\overline{h}_i$, $\overline{p}_i$ and the $\overline{x}_i$ ($i \geq 2$) all fuzzy, subject to $\overline{x}_{N+1}$ approximately zero. A numerical example is also presented for this fuzzy inventory model.

The fuzzy inventory model in 6.3.3 is under development and the authors plan to present the results at an international conference [8] in 2001.

There will be a slight change in the fuzzy mathematics in Subsections 6.3.2 and 6.3.3 that does not occur in the rest of the book. We will be working with fuzzy sets that are parts of fuzzy numbers.

In Subsections 6.3.2 and 6.3.3 the $\overline{D}_i$, $\overline{K}_i$, $\overline{h}_i$ and $\overline{p}_i$ will all be non-negative trapezoidal fuzzy numbers, the $\overline{x}_i$ ($i \geq 2$) need not be fuzzy numbers but will be non-negative, and $\overline{Z}$ will be a non-negative trapezoidal shaped fuzzy number. Other fuzzy sets are constructed which are not standard fuzzy numbers. Let

$$\overline{\mathrm{Pos}}(x) = \begin{cases} 1 & : \quad x \geq 0 \\ 0 & : \quad \text{otherwise} \end{cases}, \tag{6.46}$$

and let $\overline{N} = (a/b, c/d)$, with $a < 0 < d$. Then fuzzy sets like

$$\overline{N} \cap \overline{\mathrm{Pos}} \tag{6.47}$$

will be used. This causes no problem in fuzzy arithmetic because we will perform fuzzy arithmetic using α–cuts and interval arithmetic.

The N-period inventory problem studied in this paper is usually solved, with no fuzzy sets, using dynamic programming. So a solution method, when some of the parameters are fuzzy, might be fuzzy dynamic programming. Some work has been done, employing (fuzzy) dynamic programming, for N–period fuzzy inventory control problem ([31],[40],[53]).

The book [40] gives an excellent review of multi–stage inventory control. Both fuzzy sets and stochastic models are used. The solution method is dynamic programming. They seek to maximize the weighted sum of the membership functions which assures us that the system states, over their respective stages, are satisfactory. This is not the standard cost minimization, or profit maximization, model in inventory control. The set up cost, demand, storage cost, lost sales cost are all crisp and not fuzzy.

A recent survey of fuzzy sets applied to production management research (including inventory control, job shop scheduling, project scheduling, and forecasting) is [24].

Dynamic programming can be very computationally intensive, and even more so, for fuzzy dynamic programming. Therefore, we wished to avoid a fuzzy dynamic programming approach. Instead we employ an evolutionary algorithm.

Our general purpose evolutionary algorithm is discussed in Chapter 13 at the end of this book and in Section 6.3.4 we briefly show how it was adopted to the fuzzy inventory control problem.

However, using evolutionary algorithms has its drawbacks: we do not get any results about the properties of the solutions. The best result would be to derive "closed form" solutions (formulas for the solutions) so that we can investigate their properties. So, without such "closed form" solutions we have to be content with evolutionary algorithm solutions.

6.3.1 Demand Not Fuzzy

The objective is to find non-negative integers z_i, $1 \leq i \leq N$, to minimize

$$\lambda_1[M - A_1] + \lambda_2 u_2 + \lambda_3 A_2 \ , \tag{6.48}$$

where the A_i are defined in equations (6.42) and (6.43). $\overline{K}_i$ and $\overline{h}_i$ will be trapezoidal fuzzy numbers which implies $\overline{Z}_i$ will also be a trapezoidal fuzzy number. Let $\overline{Z} = (z_1/z_2, z_3/z_4)$. In computing A_1 and A_2 we will use $u_1 = z_1$, $u_2 = (z_2 + z_3)/2$ and $u_3 = z_4$. The only other thing we must do is to expand constraint $x_{N+1} = 0$ to involve the z_i (decision variables).

Since no shortages are allowed the constraints are:

$$\max\{0, D_k - x_k\} \leq z_k \leq \sum_{i=k}^{N} D_i - x_k \ , \tag{6.49}$$

$k = 1, 2, \ldots, N-1$, and

$$z_N = D_N - x_N \ . \tag{6.50}$$

Of course, $D_N - x_N \geq 0$ for otherwise we cannot get $x_{N+1} = 0$. These constraints can be simplified to

$$\max\{0, D_k - x_k\} \leq z_k \ , \ 1 \leq k \leq N-1 \ , \tag{6.51}$$

and

$$\sum_{i=1}^{N} z_i \leq \sum_{i=k}^{N} D_i - x_1 \ , \tag{6.52}$$

together with equation (6.50).

We summarize this fuzzy inventory control problem: (1) Given M, the λ_i, find non-negative integers z_i to minimize the expression in equation (6.48); (2) subject to the constraints in equations (6.50) – (6.52); (3) for the A_i computed in equations (6.42) and (6.43) and $\overline{Z}$, determined using α-cuts and interval arithmetic, from equations (6.40) and (6.41); and (4) for given values of x_i, $\overline{K}_i$, $c_i(z_i)$, N, D_i and $\overline{h}_i$.

Example 6.3.1.1

Now we apply our evolutionary algorithm to generate an approximate solution. We use $N = 4$ to be able to easily compute the corresponding crisp solution to compare to the result of the evolutionary algorithm.

Initial inventory (x_1) is 15 units, there are no price breaks with $c_i(z_i) = 2z_i$ all i and $\overline{K}_i$, D_i, $\overline{h}_i$ are given in Table 6.6. The decision maker decides to use

Period	D_i	$\overline{K}_i$	$\overline{h}_i$
1	76	(92/96, 100/104)	(0.8/1, 1/1.2)
2	26	(108/112, 116/120)	(0.8/0.9, 1.1/1.2)
3	90	(179/183, 187/191)	(0.6/0.8, 1.2/1.4)
4	67	(64/68, 72/76)	(0.7/1, 1/1.3)

Table 6.6: Data for Example 6.3.1.1

$\lambda_1 = \lambda_2 = 0.4$, $\lambda_3 = 0.2$ since he was more interested in maximizing A_1 and minimizing u_2, than in minimizing A_2. The value of M in equation (6.48) that we used was $M = 1000$.

The evolutionary algorithm randomly generates an initial population. A population member is

$$z = (z_1, z_2, z_3, z_4) , \tag{6.53}$$

where the z_i's are non-negative integers and they satisfy the constraints in equations (6.50) – (6.52). A subset of the fittest individuals are selected to produce the next generation. The fittest are those which produce the smallest values of the objective function (equation (6.48)). The next generation is obtained through crossover and mutation. After many generations the population tends to stabilize and we stop the evolutionary algorithm. All computations are done using α-cuts of $0, 0.1, \ldots$, 0.9, 1.0. and interval arithmetic to compute $\overline{Z}$, which is then used to get A_1 and A_2, etc.

The evolutionary algorithm solutions for $\overline{Z}$ produced $A_1 = 16.0$, $u_2 = 955.0$, and $A_2 = 16.0$. Finally, the best values found for the decision variables were $z_1 = 61$, $z_2 = 26$, $z_3 = 90$ and $z_4 = 67$.

To obtain a crisp solution we defuzzify $\overline{K}_i$ and $\overline{h}_i$ as follows: (1) if $\overline{K}_i = (k_{i1}/k_{i2}, k_{i3}/k_{i4})$, the defuzzified value is $(k_{i2} + k_{i3})/2 = k_i^*$; (2) if $\overline{h}_i = (h_{i1}/h_{i2}, h_{i3}/h_{i4})$, use $h_i^* = (h_{i2}+h_{i3})/2$. These values are $k_1^* = 98$, $k_2^* = 114$, $k_3^* = 185$, $k_4^* = 70$ and $h_i^* = 1.0$ for all i. The crisp optimal solution is $z_1 = 61$, $z_2 = 26$, $z_3 = 90$ and $z_4 = 67$. The minimum cost in the crisp solution was \$955.

Both solutions, fuzzy and crisp, had the same optimal solution for the z_i, $1 \leq i \leq 4$. This solution is not surprising because, since $c_i(z_i) = 2z_i$, all i, and $h_i^* = 1.0$, all i, it is buy to meet demand each period so as not to have any shortage costs. It appears that the evolutionary algorithm has obtained the correct solution.

6.3.2 Fuzzy Demand

What is new in this section, from the previous section, is: (1) $\overline{D}_i$, demand is fuzzy; (2) $\overline{x}_i$, $i \geq 2$, incoming inventory is fuzzy; (3) new definition of holding cost; (4) shortage costs; and (5) new definition of $\overline{x}_{N+1} \approx 0$ which specifies the constraints on the z_i. What has not changed is $\overline{C}_i(z_i)$, $\overline{h}_i$, A_1, A_2, M, λ_i and crisp starting inventory x_1.

$\overline{D}_i$ will be a non-negative trapezoidal fuzzy number for all i. The $\overline{x}_i$, $i \geq 2$, as we shall see, will also be fuzzy but the equations $\overline{x}_{i+1} = \overline{x}_i + z_i - \overline{D}_i$, $i \geq 2$, will no longer hold.

Shortages are allowed but there are no backorders. When demand exceeds supply sales are lost. Let $\overline{H}_i(z_i)$ measure how much supply exceeds demand and $\overline{P}_i(z_i)$ measure how much demand exceeds supply, for all periods i. Define

$$\overline{H}_i(z_i) = (\overline{x}_i + z_i - \overline{D}_i) \cap \overline{\text{Pos}} , \tag{6.54}$$

$$\overline{P}_i(z_i) = (\overline{D}_i - [\overline{x}_i + z_i]) \cap \overline{\text{Pos}} , \tag{6.55}$$

$1 \leq i \leq N$, for $\overline{\text{Pos}}$ defined by equation (6.46). Of course, for $i = 1$ we have $x_1 = \overline{x}_1$. Then

$$\overline{TC}_i = \overline{C}_i(z_i) + \overline{h}_i \cdot \overline{H}_i(z_i) + \overline{p}_i \cdot \overline{P}_i(z_i) , \tag{6.56}$$

and

$$\overline{Z} = \sum_{i=1}^{N} \overline{TC}_i . \tag{6.57}$$

In equation (6.56) $\overline{p}_i$ is the fuzzy penalty cost due to shortages. $\overline{Z}$ will be a trapezoidal shaped fuzzy number computed using α-cuts and interval arithmetic. Since $\overline{H}_i(z_i)$ measures how much supply exceeds demand, then

$$\overline{x}_{i+1} = \overline{H}_i(z_i) , \tag{6.58}$$

$i = 1, 2, \ldots, N-1$, which is the outgoing inventory for the i-th period.

Let us now justify equations (6.54) and (6.55). Let $\overline{D}_i = (d_{i1}/d_{i2}, d_{i3}/d_{i4})$ and $\overline{x}_i + z_i = (x_{i1}+z_i/x_{i2}+z_i, x_{i3}+z_i/x_{i4}+z_i)$. First consider the extreme case of $x_{i1} + z_i \geq d_{i4}$, or supply completely exceeds demand. Then $\overline{x}_i + z_i - \overline{D}_i \geq 0$ so $\overline{H}_i(z_i) = \overline{x}_i + z_i - \overline{D}_i = \overline{x}_{i+1}$ and $\overline{P}(z_i) = \overline{0}$ (identically zero). The other extreme case is $x_{i4} + z_i \leq d_{i1}$, or demand completely exceeds supply. Then $\overline{D}_i - (\overline{x}_i + z_i) \geq 0$, $\overline{H}_i(z_i) = \overline{x}_{i+1} = \overline{0}$ and $\overline{P}_i(z_i) = \overline{D}_i - (\overline{x}_i + z_i)$. The other cases are when $x_{i1} + z_i - d_{i4} < 0 < x_{i4} + z_i - d_{i1}$. One of these situations is shown in Figures 6.19 – 6.21, when $x_{i2} + z_i - d_{i3} < 0 < x_{i3} + z_i - d_{i2}$. Another case, if $x_{i3} + z_i - d_{12} < 0 < x_{i4} + z_i - d_{i1}$ is in Figures 6.22 – 6.24.

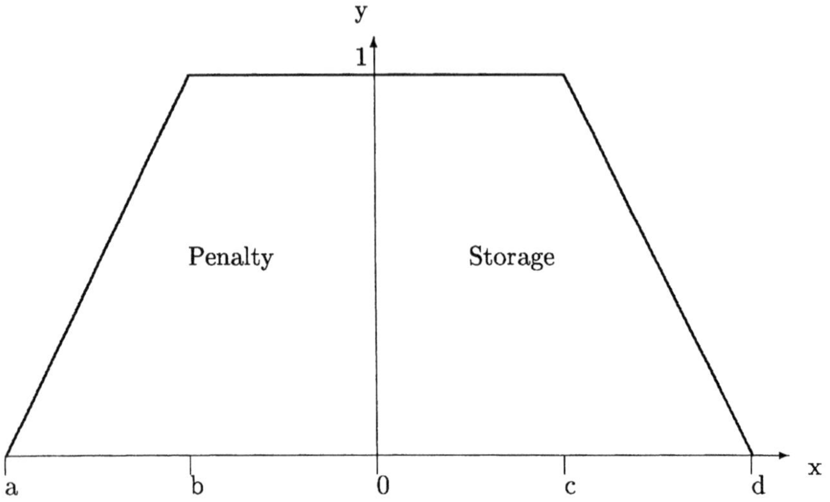

Figure 6.19: The Fuzzy Set $\overline{x}_i + z_i - \overline{D}_i$

Figures 6.19 – 6.24 show that the fuzzy sets $\overline{H}_i(z_i)$ and $\overline{P}_i(z_i)$ need not be fuzzy numbers, nor need they be normalized (have a membership value to one). However, this presents no real problem since we are using α-cuts and interval arithmetic to do fuzzy arithmetic.

The last change, from Section 6.3.1, to discuss is the constraint $\overline{x}_{N+1} \approx 0$. From the previous discussion, and Figures 6.19 – 6.24, we see that there are four cases for $\overline{x}_{N+1}$: (1) $\overline{x}_{N+1} = \overline{0}$; (2) $\overline{x}_{N+1} = \overline{H}_N(z_N)$ in Figure 6.23; (3) $\overline{x}_{N+1} = \overline{H}_N(z_N)$ in Figure 6.20 ; or (4) $\overline{x}_{N+1} = \overline{H}_N(z_N)$ shown in Figures 6.25 – 6.26. So we have $\overline{x}_{N+1}[1] = \phi$ for cases 1 and 2, $\overline{x}_{N+1}[1] = [0, t]$ in case 3 and $\overline{x}_{N+1}[1] = [s, t]$, $s > 0$, in case 4. We will say that $\overline{x}_{N+1} \approx 0$ is satisfied when $\overline{x}_{N+1}[1] = \phi$ or $\overline{x}_{N+1}[1] = [0, t]$, but it is not satisfied when $\overline{x}_{N+1}[1] = [s, t]$, $s > 0$.

We summarize this fuzzy N-period inventory control problem: (1) given M, λ_i find non-negative integers z_i to minimize the expression in equation (6.48); (2) subject to the constraint $\overline{x}_{N+1} \approx 0$; (3) for $\overline{Z}$, computed using α-cuts and interval arithmetic, given by equation (6.57) and the A_i from equation (6.42) and (6.43); and (4) for given values of x_1, $c_i(z_i)$, $\overline{K}_i$, N, $\overline{D}_i$, $\overline{h}_i$, $\overline{p}_i$.

Example 6.3.2.1

This will continue Example 6.3.1.1. Fuzzy demand $\overline{D}_i$ and fuzzy penalty cost $\overline{p}_i$ are shown in Table 6.7. The defuzzified values of the $\overline{D}_i$ in Table 6.7 will equal the crisp D_i values in Table 6.6. The defuzzified values of the $\overline{p}_i$ are all equal to 3.00. All the other parameter values are the same as in Example 6.3.1.1.

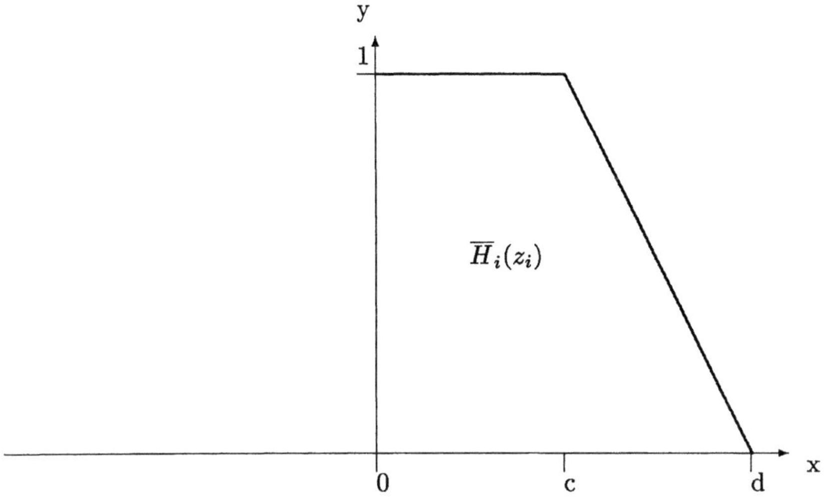

Figure 6.20: Storage from Figure 6.19

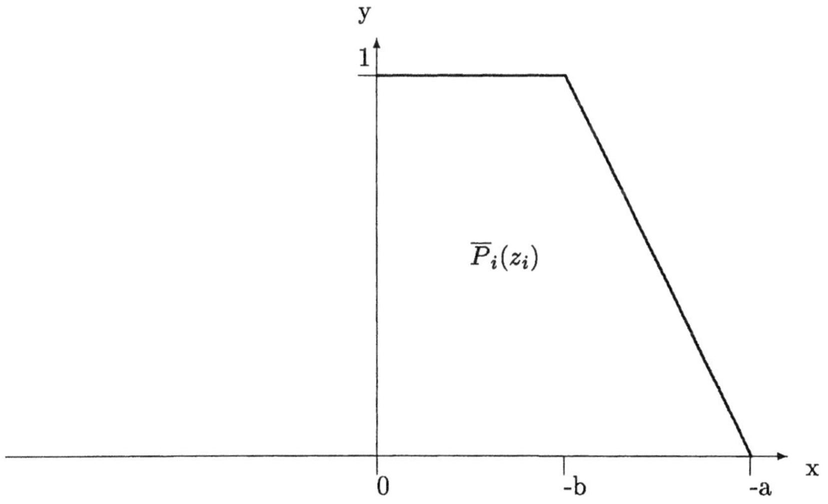

Figure 6.21: Penalty from Figure 6.19

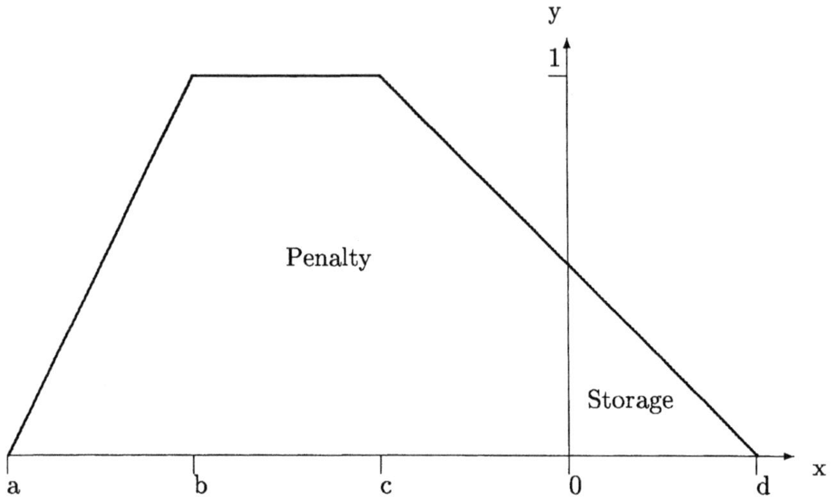

Figure 6.22: The Fuzzy Set $\overline{x}_i + z_i - \overline{D}_i$

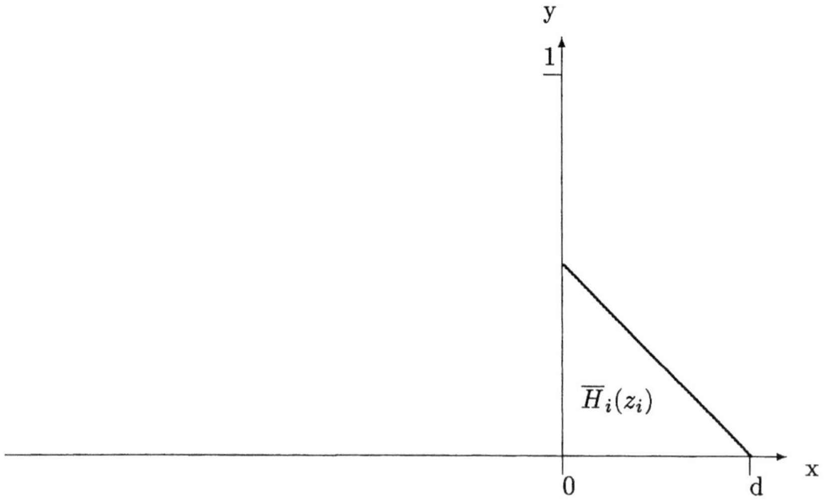

Figure 6.23: Storage from Figure 6.22

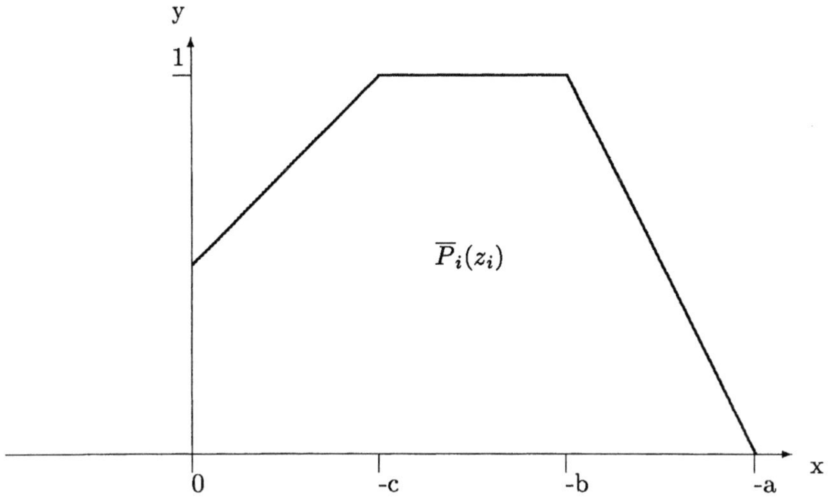

Figure 6.24: Penalty from Figure 6.22

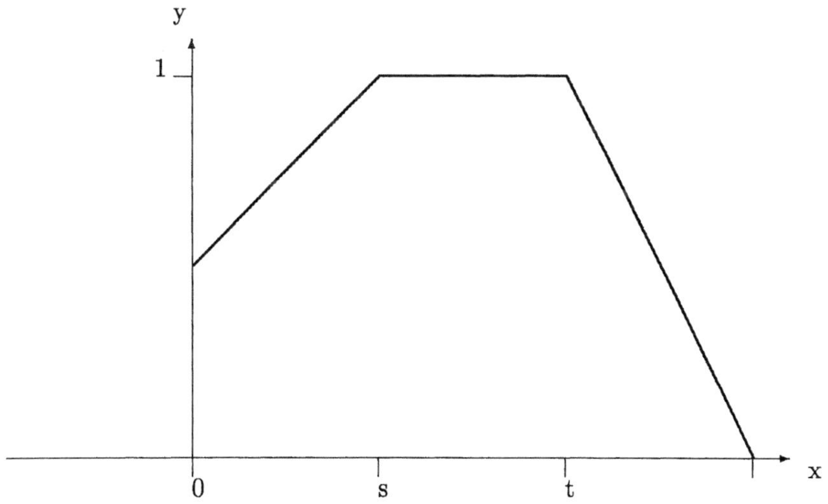

Figure 6.25: $\overline{x}_{N+1}$, One Case of Final Inventory

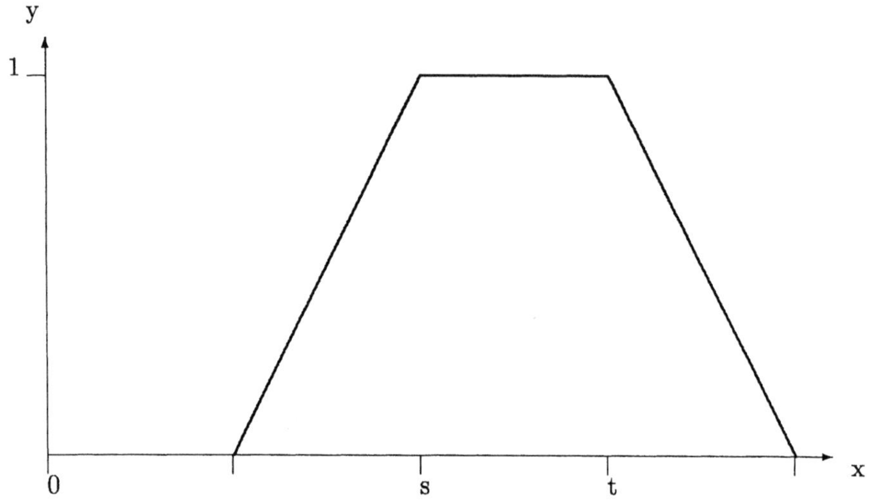

Figure 6.26: $\overline{x}_{N+1}$, Another Case of Final Inventory

Period	$\overline{D}_i$	$\overline{p}_i$
1	(70/74, 78/82)	(2.0/3.0, 3.0/4.0)
2	(24/26, 26/28)	(2.5/2.8, 3.2/3.5)
3	(86/89, 91/94)	(2.0/2.7, 3.3/4.0)
4	(64/65, 69/70)	(2.5/2.9, 3.1/3.5)

Table 6.7: Data for Example 6.3.2.1

A main problem now is for the EA to check the constraints. A population member is still $z = (z_1, z_2, z_3, z_4)$ for $z_i \geq 0$ integer. These are randomly generated in the initial population, or created through crossover and mutation, in later populations.

To check feasibility we first compute $\overline{x}_2 = \overline{H}_1(z_1)$ from z_1. Then $\overline{x}_3 = \overline{H}_2(z_2)$ from $\overline{x}_2$, and z_2 and $\overline{x}_4 = \overline{H}_3(z_3)$ from $\overline{x}_3$ and z_3. Finally, compute $\overline{x}_5$ and check $\overline{x}_5[1]$. If $\overline{x}_5[1] = \phi$ or $\overline{x}_5[1] = [0, t]$ we keep this $z = (z_1, \ldots, z_4)$. Otherwise, discard this population member and generate another one. Keep generating new population members until we obtain one that is feasible, which is kept.

The evolutionary algorithm's solution for $\overline{Z}$ produced A_i values $A_1 = 34.432$, $A_2 = 71.12$ with $u_2 = 883.10$. The core of $\overline{Z}$, where the membership values are equal to one, was the interval [859.0, 907.2]. The best values for the decision variables were $z_1 = 89$, $z_2 = 0$, $z_3 = 89$ and $z_4 = 65$.

Next, we defuzzified the fuzzy inventory problem and obtained its solution $z_1 = 61$, $z_2 = 116$, $z_3 = 0$ and $z_4 = 67$. We used the same defuzzified values of $\overline{K}_i$ and $\overline{h}_i$ as in Example 6.3.1.1, defuzzified $\overline{D}_i$ gives D_i values in Table 6.6, and defuzzified $\overline{p}_i$ is always 3.00. The minimum cost in the crisp problem was \$860, which was in the core of $\overline{Z}$. Since the optimal result to the crisp is in the core of the optimal $\overline{Z}$ we believe this is more evidence that the evolutionary algorithm is approximating the correct fuzzy solution.

6.3.3 Backorders

The major changes in this section, from the previous section, are: (1) we now allow for backorders: (2) we maximize profit instead of minimizing cost; and (3) we discount future monies back to their present value.

We still assume we buy the item, and we do not produce it ourselves, so we still use $\overline{C}_i(z_i)$ in equation (6.38). However, we will need to redefine $\overline{H}_i(z_i)$ and $\overline{P}_i(z_i)$ in equations (6.54) and (6.55), respectively.

We are going to resell the item at r\$/ unit where r is known (not fuzzy). We cannot use $\overline{x}_i + z_i - \overline{D}_i$ directly, as shown in Figures 6.19 – 6.24, because they do not show what we sell each period. Let $\overline{S}_i(z_i)$ be the (fuzzy) amount we sell during period i, $1 \leq i \leq N$. We will employ the following procedure to determine $\overline{S}_i(z_i)$: (1) if $\overline{x}_i + z_i > \overline{D}_i$, then $\overline{S}_i(z_i) = \overline{D}_i$, $\overline{H}_i(z_i) = \overline{Poss} \cap (\overline{x}_i + z_i - \overline{D}_i) = \overline{x}_{i+1}$, $\overline{P}_i(z_i) = \emptyset$; (2) if $\overline{D} > \overline{x}_i + z_i$, then $\overline{S}_i(z_i) = \overline{x}_i + z_i$, $\overline{P}_i(z_i) = (\overline{D}_i - [\overline{x}_i + z_i]) \cap \overline{Poss}$, $\overline{H}_i(z_i) = \emptyset$ and $\overline{D}_{i+1}$ will be $\overline{D}_{i+1} + \overline{P}_i(z_i)$; and (3) if $\overline{x}_i + z_i \approx \overline{D}_i$, then $\overline{S}_i(z_i) = \overline{D}_i$, $\overline{H}_i(z_i) = \overline{P}_i(z_i) = \emptyset$. $\overline{H}_i(z_i)$ is what goes to storage and $\overline{P}_i(z_i)$ is what is backordered to be added to next periods demand. The value $\overline{p}_i$ will now be the backorder cost per unit.

All we need now is a method of deciding on one of the following: $\overline{x}_i + z_i > \overline{D}_i$; $\overline{D}_i > \overline{x}_i + z_i$; $\overline{D}_i \approx \overline{x}_i + z_i$. The solutions obviously will depend on how we evaluate these inequalities and $\approx$. The method we shall use is: (1) $\overline{x}_i + z_i > \overline{D}_i$ if the core of $(\overline{x}_i + z_i - \overline{D}_i)$ lies to the right of zero; (2) $\overline{D}_i > \overline{x}_i + z_i$ when the core of $(\overline{x}_i + z_i - \overline{D}_i)$ lies to the left of zero; and (3) $\overline{x}_i + z_i \approx \overline{D}_i$ if zero belongs to the core of $(\overline{x}_i + z_i - \overline{D}_i)$. Notice that $\overline{x}_i + z_i - \overline{D}_i$ will be a trapezoidal shaped fuzzy number.

Next we may set up the profit function to be maximized. Let

$$\overline{TP}_i = r\overline{S}(z_i) - \overline{C}_i(z_i) - \overline{h}_i\, \overline{H}_i(z_i) - \overline{p}_i\, \overline{P}_i(z_i)\ , \tag{6.59}$$

and

$$\overline{Z} = \sum_{i=1}^{N} (\overline{\alpha})^{i-1} \overline{TP}_i\ . \tag{6.60}$$

The $\overline{\alpha}$ are fuzzy discount factors which bring future fuzzy monies back to their present value. As before we compute A_1 (equation (6.42)) and A_2 (equation (6.43)) and we wish to

$$\max\left[\lambda_1 A_1 + \lambda_2 u_2 + \lambda_3 (M - A_2)\right]\ , \tag{6.61}$$

for $\lambda_i > 0$, $\lambda_1 + \lambda_2 + \lambda_3 = 1$, subject to the constraint $\overline{x}_{N+1} \approx 0$ from Section 6.3.2. Also the fuzzy demand in the $(i+1)$–th period is $\overline{D}_{i+1} + \overline{P}_i(z_i)$.

Future details on this fuzzy inventory model, plus a numerical example whose approximate solution was obtained using an evolutionary algorithm, may be found in [8].

6.3.4 Evolutionary Algorithm

A member of the population is $v = (z_1, z_2, z_3, z_4, \sigma)$ for z_i a non-negative integer $1 \leq i \leq 4$ and σ the mutation rate. Initially we set $\sigma = 0.3$. Population size was always 10,000. All members of the initial population were randomly chosen subject to the constraints. The constraints were : (1) final inventory zero ($x_5 = 0$ in Section 6.3.1); or (2) final fuzzy inventory is approximately zero($\overline{x}_5 \approx 0$ in Sections 6.3.2 and 6.3.3). The fitness function was: (1) equations (6.41) and (6.48) in Section 6.3.1 ; (2) equations (6.57) and (6.48) for Section 6.3.2; and (3) equations (6.60) and (6.61) in Section 6.3.3 (even though no numerical results were reported). The fittest 3000 individuals were taken to generate the next population. There was no constraint on crossover except that the new individual satisfy the constraints. The same for mutation and in addition the mutated individual must have the z_i non-negative integers. The rest of the details on the evolutionary algorithm are in Chapter 13. In all of the examples presented, both fuzzy and crisp, we obtained the best results in less than 40 generations.

6.4 Fuzzy Queuing Theory

We consider a queuing system with multiple parallel servers, finite or infinite system capacity and calling source, but whose arrivals and service completions are governed by a possibility distribution. Changes in the number of customers in the system is modeled by a transition matrix Ω and powers of Ω, under max–min product, produce (if it converges) the steady–state possibility distribution. Results are first applied to the machine servicing problem to determine the most economical size of the repair crew. Then we look at a fuzzy queuing decision problem where we are to decide on the optimal number of servers.

Standard queuing notation is

$$(a/b/c) : (d/e/f) \; , \tag{6.62}$$

where: (1) a denotes the arrivals distribution, (2) b is the service time (or departure) distribution; (3) c is the number of parallel servers; (4) d stands for the service discipline (first come – first served, etc.); (5) e is system capacity (in queue plus in service); and (6) f is the size of the calling source (potential customers). In this section we will have

$$(a/b/c) = (G/G/c) \; , \tag{6.63}$$

where: (1) $a = G$ means that the number of arrivals, per unit time, is a (stationary) fuzzy variable whose values are restricted by some possibility distribution; and (2) $b = G$ means that the number of service completions, per unit time, is a (stationary) fuzzy variable whose values are restricted by a possibility distribution and the number of full servers. We will also consider: (1) d arbitrary; (2) $e = M \geq c$ or $e = +\infty$; and (3) $f = N \geq c$ or $f = +\infty$. We will assume that all fuzzy variables are noninteractive. We are not interested in computing the possibility distribution of waiting time and therefore d can be arbitrary.

Let X be the number of customers in the system. We wish to find

$$p_n(t) = \text{Poss}[X = n| \text{ time } = t] \ , \tag{6.64}$$

for $n = 0, 1, 2, \ldots$ and $t = 1, 2, \ldots$. One unit time, which can be one second, one minute, one hour, etc., is represented by $t = 1$. We assume that we have the initial possibility distribution

$$p(0) = (p_0(0), p_1(0), \ldots) \ . \tag{6.65}$$

We also wish to compute

$$p_n = \lim_{t \to +\infty} p_n(t) \ , \tag{6.66}$$

when this limit exists, which will be called the steady–state possibility distribution. Once we have the p_n we evaluate the system using performance measures such as: (1) possibility all servers are busy $=$ Poss$[X \geq c] = \max\{p_n | n \geq c\}$ $=$ possibility of an arriving customer ($M = +\infty$) will have to wait for service; (2) possibility of turning customers away when system capacity is finite; and (3) the possibility that at least one server is empty $= \max\{p_n | 0 \leq n < c\}$.

Initial work on applying possibility theory to queuing systems is in [47] where the author concentrates on finding the possibility distribution for waiting time. A second paper on possibilistic queuing theory is ([36],[37]) where the authors compute, using the extension principle, the usual performance measures for a queue when λ, the arrival rate, and departure rate μ are fuzzy variables. However, their formulas are very complicated and appear difficult to evaluate. For two other more recent papers on fuzzy queuing theory see [30] and [44]. We take a different approach with the arrival and service process described by fuzzy variables and we will model changes by a transition matrix.

We now need to discuss our basic assumptions on the queuing system in order to be able to model changes by a transition matrix Ω. Changes occur only at times $t = 1, 2, 3, \ldots$. Assume we are at $t = k$ and we either turn the clock off during a change in the system or we assume that the change takes zero time. First, customers who have completed service depart the system to return to the calling source. Secondly, all customers in queue fill any empty servers . Thirdly, customers who have arrived during the previous time interval are allowed into the system (up to capacity if M is finite)and immediately fill any empty servers the rest going into queue. Once system capacity is reached all other arriving customers are turned away (lost to the system) and return to the calling source. Then, if the clock was off, we start the clock until $t = k + 1$ and the system changes can again take place. The time intervals between system changes do not all have to be the same, but for simplicity we will assume that the time intervals between changes are all equal. Alternatively, we may allow customers to leave the servers early, but they do not arrive back in the calling source until the end of the time interval if the calling source is finite, but only fill the empty servers and the queue at fixed times.

The queuing system described above may be an approximation to some real system but also such systems do exist. Examples may include: (1) rides at an amusement park (seats on the ride are the servers) when customers can stay on for a second ride; (2) stores with crowd control where they wish to keep the number of customers in the store within some maximum amount and they let people into the store only at certain times; and (3) systems served by rapid transit systems where customers arrive, and depart, only when the train arrives. More generally, systems where service lasts a fixed time (apartment rentals; certificates of deposit; movies, jobs filled daily, weekly, ...; groups of pay phones which charge a fixed amount for each three minutes; ...), and customers may stay on for more "service" at the end of each time interval, may fit the above description. It will turn out that we will be dealing with a discrete time, stationary Markov process whose transitions are governed by possibility distributions instead of probability distributions. This section is based on [3] and [9].

We next need to discuss the possibility distribution associated with arrivals and departures.

6.4.1 Service

We have c servers which serve individual customers. Let D be the number of departures (service completions) per unit time and set

$$\mu(d, s) = \text{Poss}[D = d | s \text{ servers full}] \; , \tag{6.67}$$

for $s = 0, 1, \ldots, c$ and $d = 0, 1, \ldots, s$. The $\mu(d, s) \in [0, 1]$ give the possibility of d departures at the end of a time interval given that s servers were full at the start of the time interval. We may assume that the $\mu(d, s)$ are given or we can derive them from knowledge of the individual servers.

Label the servers S_1, S_2,..., S_c and suppose we are given

$$\text{Poss}[D = d_i | S_c] = \begin{cases} \mu_0 & : \text{ if } d_i = 0 \; , \\ \mu_1 & : \text{ if } d_i = 1 \; , \\ 0 & : \text{ otherwise} \; , \end{cases} \tag{6.68}$$

for each S_i, $1 \leq i \leq c$. We are assuming that we have c identical servers. All our possibility distributions will be normalized which means μ_0 or μ_1 will equal one. Also, all our fuzzy variables (possibility distributions) are assumed non–interactive (see Section 2.4). This means that the servers are non–interactive and we may find the joint distribution as follows

$$\mu(d_1, \ldots, d_s) = \min_{1 \leq i \leq s} \left\{ \text{Poss}[D = d_i | S_i] \right\} , \tag{6.69}$$

for $d_i = 0$ or 1, $1 \leq i \leq s$. Then we may compute $\mu(d, s)$ as

$$\mu(d, s) = \max\{\mu(d_1, \ldots, d_s) | d_1 + \cdots d_s = d\} , \tag{6.70}$$

for $d = 0, 1, 2, \ldots, s$, $s = 1, 2, \ldots, c$.

6.4.2 Arrivals

We are assuming individual arrivals and no bulk (group) arrivals. Let A be the number of arrivals per time unit. If the calling source is infinite, then let

$$\lambda(\alpha) = \text{Poss}[A = \alpha] , \tag{6.71}$$

for $\alpha = 0, 1, 2, \ldots$. We assume that the $\lambda(\alpha) \in [0, 1]$ are given. Next suppose that the calling source is finite ($f = N$) and let

$$\lambda(\alpha, n) = \text{Poss}[A = \alpha | n \text{ in system}] , \tag{6.72}$$

for $\alpha = 0, 1, 2, \ldots, N - n$ and $n = 0, 1, 2, \ldots, \min(M, N)$. Of course, $\lambda(0, N) = 1$. We would expect that $\lambda(\alpha, n)$ is a decreasing function of n but there is no need to make this assumption. We may derive the $\lambda(\alpha, n)$ if we know the characteristics of the customers. We will need to assume identical customers or else we get into the complication of keeping track of which customers are in the system. For each customer C_i, $1 \leq i \leq N$, let

$$\text{Poss}[A = \alpha_i | C_i] = \begin{cases} \lambda_0 & : \text{ if } \alpha_i = 0 , \\ \lambda_1 & : \text{ if } \alpha_i = 1 , \\ 0 & : \text{ otherwise} . \end{cases} \tag{6.73}$$

Next find the joint distributions

$$\mu(\alpha_1, \ldots, \alpha_l) = \min_{1 \leq i \leq l} \left\{ \text{Poss}[A = \alpha_i | C_i] \right\} , \tag{6.74}$$

for $1 \leq l \leq N$, assuming non–interactive customers. Then

$$\lambda(\alpha, n) = \sup \left\{ \mu(\alpha_1, \ldots, \alpha_l) \middle| \alpha_1 + \cdots + \alpha_l = \alpha, l = N - n \right\} , \tag{6.75}$$

for $0 \leq \alpha \leq N - n$, $n = 0, 1, \ldots, \min(M, N)$.

Once we have the $\mu(d, s)$ and $\lambda(\alpha)$, or the $\lambda(\alpha, n)$, we can proceed to construct Ω, the transition matrix, and discuss finding the steady–state distribution. However, some justification (see [36], [47]) of using possibility in place of probability to study queuing systems needs to be presented. Obviously, it is of mathematical interest to develop and compare queuing theory using possibilities to queuing theory using probabilities. Another reason is that one might choose subjective possibilities over subjective probabilities to model the uncertainties. When there is no data, or very little concrete data available on the system we are forced to employ subjective possibilities or probabilities. This situation arises when we are planning a new system, or when some information about an existing system is unavailable because it is classified, confidential, too expensive to obtain, etc. In any case let us proceed to model the arrival and departure processes with possibility distributions.

In the next section we look at finding the steady–state distribution for infinite calling source but with limited or infinite system capacity. In Section 6.4.4 we discuss the special case $N = M \geq c$, or calling source equals system capacity, and look at the machine servicing problem in order to determine the optimal size of the repair crew. The fuzzy queuing decision problem follows in Section 6.4.5 and the last section is a brief summary and our conclusions.

6.4.3 Finite or Infinite System Capacity

We first consider finite system capacity ($M \geq c$) and then generalize to infinite capacity. Let

$$\overline{\lambda}(\alpha) = \text{Poss}[A \geq \alpha] = \max_{i \geq \alpha}\{\lambda(i)\}. \tag{6.76}$$

The one step transition matrix Ω will be $(M+1) \times (M+1)$, with its rows labels $0, 1, 2, \ldots, M$ indicating the number of customers in the system during the last time interval, and its columns labeled $0, 1, 2, \ldots, M$ for the number of customers in the system during the next time interval, Let w_{ij} be an element in Ω. Many of the w_{ij}, below the main diagonal of Ω, are zero. For example, for $M = 5$ and $c = 2$ we see that $w_{3,0} = w_{4,0} = w_{4,1} = w_{5,0} = w_{5,1} = w_{5,2} = 0$. Let us consider computing a non–zero w_{ij} for $j < M$. For example

$$w_{11} = \max\left\{ \min\{\lambda(0), \mu(0,1)\}, \min\{\lambda(1), \mu(1,1)\}, \right\}, \tag{6.77}$$

because we can go from one customer in the system to one in the system only two ways: (1) no arrivals and the customer in a server does not depart; or (2) one arrival and the customer in a server departs. The possibility of the first situation is given by $\min\{\lambda(0), \mu(0,1)\}$ and $\min\{\lambda(1), \mu(1,1)\}$ is the possibility of the second event.

Let us now justify the case of max and min in equation (6.77). Our method of justification can be generalized to show that the elements in Ω^k

give the k–step possibilities, but we shall omit these details. Assuming the customers and servers are non–interactive form the joint distribution

$$\mu(\alpha, d, s) = \min\{\lambda(\alpha), \mu(d, s)\} , \tag{6.78}$$

for $\alpha = 0, 1, 2, \ldots$, $d = 0, 1, 2, \ldots, s$, $s = 1, 2, \ldots, c$. Let E_1 be the event of no arrivals and no departures and E_2 is the event of one arrival and one departure. Then $\text{Poss}(E_1) = \min\{\lambda(0), \mu(0,1)\}$, $\text{Poss}(E_2) = \min\{\lambda(1), \mu(1,1)\}$. What we want is the event $E = E_1 \cup E_2$ which describes w_{11} in Ω. But $\text{Poss}(E) = \max\{\text{Poss}(E_1), \text{Poss}(E_2)\}$ in possibility theory. Hence, $\text{Poss}(E)$ gives equation (6.77) for w_{11}. In this way we compute all the non–zero w_{ij}, $j < M$.

Now consider a $w_{i,M}$. Suppose $M = 5$ and $c = 2$ and we want the value of $w_{2,5}$. It is

$$w_{2,5} = \max\left\{ \min\{\overline{\lambda}(3), \mu(0,2)\}, \min\{\overline{\lambda}(4), \mu(1,2)\}, \min\{\overline{\lambda}(5), \mu(2,2)\} \right\} , \tag{6.79}$$

because $w_{2,5}$ is made up of: (1) three or more arrivals, no departures ($M = 5$); (2) four or more arrivals, one departure ($M = 5$); and (3) five or more arrivals, and two departures ($M = 5$). Hence, all of Ω is now determined and we can try to find the steady–state possibilities.

Ω^k will give the k–step possibilities. We evaluate $\Omega \cdot \Omega$ using standard matrix multiplication except min replaces multiplication and use max for addition. We also use max–min composition for a vector times Ω. Therefore, the k–step state possibilities are

$$p(0)\Omega^k , \tag{6.80}$$

and if Ω^k converges, the steady–state distribution is

$$\lim_{k \to +\infty} p(0)\Omega^k . \tag{6.81}$$

Let $\Omega^k \to \Pi$, as $k \to +\infty$, when the powers of Ω converge. Therefore

$$p = p(0)\Pi , \tag{6.82}$$

where $p = (p_0, p_1, \ldots)$ is the steady–state possibility distribution.

Results on the convergence of max–min powers of a fuzzy matrix may be found in ([4], [25], [28],[56]). It is known that powers of Ω converge to Ω^l, for positive integer l, or oscillate with a finite period. When Ω^k converges to Π the rows in Π need not be identical (note that in probability theory the rows in Π will be identical for Ω the one step transition matrix for an ergodic Markov chain). If powers of Ω oscillate, then we have no steady–state possibilities.

We generalize to $M = \infty$ by simply replacing $\overline{\lambda}(\alpha)$ with $\lambda(\alpha)$ in computing Ω. Of course, Ω is now infinite, but "most" of Ω will consist of zeros. We would expect $\lambda(\alpha)$ to be zero for $\alpha \geq$ some positive integer which will make rows in Ω eventually zero. Powers of Ω may converge or oscillate.

6.4.4 Machine Servicing Problem

There are c repair crews available to service N machines. We therefore have

$$(d/e/f) = (d/N/N) , \tag{6.83}$$

or $M = N \geq c$. Let R be the hourly cost of each repair crew and let B be the hourly cost of lost production per waiting machine. When a machine breaks down it is assigned a repair crew, if one is available, or it has to wait for the first free repair crew. To model this problem with a transition matrix Ω we assume repair crews are scheduled hourly and repaired machines go back into production on the hour. If this seems unreasonable, then we could take 30 minutes, or 5 minutes, for our unit of time.

The cost function, which is to be minimized, is

$$\overline{Z} = Rc + \overline{B} \cdot \overline{w}, \tag{6.84}$$

where $\overline{w}$ is the number of broken machines in the system (either in queue or being serviced). The hourly cost per disabled machine is difficult to estimate so we model it as a trapezoidal fuzzy number $\overline{B}$. The number of broken machines will be a fuzzy variable $\overline{w}$ having a discrete possibility distribution. For each value of $c = 1, 2, \ldots, N$ we find the possibility distribution of $\overline{w}$ and compute $\overline{Z}$. In this application $\overline{Z}$ need not be a fuzzy number.

Now there are two things to do: (1) discuss how we will find min $\overline{Z}$, $1 \leq c \leq N$; and (2) show how we are to compute $\overline{B}\overline{w}$ and $\overline{Z}$. Let us first consider minimizing fuzzy set $\overline{Z}$. Since $\overline{Z}$ is fuzzy we cannot directly minimize it so we will first change min $\overline{Z}$ into a crisp vector optimization problem and then into a crisp single objective minimization problem as was done in Section 6.3. Let u_2 be the centroid of $\overline{Z}$. Define

$$A_1 = \text{Poss}[x < u_2] , \tag{6.85}$$

$$A_2 = \text{Poss}[x > u_2] . \tag{6.86}$$

A_1 and A_2 are easily found as

$$A_1 = \sup\{\overline{Z}(x) | x < u_2\} , \tag{6.87}$$

$$A_2 = \sup\{\overline{Z}(x) | x > u_2\} . \tag{6.88}$$

In place of min $\overline{Z}$ we substitute

$$\min(1 - A_1, u_2, A_2) , \tag{6.89}$$

because we wish to: (1) minimize the central value of $\overline{Z}$; (2) minimize the possibility of obtaining values greater than u_2 (A_2); and (3) maximize the possibility of getting less than u_2, or min$(1 - A_1)$.

What is usually done in vector optimization problems is now to generate undominated solutions for the decision maker. One way to do this is to consider the single objective function

$$\min\{\lambda_1(1-A_1)+\lambda u_2+\lambda_3 A_2\}\ , \tag{6.90}$$

for $\lambda_i > 0$, $\lambda_1+\lambda_2+\lambda_3=1$, $c=1,2,\ldots,N$. What is known [23] is that solving this optimization problem for $\lambda_i > 0$ we obtain undominated solutions but may not be able to generate all undominated solutions this way. However, we will use this method in this section (also used in Section 6.3).

Next let us consider computing $\overline{B}\overline{w}$ and $\overline{Z}$. In $\overline{B}\overline{w}$, $\overline{B}$ is a trapezoidal fuzzy number, and $\overline{w}$ is a discrete fuzzy set. Suppose $N=3$ so $\overline{w}$, for a certain value of $c \in \{1,2,3\}$, could be

$$\overline{w}=\{\frac{0.3}{0},\frac{0.8}{1},\frac{1.0}{2},\frac{0.5}{3}\}\ . \tag{6.91}$$

Once we find $\overline{B}\overline{w}$ just add the crisp constant Rc to get $\overline{Z}$. We would like to use α–cuts and interval arithmetic to get $\overline{B}\overline{w}$ but α–cuts of $\overline{w}$ are not intervals. So, we must use the extension principle. If $\overline{C}=\overline{B}\overline{w}$, then

$$\overline{C}(z)=\max\{\min\{\overline{B}(x),\overline{w}(y)\}|xy=z\}\ . \tag{6.92}$$

But this computation is not too difficult because $\overline{w}(y)\neq 0$ only for $y=0,1,2,3$ if $N=3$. We can easily see that, using the $\overline{w}$ given above, that $\overline{C}$ is

$$\min\{0\cdot\overline{B},0.3\}\cup\min\{1\cdot\overline{B},0.8\}\cup\min\{2\cdot\overline{B},1\}\cup\min\{3\cdot\overline{B},0.5\} \tag{6.93}$$

where "$\cup$" denotes max. A graph of $\overline{C}(z)$ for $c=1$ is shown in Figure 6.27.

Example 6.4.4.1

Let $N=3$ and $R=\$\,50$, $\overline{B}=(100/120,130/150)$. For arrivals we choose: (1) $\lambda(0,0)=0.2$, $\lambda(1,0)=0.5$, $\lambda(2,0)=1.0$, $\lambda(3,0)=0.3$; (2) $\lambda(0,1)=0.4$, $\lambda(1,1)=1.0$, $\lambda(2,1)=0.6$; (3) $\lambda(0,2)=1.0$, $\lambda(1,2)=0.5$; and (4) $\lambda(0,3)=1.0$. For the servers we have: (1) $\mu(0,1)=1.0$, $\mu(1,1)=0.5$; (2) $\mu(0,2)=0.6$, $\mu(1,2)=1.0$, $\mu(2,2)=0.4$; and (3) $\mu(0,2)=0.2$, $\mu(1,3)=1.0$, $\mu(2,3)=0.8$, $\mu(3,3)=0.2$. Calculate Ω_i, $i=1,2,3$ as shown in Table 6.8. .

Next we determine if the Ω_i^k converge. We found that Ω_1^k converges to Π_1 but the sequences Ω_i^k oscillate, $i=2,3$. First consider Π_1. If initially no machines are broken down, so that $p(0)=(1,0,0,0)$, then we get the steady state possibility $p(0)\Pi_1=(0.5,0.5,1.0,0.5)$. This gives the steady state possibility distribution for $\overline{w}$ when $c=1$. From this compute $\overline{Z}$ and the objective function in equation (6.90) using $\lambda_1=0.4$, $\lambda_2=0.4$, $\lambda_3=0.2$. The value of the objective function for $c=1$ is 115.63.

Now for $c=2$ and $c=3$. First consider $c=2$. It turns out that the sequence Ω_2^k oscillates between 4×4 matrices S_{21} and S_{22}. Compute

		Future State			
Repair crews	Prev. state	0	1	2	3
c=1	0	$\lambda(0,0)$	$\lambda(1,0)$	$\lambda(2,0)$	$\lambda(3,0)$
	1	$\lambda(0,1)\mu(1,1)$	$\lambda(0,1)\mu(0,1)$ $+\lambda(1,1)\mu(1,1)$	$\lambda(1,1)\mu(0,1)$ $+\lambda(2,1)\mu(1,1)$	$\lambda(2,1)\mu(0,1)$
Ω_1	2	0	$\lambda(0,2)\mu(1,1)$	$\lambda(0,2)\mu(0,1)$ $+\lambda(1,2)\mu(1,1)$	$\lambda(1,2)\mu(0,1)$
	3	0	0	$\mu(1,1)$	$\mu(0,1)$
c=2	0	$\lambda(0,0)$	$\lambda(1,0)$	$\lambda(2,0)$	$\lambda(3,0)$
	1	$\lambda(0,1)\mu(1,1)$	$\lambda(0,1)\mu(0,1)$ $+\lambda(1,1)\mu(1,1)$	$\lambda(1,1)\mu(0,1)$ $+\lambda(2,1)\mu(1,1)$	$\lambda(2,1)\mu(0,1)$
Ω_2	2	$\lambda(0,2)\mu(2,2)$	$\lambda(0,2)\mu(1,2)$ $+\lambda(1,2)\mu(2,2)$	$\lambda(0,2)\mu(0,2)$ $+\lambda(1,2)\mu(1,2)$	$\lambda(1,2)\mu(0,2)$
	3	0	$\mu(2,2)$	$\mu(1,2)$	$\mu(0,2)$
c=3	0	$\lambda(0,0)$	$\lambda(1,0)$	$\lambda(2,0)$	$\lambda(3,0)$
	1	$\lambda(0,1)\mu(1,1)$	$\lambda(0,1)\mu(0,1)$ $+\lambda(1,1)\mu(1,1)$	$\lambda(1,1)\mu(0,1)$ $+\lambda(2,1)\mu(1,1)$	$\lambda(2,1)\mu(0,1)$
Ω_3	2	$\lambda(0,2)\mu(2,2)$	$\lambda(0,2)\mu(1,2)$ $+\lambda(1,2)\mu(2,2)$	$\lambda(0,2)\mu(0,2)$ $+\lambda(1,2)\mu(1,2)$	$\lambda(1,2)\mu(0,2)$
	3	$\mu(3,3)$	$\mu(2,3)$	$\mu(1,3)$	$\mu(0,3)$

Table 6.8: The Transition Matrices in Example 6.4.4.1

$p(0)S_{2j} = p^{2j}$, $j = 1,2$. This gives two possibility distributions for $\overline{w}$. Using p^{2j} find the value of $\overline{Z}$, say $\overline{Z}_{2j}$, and then the value (same λ_i) of the objective function, say O_{2j}, $j = 1,2$. Since, after some power of Ω_2, we get S_{21}, S_{22}, S_{21}, S_{22},... forever, the S_{2j} are "equally likely", $j = 1,2$. For this reason we take the average of the O_{2j} for the objective function value of $c = 2$. The value of the objective function is 125.53 for $c = 2$.

If $c = 3$, then we found that the sequence Ω_3^k eventually oscillates between S_{31}, and S_{32}. As above we compute O_{3j}, $j = 1,2$, and then their average for the value of the objective function. The value of the objective function is 130.92 for $c = 3$.

Hence, the optimal number of repair crews is $c^* = 1$.

6.4.5 Fuzzy Queuing Decision Problem

In this queuing problem we have finite calling capacity ($M < \infty$) with finite calling source ($f = N > M \geq c$), and we wish to find the optimal number of servers c to minimize the total cost composed of the cost of the c servers plus the cost of lost customers. We want to minimize $\overline{Z}$ where

$$\overline{Z} = C_1 c + \overline{C}_2 \, \overline{L} \,, \tag{6.94}$$

where C_1 is the cost per server in \$/unit time, $\overline{C}_2$ = cost (\$/unit time) per lost customer (turned away due to finite capacity), and $\overline{L}$ is the number of lost customers. The constant $\overline{C}_2$ is a trapezoidal fuzzy number due to the fact

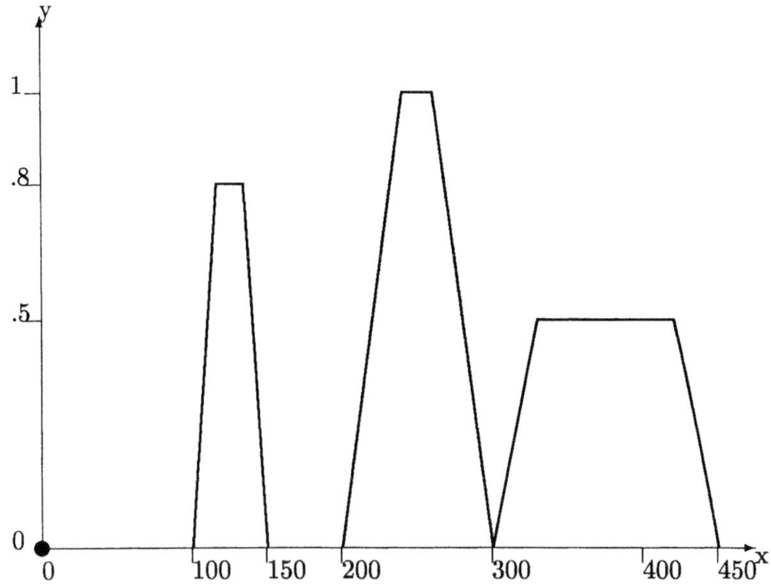

Figure 6.27: The Fuzzy Set $\overline{B}\ \overline{w}$ in Example 6.4.4.1

that the price tag for lost customers is uncertain and hard to estimate. $\overline{L}$ is a discrete fuzzy variable having a possibility distribution giving Poss$[\overline{L}=0]$, Poss$[\overline{L}=1]$, etc.

To find $\overline{L}$ we first use $M=N$. Given values for $\lambda(\alpha)$ and $\mu(d,s)$ construct Ω and, if it converges, its limit Π. Given an initial possibility distribution $p(0)$ we then compute

$$p(0)\Pi = (p_0, p_1, \ldots, p_M, p_{M+1}, \ldots, p_N)\ . \tag{6.95}$$

Now since we actually have finite capacity M the p_i, $i \geq M+1$, give the possibilities for lost customers. That is, Poss$[\overline{L}=0] = \max_{1 \leq i \leq M} p_i$, Poss$[\overline{L}=1] = p_{M+1}$, etc. This is the possibility distribution for $\overline{L}$ the number of lost customers.

Now, as in Section 6.4.4 we compute $\overline{Z}$, A_1, u_2, A_2 and the objective function , equation (6.90), for $c=1,2,3,\ldots$ and choose the optimal c^*.

Example 6.4.5.1

Let C_1 = \$17/hour and $\overline{C}_2$ is approximately \$100/hour, or $\overline{C}_2$ = (80/100, 100/120) a triangular fuzzy number. For the $\mu(i,j)$, $0 \leq i \leq j = 1,2,3$ we use the data from Example 6.4.4.1. We start with capacity N to find the possibility distribution for $\overline{L}$. For arrivals let $\lambda(0)=0.3$, $\lambda(1)=1$, $\lambda(2)=0.6$ and $\lambda(\alpha)=0$, $\alpha \geq 3$. Let $N=10$, $M=5$ and $c=1,2$, or 3.

We wish to find Ω_i for $i = 1,2,3$. These will be 11×11 matrices but they contain a definite pattern of numbers so that we can easily compute $\lim_{k\to\infty} \Omega_i^k = \Pi_i$, $i = 1,2,3$. All sequences converge. The transition matrices are

$$\Omega_1 = \begin{pmatrix} 0.3 & 1.0 & 0.6 & 0.0 & 0.0 & 0.0 & 0.0 & \cdots \\ 0.3 & 1.0 & 0.6 & 0.5 & 0.0 & 0.0 & 0.0 & \cdots \\ 0.0 & 0.3 & 1.0 & 0.6 & 0.5 & 0.0 & 0.0 & \cdots \\ 0.0 & 0.0 & 0.3 & 1.0 & 0.6 & 0.5 & 0.0 & \cdots \\ \vdots & \vdots & & & & & & \end{pmatrix}, \tag{6.96}$$

$$\Omega_2 = \begin{pmatrix} 0.3 & 1.0 & 0.6 & 0.0 & 0.0 & 0.0 & 0.0 & \cdots \\ 0.3 & 0.5 & 1.0 & 0.6 & 0.0 & 0.0 & 0.0 & \cdots \\ 0.3 & 0.4 & 1.0 & 0.6 & 0.6 & 0.0 & 0.0 & \cdots \\ 0.0 & 0.3 & 0.4 & 1.0 & 0.6 & 0.6 & 0.0 & \cdots \\ 0.0 & 0.0 & 0.3 & 0.4 & 1.0 & 0.6 & 0.6 & \cdots \\ \vdots & \vdots & & & & & & \end{pmatrix}, \tag{6.97}$$

$$\Omega_3 = \begin{pmatrix} 0.3 & 1.0 & 0.6 & 0.0 & 0.0 & 0.0 & 0.0 & \cdots \\ 0.3 & 0.5 & 1.0 & 0.6 & 0.0 & 0.0 & 0.0 & \cdots \\ 0.3 & 0.4 & 1.0 & 0.6 & 0.6 & 0.0 & 0.0 & \cdots \\ 0.2 & 0.3 & 0.8 & 1.0 & 0.6 & 0.2 & 0.0 & \cdots \\ 0.0 & 0.2 & 0.3 & 0.8 & 1.0 & 0.6 & 0.2 & \cdots \\ \vdots & \vdots & & & & & & \end{pmatrix}. \tag{6.98}$$

We do not need to display all of the Π_i because all we need is $p(0)\Pi_i$, $i = 1,2,3$, for $\overline{L}_i$. We found for $p(0) = (1,0,0,\ldots)$, or system initially empty, that

$$p(0)\Pi_1 = (0.3, 1.0, 0.6, 0.6, 0.6, 0.6, 0.6, 0.6, 0.6, 0.6, 0.6)\,, \tag{6.99}$$

$$p(0)\Pi_2 = (0.3, 0.5, 1.0, 0.6, 0.6, 0.6, 0.6, 0.6, 0.6, 0.6, 0.6), \tag{6.100}$$

$$p(0)\Pi_3 = (0.3, 0.5, 1.0, 0.6, 0.6, 0.6, 0.6, 0.6, 0.6, 0.6, 0.6)\,. \tag{6.101}$$

We then find the possibility distribution of $\overline{L}_i$ for $i = 1,2,3$. The results are

$$\overline{L}_1 : \left\{ \frac{1.0}{0}, \frac{0.6}{1}, \frac{0.6}{2}, \frac{0.6}{3}, \frac{0.6}{4}, \frac{0.6}{5} \right\}, \tag{6.102}$$

$$\overline{L}_2 : \left\{ \frac{1.0}{0}, \frac{0.6}{1}, \frac{0.6}{2}, \frac{0.6}{3}, \frac{0.6}{4}, \frac{0.6}{5} \right\}, \tag{6.103}$$

$$\overline{L}_3 : \left\{ \frac{1.0}{0}, \frac{0.6}{1}, \frac{0.6}{2}, \frac{0.6}{3}, \frac{0.6}{4}, \frac{0.6}{5} \right\}. \tag{6.104}$$

Now we compute $\overline{C}_2\,\overline{L}_i$, as in Example 6.4.4.1, find A_1, u_2, A_2 for $\overline{Z}_i$ and using $\lambda_1 = 0.4$, $\lambda_2 = 0.4$, $\lambda_3 = 0.2$ compute equation (6.90). It is obvious that the optimal c value is $c^* = 1$ because $\overline{L}_1 = \overline{L}_2 = \overline{L}_3$ so that $\overline{C}_2\,\overline{L}_i$ are all the same, $i = 1, 2, 3$. Then the A_1 and A_2 values are all the same for $c = 1, 2, 3$ but the centroids increase as c increases because we add $17c$ to $\overline{C}_2\,\overline{L}_i$ to get the $\overline{Z}$ values.

Notice that for $f = N = \infty$ the Ω_i are infinite matrices and $p(0)\Pi_i$ can have all values not zero so that $\overline{C}_2\overline{L}$ is an unbounded fuzzy subset of $[0, \infty)$. In this case we cannot find the centroid of $\overline{C}_2\overline{L}$ and are unable to apply our solution method.

6.4.6 Summary and Conclusions

We first surveyed the basic results in fuzzy queuing theory and then adapted our method of solving max/min $\overline{Z}_k$, decision variable k in feasible set $\mathcal{F}$, from Section 6.3, to two application problems. Of course other procedures for handling max/min $\overline{Z}_k$, $k \in \mathcal{F}$, can be devised but we think our method is quite reasonable.

Future research could be concerned with applying possibility theory to other areas of queuing theory and also to other areas of queuing decision problems in the design of optimal queuing systems.

We have also applied possibility theory to Markov chains ([11]), but we have not included these results in this book.

6.5 Fuzzy Network Analysis

Fuzzy network analysis would include the shortest route problem, the max-flow problem, the min-cost capacitated flow problem and PERT. Fuzzy PERT was discussed in Section 6.2 and the max-flow problem is contained within the min-cost capacitated flow problem. In this section we will use an evolutionary algorithm to compute approximate solutions to the fuzzy shortest path problem and the fuzzy min-cost capacitated flow problem. For another approach see [41].

In the next section we discuss what we mean by a solution to the fuzzy shortest path problem and present an example with our evolutionary algorithm solution. Section 6.5.2 contains the fuzzy min-cost capacitated flow problem, a discussion of its solution, and an example. How we adapted the evolutionary algorithm to the fuzzy optimization problems of this section is in Section 6.5.3 and a brief summary and our conclusions are in Section 6.5.4. This section is based on [6].

If $\overline{M}$ and $\overline{N}$ are two triangular shaped fuzzy numbers we will need to define the meaning of $\overline{M} \leq \overline{N}$, $\overline{M} < \overline{N}$ and $\overline{M} \approx \overline{N}$. This was presented in Section 4.4, but for completeness we will repeat it here. The only thing that really matters is that $\leq$, $<$ and $\approx$ can be easily incorporated into the evolutionary algorithm. Of course, by choosing different definitions of $\leq$, $<$ and $\approx$ you can obtain different solutions to fuzzy optimization problems.

There have been numerous definitions proposed for $\leq$ between fuzzy numbers [15]. We will start with $<$ between two fuzzy numbers. Define ([1],[33] p.407)

$$v(\overline{M} \leq \overline{N}) = sup\{min(\overline{M}(x), \overline{N}(y)) | x \leq y\}, \tag{6.105}$$

which is the fuzzification of crisp $\leq$. We write $\overline{M} < \overline{N}$ when $v(\overline{M} \leq \overline{N}) = 1$ and $v(\overline{N}, \overline{M}) < \theta$ for some fixed $\theta \in (0, 1]$. In this section we will use $\theta = 0.8$. Also, we write $\overline{M} \approx \overline{N}$ when $\overline{M} < \overline{N}$ is not true and $\overline{N} < \overline{M}$ is false. In other words, if

$$min(v(\overline{M}, \overline{N}), v(\overline{N}, \overline{M})) \geq \theta, \tag{6.106}$$

then $\overline{M} \approx \overline{N}$. Finally we say $\overline{M} \leq \overline{N}$ when $\overline{M} < \overline{N}$ or $\overline{M} \approx \overline{N}$. These are the definitions of $\leq$, $<$ and $\approx$ that we shall use in the next two sections. An easy way to evaluate $v(\overline{M} \leq \overline{N})$ for $\overline{M}$ $(\overline{N})$ a triangular shaped fuzzy number with vertex at b_1 (b_2) is

$$v(\overline{M}, \overline{N}) = \begin{cases} 1, & b_1 \leq b_2, \\ ht(\overline{M}, \overline{N}), & b_2 < b_1, \end{cases} \tag{6.107}$$

where $ht(\overline{M}, \overline{N})$ is the height of the intersection of $\overline{M}$ and $\overline{N}$

$$ht(\overline{M}, \overline{N}) = sup_x(min(\overline{M}(x), \overline{N}(x))). \tag{6.108}$$

Let $\mathcal{F}$ be a collection of triangular shaped fuzzy numbers. $\mathcal{F}$ could be the solution set to a fuzzy optimization problem. Our fuzzy optimization problems in this section will all be minimization problems so we say $\overline{N} \in \mathcal{F}$ is undominated if there is no $\overline{M} \in \mathcal{F}$ so that $\overline{M} < \overline{N}$. Let H_1 be all the undominated $\overline{N}$ in $\mathcal{F}$. Then H_2 is all the undominated $\overline{N}$ in $\mathcal{F} - H_1$, etc. In our application in the next section $\mathcal{F}$ is finite so we are assured that H_1 is non-empty. The solution set $\mathcal{F}$ becomes partitioned into undominated sets $H_1, H_2, \cdots$. Properties of $\leq$, $<$, $\approx$ and the H_i have been studied in [1]. We will use these definitions of $\leq$, $<$ and $\approx$ again in Section 10.3. This definition of $\leq$ is obviously different from those in Section 6.1.2.

Our first solution procedure is to find H_1 in the fuzzy shortest route problem. Our second solution method, for the fuzzy min-cost capacitated flow problem, requires the minimization of $\overline{Z}$, a triangular shaped fuzzy number, and then we follow the procedures in Sections 6.3 and 6.4 .

6.5.1 Fuzzy Shortest Route

Consider a network having n nodes linked by arcs. We allow all nodes to be connected by an arc. All arcs will be two-way streets (can travel in both directions). Define $\overline{a}_{ij}$ to be the fuzzy distance (or cost) of traveling from node i to node j, $1 \leq i, j \leq n$. Set $\overline{a}_{ii} = 0$ for $1 \leq i \leq n$. If there is no arc connecting node i to node j we set $\overline{a}_{ij} = M$, M is a large positive number.

Since this is a minimization problem the optimal solution will not select a path from i to j if $\overline{a}_{ij} = M$. We wish to find paths from node 1 to node n with smallest fuzzy distance.

For a survey of the literature on the fuzzy shortest path problem see the papers [6],[29] and [45] and the references in these papers. Our solution concept, outlined below, is different from those used by other authors.

Define a path of length $K+1$, $1 \leq K \leq n-1$, to be $(1, i_1, \cdots, i_K, n)$ where $i_1, i_2, \cdots, i_K$ are distinct numbers in the set $\{2, \cdots, n-1\}$. A path of length 1 is $(1, n)$ and a path of length 2 is $(1, i_1, n)$ for $2 \leq i_1 \leq n-1$. Let Ω be all paths from 1 to n. If $\Gamma \in \Omega$, then define $D(\Gamma) = \overline{a}_{1,i_1} + \overline{a}_{i_1,i_2} + \cdots + \overline{a}_{i_K,n}$. If all the $\overline{a}_{ij}$ are triangular fuzzy numbers, or real numbers in $[0, M]$, then $D(\Gamma)$ is also a triangular fuzzy number. If some of the $\overline{a}_{ij}$ are triangular shaped fuzzy numbers, then so is $D(\Gamma)$. Our solution concept is to find all $\Gamma \in \Omega$ producing $D(\Gamma)$ in H_1.

To initiate our evolutionary algorithm we choose a number of paths $\Gamma_1, \Gamma_2, \cdots, \Gamma_N$ in Ω and compute their fuzzy distances $D(\Gamma_i)$, $1 \leq i \leq N$. Since all the $\overline{a}_{ij}$ will be triangular fuzzy numbers, or real numbers, we can also use the vertices of the $\overline{a}_{ij}$ (real numbers) to obtain a crisp solution. Let a_{ij2} be the vertex point of triangular fuzzy number $\overline{a}_{ij}$, $a_{ii2} = 0$ all i and $a_{ij2} = M$ if there is no path from i to j. Solve the crisp problem using the distances a_{ij2} producing optimal (shortest distance) paths $\Gamma_1^*, \cdots, \Gamma_P^*$. The crisp solution can be used to initialize the evolutionary algorithm.

Crossover and mutation are straightforward in that they just choose new paths through the network. To start the initial population evolving we need a fitness function. Suppose our initial population consists of 50 paths $\Gamma_1, \cdots, \Gamma_{50}$. We will choose 20 of the "best" paths for reproduction and mutation. Let $D(\Gamma_i) = \overline{Q}_i$ where $\overline{Q}_i$ is a triangular shaped fuzzy number with vertex at q_{i2}, $1 \leq i \leq 50$. We order the $\overline{Q}_i$ according to their vertex values q_{i2} and choose the 20 paths with the smallest q_{i2} values to generate the next population. We do not want to pick that path with the smallest vertex value because then we would have only one individual to produce the next population of size 50.

After many generations the population tends to stabilize and we stop the evolutionary algorithm. Then we choose the undominated values in the last population to put in H_1 and record their corresponding paths.

Example 6.5.1.1

Let $n = 6$ and the fuzzy distances are shown in Table 6.9. The results of our evolutionary algorithm are displayed in Table 6.10. Notice that we used triangular fuzzy numbers for all the distances in Table 6.9. The optimal, undominated, fuzzy distances given in Table 6.10 will then be all triangular fuzzy numbers.

In the first test run the population size was set at 50. The fittest 10 were

	1	2	3	4	5	6
1	0	(1/3/4)	(4/6/8)	(11/13/16)	(13/15/18)	M
2	(2/4/5)	0	(3/5/7)	(6/8/9)	(10/12/15)	(14/19/23)
3	(1/3/4)	(3/4/6)	0	(3/4/6)	M	(11/14/18)
4	(9/11/14)	(5/8/10)	(5/7/10)	0	(1/2/3)	(7/9/12)
5	(11/15/18)	(8/10/13)	M	(5/6/7)	(0	(5/6/8)
6	M	(16/20/26)	(8/10/13)	(7/9/11)	(2/3/4)	0

Table 6.9: Fuzzy Distances Between the Nodes in Example 6.5.1.1

Path	Fuzzy Distance
1–3–4–5–6	(13/18/25)
1–2–4–5–6	(13/19/24)
1–3–4–6	(14/19/26)
1–2–4–6	(14/20/25)
1–3–6	(15/20/26)
1–2–3–4–5–6	(13/20/28)
1–5–6	(18/21/26)

Table 6.10: Undominated Fuzzy Shortest Routes in Example 6.5.1.1

chosen to build the next generation. After one single iteration the shortest route $(1, 3, 4, 5, 6)$ with fuzzy distance (13/18/25) was found (see Table 6.10). We stared the evolutionary algorithm 10 more times with initial population of 50 and obtained a fuzzy shortest route after at most two generations. In a second test run we set the initial population size at 30 where the best 5 paths were selected to produce the next generation. The settings for the second test run were repeated ten times and the fuzzy shortest path $(1, 3, 4, 6)$ was gotten in the fourth generation. In this way we constructed Table 6.10. The solution to the crisp problem, using the vertices of the triangular fuzzy numbers, is 18. The algorithm is quite fast and could be used, without too much computing time, for networks with n up to 100.

6.5.2 Fuzzy Min-Cost Capacitated Network

The min-cost capacitated flow network problem generalizes the transportation, transshipment and max-flow problems [54]. Some authors ([12],[50]) have worked on the fuzzy max-flow problem, the fuzzy transportation problem and the fuzzy assignment problem. This section is based on [6]. Let us first discuss the crisp problem and then fuzzify it.

There are n nodes, some of which are "sources" which accept input , and some others are "sinks" which produce output. Let I_i be the input to node i and set $I_i = 0$ if node i is not a source node. Let O_i be the output from node i and set $O_i = 0$ if node i is not a sink node.

We again assume that all nodes are connected by an arc which are now all one-way streets. All arcs have a unit cost c_{ij} associated with them. We set $c_{ii} = 0$ all i and if there is no arc from node i to node j we make $c_{ij} = M$, M a large positive number, so that the optimal solution to the minimization problem will not choose an arc with $c_{ij} = M$.

Let x_{ij} be the flow from node i to node j. For example, x_{ij} may be gallons of oil. The x_{ij} flow along the arc from i to j at unit cost c_{ij}. Also, all arcs have a minimum capacity L_{ij} and a maximum capacity U_{ij}. Now an L_{ij} can be zero (no minimum capacity) and a U_{ij} can be infinite (no maximum capacity). If there is no maximum capacity we let $U_{ij} = M$.

So, the optimization problem is to minimize the total cost

$$minZ = \sum_{i=1}^{n} \sum_{j=1}^{n} c_{ij} x_{ij}, \tag{6.109}$$

subject to capacity constraints

$$L_{ij} \leq x_{ij} \leq U_{ij}, \tag{6.110}$$

all i, j, plus the conservation of flow constraints

$$I_i + \sum_{k=1}^{n} x_{ki} = \sum_{k=1}^{n} x_{ik} + O_i, \tag{6.111}$$

for all nodes $i = 1, 2, \cdots, n$. The conservation of flow constraint says that the total flow into a node must equal to the total flow out of that node. For the problem to be feasible we need $\sum_{i=1}^{n} I_i = \sum_{i=1}^{n} O_i$.

The fuzzification has all the $\overline{C}_{ij}$, $\overline{L}_{ij}$, $\overline{U}_{ij}$, $\overline{I}_i$ and $\overline{O}_i$ triangular (shaped) fuzzy numbers or real numbers in $[0, M]$. Let us assume they are all triangular fuzzy numbers and the fuzzy flows $\overline{X}_{ij}$ are also triangular fuzzy numbers. The fuzzy optimization problem is

$$min\overline{Z} = \sum_{i=1}^{n} \sum_{j=1}^{n} \overline{C}_{ij} \overline{X}_{ij}, \tag{6.112}$$

subject to

$$\overline{L}_{ij} \leq \overline{X}_{ij} \leq \overline{U}_{ij}, \tag{6.113}$$

for all i, j and

$$\overline{I}_i + \sum_{k=1}^{n} \overline{X}_{ki} \approx \sum_{k=1}^{n} \overline{X}_{ik} + \overline{O}_i, \tag{6.114}$$

for $1 \leq i \leq n$. In equation (6.113) we have some special cases: (1) if $\overline{L}_{ij} = 0$, then it is $0 \leq \overline{X}_{ij} \leq \overline{U}_{ij}$; (2) if $\overline{U}_{ij} = M$, then it is $\overline{L}_{ij} \leq \overline{X}_{ij}$; and (3) if $\overline{L}_{ij} = 0$ and $\overline{U}_{ij} = M$, then it is $0 \leq \overline{X}_{ij}$. In equation (6.114) we cannot, because of fuzzy arithmetic, demand exact equality as in the crisp case, so we use $\approx$. We also assume that $\sum_{i=1}^{n} \overline{I}_i \approx \sum_{i=1}^{n} \overline{O}_i$. We are using the definitions of $\leq$, $<$ and $\approx$ between fuzzy numbers given at the beginning of this section. However, if one of the fuzzy numbers is a real number the comparisons become much easier. If $\overline{X}$ is a triangular (shaped) fuzzy number with base on the interval $[x_1, x_2]$, then $\overline{X} \geq 0$ holds when $x_1 \geq 0$ and $\overline{X} \leq M$ if $x_2 \leq M$. Also, we will not allow any negative flows through the network, so we add the extra constraint that $\overline{X}_{ij} \geq 0$.

$\overline{Z}$ will be a triangular shaped fuzzy number so let the vertex be at u_2 and base be the interval $[u_1, u_3]$. Now compute A_1 as in equation (6.42) and A_2 as in equation (6.43) and then min$\overline{Z}$ becomes

$$min\{\lambda_1(M^* - A_1) + \lambda_2 u_2 + \lambda_3 A_2\}, \tag{6.115}$$

for M^* a large positive number, $\lambda_i \geq 0$ with sum one, subject to the constraints given in equation (6.113) and (6.114).

In the evolutionary algorithm an individual will be a feasible flow vector $(\overline{X}_{11}, \overline{X}_{12}, \cdots)$. By feasible we mean that equations (6.113) and (6.114) are satisfied. The fitness function is the value of equation (6.115). Crossover and mutation just generate new feasible flow vectors. We place into the initial population the fuzzification of the crisp optimal solution. The crisp problem is generated by using the vertex values of all the triangular fuzzy numbers.

Example 6.5.2.1

Our example is shown in Figure 6.28. For all arcs not shown we set $\overline{C}_{ij} = M$. If no pair $(\overline{L}_{ij}, \overline{U}_{ij})$ is specified for an arc it means that it is equal to $(0, M)$, or no restrictions. Nodes 1 and 2 are source nodes and the sink nodes are 4 and 5. The values of the fuzzy numbers are: (1) $\overline{I}_1 = (35/40/45)$ and $\overline{I}_2 = (45/50/55)$; (2) $\overline{O}_4 = (25/30/35)$ and $\overline{O}_5 = (55/60/65)$: (3) $\overline{C}_{12} = (1/3/5)$, $\overline{C}_{13} = (6/7/8)$, $\overline{C}_{14} = (4/5/6)$, $\overline{C}_{23} = (1/2/3)$, $\overline{C}_{25} = (0.5/1/2)$, $\overline{C}_{35} = (6/7/8)$, $\overline{C}_{45} = (2/4/6)$; and (4) $\overline{U}_{13} = (8/10/11)$, $\overline{U}_{14} = (30/35/40)$, $\overline{U}_{23} = (50/60/70)$, and $\overline{U}_{25} = (28/30/34)$. Using the vertices the crisp optimal solution is $x_{12} = 5$, $x_{13} = 0$, $x_{14} = 35$, $x_{23} = 25$, $x_{25} = 30$, $x_{35} = 25$ and $x_{45} = 5$ with minimum cost of 490. In equation (6.115) we used $\lambda_1 = 0.4 = \lambda_2$ and $\lambda_3 = 0.2$ since the decision maker was more interested in maximizing A_1 and minimizing u_2 than in minimizing A_2. The optimal solution for $\overline{Z}$ is shown in Figure 6.29 with the corresponding values for the $\overline{X}_{ij}$ in Table 6.11. Notice that for this type of fuzzy optimization problem we do not necessarily get the optimal values to the crisp problem to be at the vertices of optimal $\overline{Z}$ or the optimal $\overline{X}_{ij}$.

The algorithm is not fast because it spends a lot of time checking the constraints. For much larger n we would need to construct special algorithms just for checking the constraints.

For this problem we used a population size of 1000 where the fittest 130 were selected to generate the next generation. Two different mutation rates

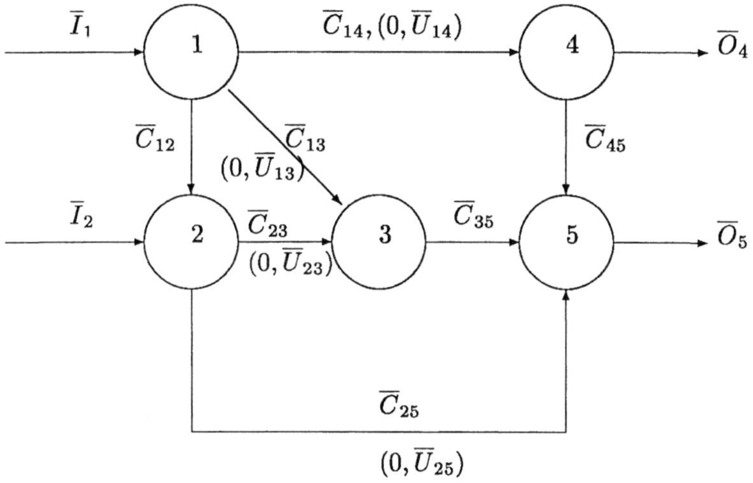

Figure 6.28: Fuzzy Min-Cost Capacitated Flow Problem in Example 6.5.2.1

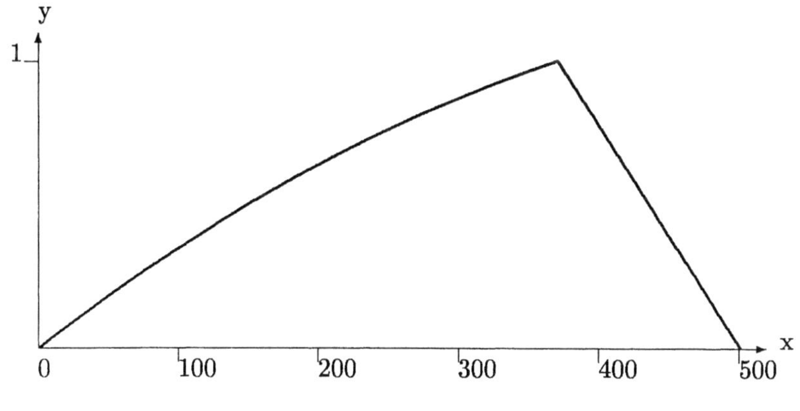

Figure 6.29: Optimal Fuzzy Cost in Example 6.5.2.1

$\overline{X}_{12}$	(0.0041/4.8012/4.8168)
$\overline{X}_{13}$	(0.0035/0.0052/0.0112)
$\overline{X}_{14}$	(0.0020/34.1925/34.1955)
$\overline{X}_{23}$	(0.0000/13.4586/13.4641)
$\overline{X}_{25}$	(0.0034/38.4981/42.9179)
$\overline{X}_{35}$	(0.0003/13.1353/13.1410)
$\overline{X}_{45}$	(0.0000/6.4814/6.4830)

Table 6.11: Optimal Fuzzy Flows in Example 6.5.2.1

were used, one for the vertex values of the fuzzy numbers and another rate for the left and right endpoints of the base of the fuzzy numbers. After 496 generations we obtained 280.12735 for the value of the objective function (equation (6.115)), where the area A_1=200.74106, the vertex of $\overline{Z}$ was u_2=368.68981 and the area A_2=64.73927.

6.5.3 Evolutionary Algorithm

We first discuss how our evolutionary algorithm was adapted for the fuzzy shortest route problem. Some of the details were already discussed in Section 6.5.1 and other details are in Chapter 13. Population size was 50 and a population member was $p = (p_1, \cdots, p_n)$ where $p_1 = 1$, $p_n = n$ and $p_i \in \{1, \cdots, n\}$, $2 \leq i \leq n-1$. For example, if $n = 6$ and $p = (1,3,3,5,4,6)$, then this is the same as $(1,3,5,4,6)$ since the distance from node 3 to node 3 is zero. The initial population was randomly generated. The fuzzy distance is computed for each individual (path) and the vertex of this fuzzy number is used to select 10 individuals, with the smallest vertex value, to generate the next population. Crossover is straight forward but mutation needs to be changed. After crossover an integer τ is randomly generated in $\{2, \cdots, n-2\}$ and then the positions p_τ and $p_{\tau+1}$ are interchanged producing an individual in the next population. In this way we guarantee that a mutated element satisfies the constraint $p_1 = 1$, $p_n = n$ and $p_i \in \{1, \cdots, n\}$ for $2 \leq i \leq n-1$.

Now we discuss the evolutionary algorithm for the fuzzy min-cost capacitated flow problem. Much of the description was already given at the end of Example 6.5.2.1. We discuss the algorithm only for the Example in Section 6.5.2. First let $\overline{Y}_1 = \overline{X}_{12}$, $\cdots$, $\overline{Y}_7 = \overline{X}_{45}$ and set $\overline{Y}_i = (y_{i1}/y_{i2}/y_{i3})$. Next define $l_i = y_{i2} - y_{i1}$ and $r_i = y_{i3} - y_{i2}$ all i. We then represent each $\overline{Y}_i$ as (l_i, y_{i2}, r_i). The advantage of this representation is that after crossover fuzzy numbers are obtained. So a population member is $p = (p_1, p_2, p_3, \cdots, p_{21}, \sigma)$ where $p_1 = l_1$, $p_2 = y_{12}$, $p_3 = r_1$, $\cdots$, $p_{21} = r_7$. The mutation rate is σ. The fitness function is given in equation (6.115). The constraints are in equations (6.113) and (6.114). The only extra constraint on mutation is that it produces positive numbers.

6.5.4 Summary and Conclusions

This section showed that an evolutionary algorithm can be applied to produce approximate solutions to the fuzzification of two important problems in operations research : the shortest path problem and the min-cost capacitated flow problem. We proposed two very different solution concepts for these two problems. The algorithm runs very fast on the fuzzy shortest route problem but is considerably slower on the fuzzy min-cost capacitated flow problem (due to checking many constraints). But the use of a directed search procedure, like the evolutionary algorithm, is promising to obtain approximate solutions to other fuzzy network problems.

6.6 Summary and Conclusions

As we mentioned at the start of this chapter there are many areas of operations research that can be fuzzified. There have been many applications of fuzzy sets and fuzzy logic to operations research and there promise to be many more. We only discussed five applications : (1) linear programming; (2) PERT; (3) inventory control; (4) queuing theory; and (5) network analysis. We look forward to the expanding applications of fuzzy logic/sets to operations research.

Bibliography

[1] J.J. Buckley: Fuzzy Hierarchical Analysis, Fuzzy Sets and Systems, 17 (1985), pp. 233–247.

[2] J.J. Buckley: Fuzzy PERT, in: G.W. Evans, W. Karwowski and M.R. Wilhelm (eds.), Applications of Fuzzy Set Methodologies in Industrial Engineering, Elsevier Science, Amsterdam, 1989, pp. 103–114.

[3] J.J. Buckley: Elementary Queueing Theory Based on Possibility Theory, Fuzzy Sets and Systems, 37 (1990), pp. 43–52.

[4] J.J. Buckley: Note on: Convergence of Powers of a Fuzzy Matrix, Fuzzy Sets and Systems, 121 (2001), pp. 363–364.

[5] J.J. Buckley and T. Feuring: Evolutionary Algorithm Solution to Fuzzy Problems: Fuzzy Linear Programming, Fuzzy Sets and Systems, 109 (2000), pp. 35–53

[6] J.J. Buckley, T. Feuring and Y. Hayashi: Solving Fuzzy Problems: Operations Research, J. Advanced Computational Intelligence, 3 (1999), pp. 171–176.

[7] J.J. Buckley, T. Feuring and Y. Hayashi: Solving Fuzzy Problems in Operations Research: Fuzzy Inventory Control, Proc. FUZZ-IEEE 2000, May 7–10, 2000, San Antonio, Texas, pp. 352–357.

[8] J.J. Buckley, T. Feuring and Y. Hayashi: Solving Fuzzy Problems in Operations Research: Inventory Control, Proc. IFSA/NAFIPS 2001, July 25-28, 2001, Vancouver, Canada, pp. 2704–2707.

[9] J.J. Buckley, T. Feuring and H. Hayashi: Fuzzy Queueing Theory Revisited, Int. J. Uncertainty, Fuzziness and Knowledge-Based Systems. To appear.

[10] J.J. Buckley, T. Feuring and Y. Hayashi: Multiobjective Fully Fuzzified Linear Programming, Int. J. Uncertainty, Fuzziness and Knowledge-Based Systems. To appear.

[11] J.J. Buckley, T. Feuring and Y. Hayashi: Fuzzy Markov Chains, Proc. IFSA/NAFIPS 2001, July 25-28, 2001, Vancouver, Canada, pp. 2708–2711.

[12] S. Chanas and P. Zielinski: Criticality in the Network with Imprecise Activity Times. Proc. IPUM 2000, July 3-7, 2000, Madrid, Spain, pp. 1233–1238.

[13] S. Chanas and P. Zielinski: Critical Path Analysis in the Network with Fuzzy Activity Times, Fuzzy Sets and Systems, 122 (2001), pp. 195–204.

[14] S.-C. Chang: Fuzzy Production Inventory for Fuzzy Product Quantity with Triangular Fuzzy Numbers, Fuzzy Sets and Systems, 107 (1999), pp. 37–57.

[15] S.J. Chen and C.-L. Hwang: Fuzzy Multiple Attribute Decision Making: Methods and Applications, Springer-Verlag, Heidelberg, 1992.

[16] E.W. Davis, J.J. Moder and C.R. Phillps: Project Management with CPM, PERT and Precedence Diagramming, Third Edition, Van Nostrand Reinhold, N.Y., N.Y., 1983.

[17] M. Delgado, J. Kacprzyk, J.-L. Verdegay and M.A. Vila (eds.): Fuzzy Optimization, Physica-Verlag, Heidelberg, 1994.

[18] EUROFUSE Workshop on Scheduling and Planning, European Working Group on Fuzzy Sets, March 31–April 1, 2000, Mons, Belgium.

[19] G. Evens, W. Karwowski and M.R. Wilhelm (eds.): Applications of Fuzzy Set Methodologies in Industrial Engineering, Elsevier, Amsterdam, 1989.

[20] H. Fargier, V. Galvagnon and D. Dubois: Fuzzy PERT in Series-Parallel Graphs, Proc. FUZZ-IEEE 2000, San Antonio, Texas, May 7–10, 2000, pp. 717–722.

[21] Fuzzy Sets and Systems, Elsevier Science, Amsterdam: Special Issue on Fuzzy Mathematical Programming, Vol. 111 (2000).

[22] M. Gen, Y. Tsujiomura and D. Zheng: An Application of Fuzzy Set Theory to Inventory Control Problems, Computers in Industrial Engineering, 33 (1997), pp. 553–556.

[23] A.M. Geoffrion: Proper Efficiency and Theory of Vector Maximization, J. Math. Analysis and Appl., 22 (1968), pp. 618–630.

[24] A.L. Guiffrida and R. Nagi: Fuzzy Set Theory Applications in Production Management Research: A Literature Survey, J. Intelligent Manufacturing, 9 (1998), pp. 39–56.

[25] S.-M. Guu, H.-H. Chen and C.-T. Pang: Convergence of Products of Fuzzy Matrices, Fuzzy Sets and Systems, 121 (2001), pp. 203–207.

[26] F. Herrera and J.L. Verdegay: Fuzzy Sets and Operations Research: Prespectives, Fuzzy Sets and Systems, 90 (1997), pp. 207–218.

[27] IFSA/NAFIPS 2001, Special Sessions on "Applications of Fuzzy Sets/Logic to Operations Research ", July 25–28, 2001, Vancouver, Canada.

[28] T. Imai, Y. Okahara and M. Miyakoshi: The Period of Powers of a Fuzzy Matrix, Fuzzy Sets and Systems, 109 (2000), pp. 405–408.

[29] T. Itoh and H. Ishii: A Model of Fuzzy Shortest Path by Possibility Measure, Japanese J. of Fuzzy Theory and Systems, 8 (1996), pp. 977–990.

[30] J.B. Jo, G. Yamazaki, Y. Tsujimura and M. Gen: Performance of a Fuzzy Queueing Network, Proc. IFSA Conf., Seoul, Korea, 1993, pp. 580–583.

[31] J. Kacprzyk: Multistage Decision-Making under Fuzziness, Verlag TÜV Reinland, Köln, Germany, 1983.

[32] H. Katagiri and H. Ishii: Some Inventory Problems with Fuzzy Shortage Cost, Fuzzy Sets and Systems, 111 (2000), pp. 87–97.

[33] G.J. Klir and B. Yuan: Fuzzy Sets and Fuzzy Logic, Prentice Hall, Upper Saddle River, N.J., 1995.

[34] Y.-J. Lai and C.-L. Hwang: Fuzzy Mathematical Programming, Springer-Verlag, Heidelberg, 1992.

[35] H.-M. Lee and J.-S. Yao: Economic Order Quantity in Fuzzy Sense for Inventory without Backorder Model, Fuzzy Sets and Systems, 105 (1999), pp. 13–31.

[36] R.J. Li and E.S. Lee: Analysis of Fuzzy Queues, Proc. NAFIPS 1998, San Francisco, June 8-10, 1988, pp. 158–162.

[37] R.J. Li and E.S. Lee: Analysis of Fuzzy Queues, Computers and Mathematics with Applications, 17 (1989), pp. 1143–1147.

[38] D.-C. Lin and J.-S. Tao: Fuzzy Economic Production for Production Inventory, Fuzzy Sets and Systems, 111 (2000), pp. 465–495.

[39] B. Liu: Uncertain Programming, Wiley, N.Y., N.Y., 1999.

[40] B.L. Liu and A.O. Esogbue: Decision Criteria and Optimal Inventory Processes, Kluwer Academic Publishers, Norwell, Mass., 1999.

[41] D.S. Malik and J.N. Mordeson: Fuzzy Discrete Structures, Physica - Verlag, Heidelberg, Germany, 2000.

[42] M.A.S. Monfared and S.J. Steiner: Fuzzy Adaptive Scheduling and Control Systems, Fuzzy Sets and Systems, 115 (2000), pp. 231–246.

[43] S.H. Nasution: Techniques and Applications of Fuzzy Theory to Critical Path Methods, in: C.Leondes (ed.), Fuzzy Theory Systems: Techniques and Applications, Vol. 4, Academic Press, San Diego, Cal., 1999, pp. 1561–1597.

[44] D.S. Negi and E.S. Lee: Analysis and Simulation of Fuzzy Queues, Fuzzy Sets and Systems, 46 (1992), pp. 321–330.

[45] S. Okada and T. Soper: A Shortest Path Problem on a Network with Fuzzy Arc Lengths, Fuzzy Sets and Systems, 109 (2000), pp. 129–140.

[46] Operations Research (the journal), Institute for Operations Research and Management Sciences, Linthicum, Maryland, USA.

[47] H.M. Prade: An Outline of Fuzzy or Possibilistic Models for Queueing Systems, in: P.P. Wang and S.K. Chang (eds.), Fuzzy Sets, Plenum Press, N.Y., N.Y., 1980, pp. 147–153.

[48] M. Sakawa: Fuzzy Sets and Interactive Multiobjective Optimization, Plenum, N.Y., 1993.

[49] M. Sakawa: Lage Scale Interactive Fuzzy Multiobjective Programming, Springer-Verlag, Heidelberg, 2000.

[50] H.-S. Shih and E.S. Lee: Fuzzy Multi-Level Minimum Cost Flow Problems, Fuzzy Sets and Systems, 107 (1999), pp. 159–176.

[51] R. Slowinski (ed.): Fuzzy Sets in Decision Analysis, Operations Research and Statistics, Kluwer, Boston, 1998.

[52] R. Slowinski and M. Hapke (eds.): Scheduling Under Fuzziness, Springer-Verlag, Heidelberg, 2000.

[53] G. Sommer: Fuzzy Inventory Scheduling in Applied Systems, in; G.E.Lasker (ed.), Applied Systems and Cybernetics, Vol. VI, Pergamon Press, N.Y., 1981, pp. 3052–3062.

[54] H.A. Taha: Operations Research, Fifth Edition, MacMillan, N.Y., N.Y., 1992.

[55] H. Tanaka and P. Guo: Possibilistic Data Analysis for Operations Research, Springer-Verlag, Heidelberg, 1999.

[56] M.G. Thomason: Convergence of Powers of a Fuzzy Matrix, J. Math. Analysis and Appl., 57 (1977), pp. 476–480.

[57] J.-S. Yao and H.-M. Lee: Fuzzy Inventory with or without Backorder Quantity with Trapezoidal Fuzzy Numbers, Fuzzy Sets and Systems, 105 (1999), pp. 311–337.

[58] J.-S. Yao and F.T. Lin: Fuzzy Critical Path Method Based on Signed Distance Ranking of Fuzzy Numbers, IEEE Trans. Systems, Man and Cybernetics, Series A, 30 (2000), pp. 76–82.

[59] J.-S. Yao and F.T. Lin: Corrections to "Fuzzy Critical Path Method Based on Signed Distance Ranking of Fuzzy Numbers", IEEE Trans. Systems, Man and Cybernetics, Series A, 30 (2000), pp. 230.

[60] J.-S. Yao and J.-S. Su: Fuzzy Inventory with Backorder for Fuzzy Total Demand Based on Interval–Valued Fuzzy Set, European J. Operational Research, 124 (2000), pp. 390–408.

[61] J.-S. Yao, S.-C. Shang and J.-S. Su: Fuzzy Inventory without Backorder for Fuzzy Order Quantity and Fuzzy Total Demand Quantity, Computers and Operations Research, 27 (2000), pp. 935–962.

[62] H.-J. Zimmermann(ed.): Practical Applications of Fuzzy Technology, Kluwer, Boston, 1999.

Chapter 7

Fuzzy Differential Equations

We will concentrate on the second order, linear, constant coefficient ordinary differential equation

$$y'' + ay' + by = g(x) \ , \tag{7.1}$$

for x in interval I. I can be $[0, T]$, for $T > 0$ or I can be $[0, \infty)$. The initial conditions are $y(0) = \gamma_0$, $y'(0) = \gamma_1$. We assume g is continuous on I.

In the next section we consider solutions to equation (7.1) for fuzzy initial conditions $y(0) = \overline{\gamma}_0$, $y'(0) = \overline{\gamma}_1$, for triangular fuzzy numbers $\overline{\gamma}_0$ and $\overline{\gamma}_1$. When there is uncertainty about how the system, governed by equation (7.1), starts off, we model that uncertainty using fuzzy numbers $\overline{\gamma}_0$ and $\overline{\gamma}_1$. Also in Section 7.1 there are three applications: (1) an electrical circuit; (2) a vibrating mass; and (3) dynamic supply and demand. In Section 7.2 we allow a and b to be fuzzy but with crisp initial conditions. There is no general theory for the case of a and b fuzzy so in that section we only investigate two examples: (1) a fuzzy, $b = 0$; and (2) $a = 0$, b fuzzy. In both cases we start off with a homogeneous equation.

We follow the same theme as in Chapters 3 and 4: three solutions $\overline{Y}_c$, $\overline{Y}_e$ and $\overline{Y}_I$. If we fuzzify the crisp equation (7.1) and solve, we are attempting to get the "classical" solution $\overline{Y}_c$. When we first solve equation (7.1) and then fuzzify the crisp solution we obtain $\overline{Y}_e$ (extension principle) and $\overline{Y}_I$ (α–cuts and interval arithmetic).

We will be using triangular, and triangular shaped, fuzzy numbers in this chapter. Section 7.1 is based on the papers ([1], [2]).

7.1 Fuzzy Initial Conditions

Let $\overline{Y}_c(x)$ be the classical solution. $\overline{Y}_c(x)$ is a triangular shaped fuzzy number for all x in I. Let $\overline{Y}_c(x) = [y_1(x,\alpha), y_2(x,\alpha)]$, for $x \in I$, $\alpha \in [0,1]$. We assume $y_i(x,\alpha)$ has first and second derivatives with respect to x and we write these as $y_i'(x,\alpha)$, $y_i''(x,\alpha)$, $i = 1,2$. Substitute the α-cuts of $\overline{Y}_c(x)$ into equation (7.1) giving

$$[y_1''(x,\alpha), y_2''(x,\alpha)] + a[y_1'(x,\alpha), y_2'(x,\alpha)] + b[y_1(x,\alpha), y_2(x,\alpha)] = [g(x), g(x)] , \tag{7.2}$$

to solve for $y_i(x,\alpha)$. In equation (7.2) we do interval arithmetic to obtain two differential equations to solve for $y_1(x,\alpha)$ and $y_2(x,\alpha)$. The initial conditions are $y_1(0,\alpha) = \gamma_{01}(\alpha)$, $y_2(0,\alpha) = \gamma_{02}(\alpha)$, $y_1'(0,\alpha) = \gamma_{11}(\alpha)$, $y_2'(0,\alpha) = \gamma_{12}(\alpha)$ where $\overline{\gamma}_0[\alpha] = [\gamma_{01}(\alpha), \gamma_{02}(\alpha)]$, $\overline{\gamma}_1[\alpha] = [\gamma_{11}(\alpha), \gamma_{12}(\alpha)]$.

If these solutions give intervals $[y_1(x,\alpha), y_2(x,\alpha)]$ defining a fuzzy number $\overline{Y}_c(x)$, for each x in I, then the classical solution exists and is $\overline{Y}_c$. Notice that for each fixed x in I, $[y_1(x,\alpha), y_2(x,\alpha)]$ a function of α in $[0,1]$, must define a triangular shaped fuzzy number for $\overline{Y}_c$ to exist. Assuming $y_i(x,\alpha)$ is also differentiable on α, what we need to get $\overline{Y}_c(x)$ a solution is $\partial/\partial\alpha[y_1(x,\alpha)] > 0$, $\partial/\partial\alpha[y_2(x,\alpha)] < 0$ and $y_1(x,1) = y_2(x,1)$, for all x in I.

Example 7.1.1

Consider the differential equation

$$y'' + 4y' + 3y = 0 , \tag{7.3}$$

for $y(0) = 1$, $y'(0) = 0$. Fuzzify the initial conditions producing $\overline{\gamma}_0 = (0/1/2)$, $\overline{\gamma}_1 = (-1/0/1)$. Equation 7.2 becomes

$$y_i''(x,\alpha) + 4y_i'(x,\alpha) + 3y_i(x,\alpha) = 0 , \tag{7.4}$$

for $i = 1,2$. The solution is

$$y_1(x,\alpha) = c_1e^{-x} + c_2e^{-3x} , \tag{7.5}$$

$$y_2(x,\alpha) = c_3e^{-x} + c_4e^{-3x} . \tag{7.6}$$

Using the initial conditions to obtain values for the constants c_i we solve

$$\alpha = c_1 + c_2 , \tag{7.7}$$

$$-1+\alpha = -c_1 - 3c_2 , \tag{7.8}$$

$$2-\alpha = c_3 + c_4 , \tag{7.9}$$

$$1-\alpha = -c_3 - 3c_4 , \tag{7.10}$$

giving

$$y_1(x,\alpha) = \left(-\frac{1}{2}+2\alpha\right)e^{-x} + \left(\frac{1}{2}-\alpha\right)e^{-3x} , \tag{7.11}$$

$$y_2(x,\alpha) = \left(\frac{7}{2}-2\alpha\right)e^{-x} + \left(-\frac{3}{2}+\alpha\right)e^{-3x} . \tag{7.12}$$

Computing $\partial/\partial\alpha[y_1(x,\alpha)]$, $\partial/\partial\alpha[y_2(x,\alpha)]$ and $y_1(x,1)$, $y_2(x,1)$ we see that equations (7.11) and (7.12) do define a fuzzy number for all $x \geq 0$ so

$$\overline{Y}_c[\alpha] = [y_1(x,\alpha), y_2(x,\alpha)] , \tag{7.13}$$

from equations (7.11) and (7.12).

Notice that

$$\lim_{x\to\infty} \overline{Y}_c(x) = 0 , \tag{7.14}$$

or the fuzziness disappears as x grows larger.

Example 7.1.2

We start with

$$y'' + y = 0 , \tag{7.15}$$

$y(0) = 1$, $y'(0) = 0$. The fuzzy initial conditions are the same as in Example 7.1.1. Substituting α-cuts we have to solve

$$y_i''(x,\alpha) + y_i(x,\alpha) = 0, \tag{7.16}$$

for $i = 1, 2$. The solutions are

$$y_1(x,\alpha) = c_1 cos(x) + c_2 sin(x), \tag{7.17}$$

$$y_2(x,\alpha) = c_3 cos(x) + c_4 sin(x). \tag{7.18}$$

The initial conditions imply

$$c_1 = \alpha, \tag{7.19}$$

$$c_2 = -1+\alpha, \tag{7.20}$$

$$c_3 = 2-\alpha, \tag{7.21}$$

$$c_4 = 1-\alpha. \tag{7.22}$$

We therefore find that

$$\frac{\partial y_1}{\partial \alpha} = cos(x) + sin(x), \tag{7.23}$$

$$\frac{\partial y_2}{\partial \alpha} = -cos(x) - sin(x). \tag{7.24}$$

So, for some x we get $\frac{\partial y_1}{\partial \alpha} < 0$ and $\frac{\partial y_2}{\partial \alpha} > 0$. Hence, $\overline{Y}_c(x)$ does not exist on I, assuming $T > 3\pi/4$.

We are familiar with this situation from Chapters 3 and 4: sometimes the classical solution exists and sometimes is does not exist. We prefer the classical solution, to the others discussed below, when it exists.

To obtain the other two solutions $\overline{Y}_e(x)$ and $\overline{Y}_I(x)$ we first need to get the crisp solution. The crisp solution may be written

$$y(x) = c_1 y_1(x) + c_2 y_2(x) + G(x) \ , \tag{7.25}$$

where $y_1(x)$ and $y_2(x)$ are two linearly independent solutions to the homogeneous equation, $G(x)$ is any solution to equation (7.1), and c_1, c_2 are arbitrary constants. The initial conditions are now used to specify unique values for c_1 and c_2. Since we are to fuzzify γ_0 and γ_1 we must show the exact dependence of the crisp solution on the initial conditions. Solving

$$\gamma_0 = c_1 y_1(0) + c_2 y_2(0) + G(0) \ , \tag{7.26}$$

$$\gamma_1 = c_1 y_1'(0) + c_2 y_2'(0) + G'(0) \ , \tag{7.27}$$

for c_1 and c_2 we obtain

$$c_1 = f_1(\gamma_0, \gamma_1) + h_1 \ , \tag{7.28}$$

$$c_2 = f_2(\gamma_0, \gamma_1) + h_2 \ , \tag{7.29}$$

where

$$f_1(\gamma_0, \gamma_1) = \frac{\gamma_0 y_2'(0) - \gamma_1 y_2(0)}{\Delta} \ , \tag{7.30}$$

$$f_2(\gamma_0, \gamma_1) = \frac{\gamma_1 y_1(0) - \gamma_0 y_1'(0)}{\Delta} \ , \tag{7.31}$$

$$h_1 = \frac{G'(0) y_2(0) - G(0) y_2'(0)}{\Delta} \ , \tag{7.32}$$

$$h_2 = \frac{G(0) y_1'(0) - G'(0) y_1(0)}{\Delta} \ , \tag{7.33}$$

$$\Delta = y_1(0) y_2'(0) - y_2(0) y_1'(0) \ . \tag{7.34}$$

So, the unique solution may be written

$$y(x) = f_1(\gamma_0, \gamma_1) y_1(x) + f_2(\gamma_0, \gamma_1) y_2(x) + \Phi(x) \ , \tag{7.35}$$

for

$$\Phi(x) = h_1 y_1(x) + h_2 y_2(x) + G(x) \ . \tag{7.36}$$

We have split the unique solution into two parts: (1) the first part contains the γ_0 and γ_1 values; and (2) the second part Φ is independent of the initial conditions. Also, notice that $\Delta \neq 0$ because it is the Wronskian of the homogeneous solutions evaluated at $x = 0$.

To obtain $\overline{Y}_e(x)$ we substitute $\overline{\gamma}_0$ for γ_0, $\overline{\gamma}_1$ for γ_1, in equation (7.35) and evaluate using the extension principle. $\overline{Y}_e(x)$ will be a triangular shaped fuzzy number with α–cuts $[y_{e1}(x,\alpha), y_{e2}(x,\alpha)]$ given by

$$y_{e1}(x,\alpha) \tag{7.37}$$
$$= \min\{f_1(\gamma_0,\gamma_1)y_1(x) + f_2(\gamma_0,\gamma_1)y_2(x) + \Phi(x) | \gamma_0 \in \overline{\gamma}_0[\alpha], \gamma_1 \in \overline{\gamma}_1[\alpha]\} ,$$
$$y_{e2}(x,\alpha) \tag{7.38}$$
$$= \max\{f_1(\gamma_0,\gamma_1)y_1(x) + f_2(\gamma_0,\gamma_1)y_2(x) + \Phi(x) | \gamma_0 \in \overline{\gamma}_0[\alpha], \gamma_1 \in \overline{\gamma}_1[\alpha]\} ,$$

$0 \leq \alpha \leq 1$, $x \in I$.

In this chapter we will demand more of $\overline{Y}_e(x)$ before we will call it a solution to the fuzzy initial value problem. Assuming $y_{ei}(x,\alpha)$ have first and second derivatives on x for each $\alpha \in [0,1]$, we say $\overline{Y}_e(x)$ is a solution if the $y_{ei}(x,\alpha)$ satisfy the original differential equation (7.1) and the initial conditions $y_{e1}(0,\alpha) = \gamma_{01}(\alpha)$, $y_{e2}(0,\alpha) = \gamma_{02}(\alpha)$, $y'_{e1}(0,\alpha) = \gamma_{11}(\alpha)$, $y'_{e2}(0,\alpha) = \gamma_{12}(\alpha)$, for all α. It is easy to see that the initial conditions will be satisfied, but the $y_{ei}(x,\alpha)$ will satisfy the differential equation if the *interval condition* holds.

In the *interval condition* let us first define the intervals. If $I = [0,T]$, then the intervals are I_k, $1 \leq k \leq K$ for $I_k = [\delta_{k-1}, \delta_k]$, $0 = \delta_0 < \delta_1 < \cdots < \delta_K = T$. If $I = [0,\infty)$, then $0 = \delta_0 < \delta_1 < \cdots$ with $\delta_k \to \infty$ and $I_k = [\delta_{k-1}, \delta_k]$, $1 \leq k < \infty$. The *interval condition* holds if for each interval I_k there exist $\gamma_i^* \in \overline{\gamma}_i[\alpha]$, $\gamma_i^{**} \in \overline{\gamma}_i[\alpha]$, $i = 0, 1$ so that

$$y_{e1}(x,\alpha) = f_1(\gamma_0^*,\gamma_1^*)y_1(x) + f_2(\gamma_0^*,\gamma_1^*)y_2(x) + \Phi(x) , \tag{7.39}$$
$$y_{e2}(x,\alpha) = f_1(\gamma_0^{**},\gamma_1^{**})y_1(x) + f_2(\gamma_0^{**},\gamma_1^{**})y_2(x) + \Phi(x) , \tag{7.40}$$

for all x in I_k, all k. What this means is that f_1 and f_2 are independent of x on each interval I_k, for all α in $[0,1]$. So, on each I_k, f_1 and f_2 are just constants for each fixed α in $[0,1]$ and the $y_{ei}(x,\alpha)$ satisfy the differential equation, $i = 1, 2$.

Theorem 7.1 *The interval condition holds and $\overline{Y}_e(x)$ is a solution to the fuzzy initial value problem.*

Proof 7.1 *See [2].*

The solution $\overline{Y}_I(x)$ is to substitute intervals $[\gamma_{01}(\alpha), \gamma_{02}(\alpha)]$ and $[\gamma_{11}(\alpha), \gamma_{12}(\alpha)]$ for γ_0 and γ_1, respectively, in equation (7.35) and evaluate using interval arithmetic. We also demand that $\overline{Y}_I(x)$ satisfy the differential equation and initial conditions to be called a solution to the fuzzy initial value problem.

Theorem 7.2 *Assume $a > 0$, $b > 0$. If $\overline{Y}_c(x)$ exists, then $\overline{Y}_c(x) \leq \overline{Y}_e(x)$. We always have $\overline{Y}_e(x) \leq \overline{Y}_I(x)$.*

Proof 7.2 *We assumed $a > 0$ and $b > 0$ so that solving for $\overline{Y}_c(x)$ will be easy. Let $\overline{Y}_c(x)[\alpha] = [y_1(x,\alpha), y_2(x,\alpha)]$. We know that*

$$y_i''(x,\alpha) + a y_i'(x,\alpha) + b y_i(x,\alpha) = g(x) \ , \tag{7.41}$$

for $x \in I$, $\alpha \in [0,1]$, $i = 1,2$. But the $y_i(x,\alpha)$ also satisfy the initial conditions $y_1(0,\alpha) = \gamma_{01}(\alpha)$, $y_2(0,\alpha) = \gamma_{02}(\alpha)$, $y_1'(0,\alpha) = \gamma_{11}(\alpha)$, $y_2'(0,\alpha) = \gamma_{12}(\alpha)$. So, by equation (7.35)

$$y_1(x,\alpha) = f_1(\gamma_{01}(\alpha), \gamma_{11}(\alpha)) y_1(x) + f_2(\gamma_{01}(\alpha), \gamma_{11}(\alpha)) + \Phi(x) \ , \tag{7.42}$$

since the right hand side of this equation is the unique solution. Hence, by equation (7.37) we see

$$y_{e1}(x,\alpha) \le y_1(x,\alpha) \ . \tag{7.43}$$

Similarly, we get

$$y_2(x,\alpha) \le y_{e2}(x,\alpha) \ , \tag{7.44}$$

from equation (7.38). The result $\overline{Y}_c(x) \le \overline{Y}_e(x)$, $x \in I$, follows.

We now consider three applications.

7.1.1 Electrical Circuit

Consider the electrical circuit shown in Figure 7.1 where $L = 1h$, $R = 2\Omega$, $C = 0.25f$ and $E(t) = 20 \cos t$. If Q is the charge on the capacitor at time $t > 0$, then

$$Q'' + 2Q' + 4Q = 50 \cos t \ , \tag{7.45}$$

for $Q(0) = \gamma_0$, $Q'(0) = \gamma_1$. The unique solution is

$$Q(t) = \gamma_0 e^{-t} \cos(\sqrt{3}t) + \frac{\gamma_0 + \gamma_1}{\sqrt{3}} e^{-t} \sin(\sqrt{3}t) + \Phi(t) \ . \tag{7.46}$$

The initial conditions being uncertain are modeled as fuzzy numbers $\overline{\gamma}_0 = (4/5/6)$, $\overline{\gamma}_1 = (0/1/2)$. We are first finding $\overline{Q}_e(t)$. Equations (7.37) and (7.38) must be evaluated. To do this we rewrite the unique solution as

$$Q(t) = \gamma_0 \frac{2}{\sqrt{3}} e^{-t} \sin(\sqrt{3}t + \frac{\pi}{3}) + \gamma_1 \frac{e^{-t}}{\sqrt{3}} \sin(\sqrt{3}t) + \Phi(t) \ . \tag{7.47}$$

Now we can see the intervals are $I_1 = [0, \delta_1]$, $I_2 = [\delta_1, \delta_2]$, $I_3 = [\delta_2, \delta_3]$ for $\delta_1 = (2/3)(\pi/\sqrt{3})$, $\delta_2 = \pi/\sqrt{3}$, $\delta_3 = (5/3)(\pi/\sqrt{3})$, $\delta_4 = 2\pi/\sqrt{3}$, ... We can see from Table 7.1 how to evaluate equations (7.37) and (7.38). The results for $q_{e1}(t,\alpha)$ are: (1) use $\gamma_{01}(\alpha)$ for γ_0 and $\gamma_{11}(\alpha)$ for γ_1 on I_1; (2) use $\gamma_{02}(\alpha)$ for γ_0 and $\gamma_{11}(\alpha)$ for γ_1 on I_2; (3) use $\gamma_{02}(\alpha)$ for γ_0 and $\gamma_{12}(\alpha)$ for γ_1 on I_3; (4) use $\gamma_{01}(\alpha)$ for γ_0 and $\gamma_{12}(\alpha)$ for γ_1 on I_4, etc. We obtain $q_{e2}(t,\alpha)$ by interchanging $\gamma_{01}(\alpha)$ and $\gamma_{02}(\alpha)$ and interchanging $\gamma_{11}(\alpha)$ and $\gamma_{12}(\alpha)$ in $q_{e1}(t,\alpha)$. In this way we find the α–cuts of $\overline{Q}_e(t)$ which are $[q_{e1}(t,\alpha), q_{e2}(t,\alpha)]$.

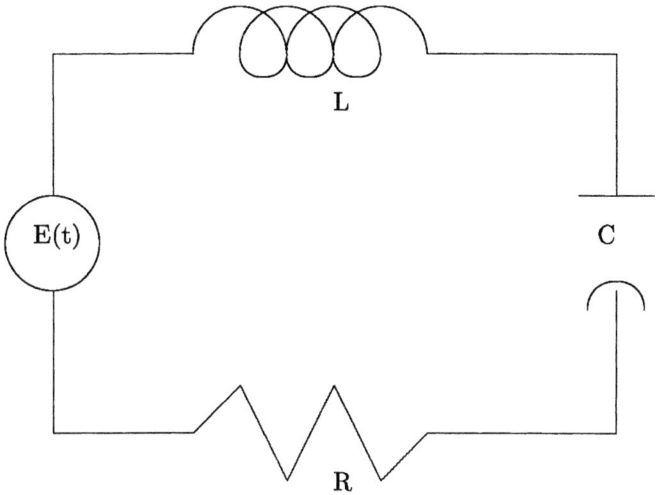

Figure 7.1: Electrical Circuit in Application 7.1.1

Interval	$\frac{\partial Q}{\partial \gamma_0}$	$\frac{\partial Q}{\partial \gamma_1}$
I_1	+	+
I_2	−	+
I_3	−	−
I_4	+	−
I_5	+	+

Table 7.1: Finding $\overline{Q}_e(t)$ in Application 7.1.1

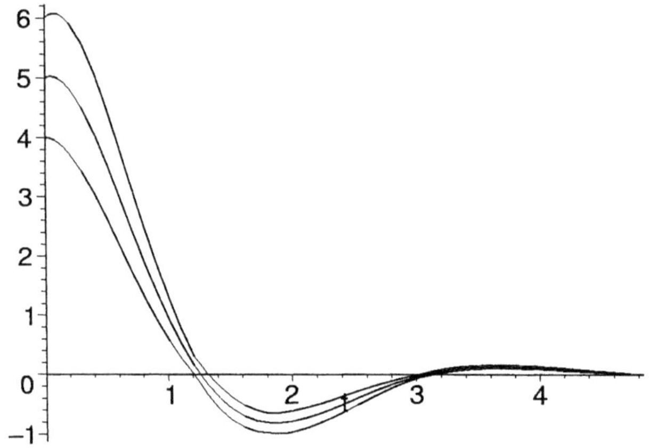

Figure 7.2: Extension Principle Solution for the Fuzzy Part of $\overline{Q}_e(t)$ in the Electrical Circuit

The graph of $\overline{Q}_e(t)$, without $\Phi(t)$, for $\alpha = 0$ and $\alpha = 1$ cuts, is in Figure 7.2 (the two outside curves give the $\alpha = 0$ cut and the center curve is the $\alpha = 1$ cut). We see that

$$\lim_{t \to \infty} \overline{Q}_e(t) = G(t) , \tag{7.48}$$

$$G(t) = \frac{100}{13} \sin t + \frac{150}{13} \cos t , \tag{7.49}$$

and the fuzzy initial conditions disappears as $t \to \infty$. Figure 7.2 only has the fuzzy part of $\overline{Q}_e(t)$ which we see goes to zero. $\Phi(t)$ has two parts: (1) a transient part that goes to zero; and (2) a steady state part which is the $G(t)$ in equation (7.49).

Now we argue that $\overline{Q}_c(t)$, the classical solution, does not exist in this application. We solve equation (7.2), for $a = 2$ and $b = 4$ and $g(t) = 50 \cos t$, for the $y_i(t, \alpha)$, $i = 1, 2$ and get

$$y_1(t,\alpha) = c_1 e^{-t} \cos(\sqrt{3}t) + c_2 e^{-t} \sin(\sqrt{3}t) + G(t) , \tag{7.50}$$

$$y_2(t,\alpha) = c_3 e^{-t} \cos(\sqrt{3}t) + c_4 e^{-t} \sin(\sqrt{3}t) + G(t) . \tag{7.51}$$

Next we use $\overline{\gamma}_0[\alpha] = [4 + \alpha, 6 - \alpha]$, $\gamma_1[\alpha] = [\alpha, 2 - \alpha]$ as in equation (7.7) –

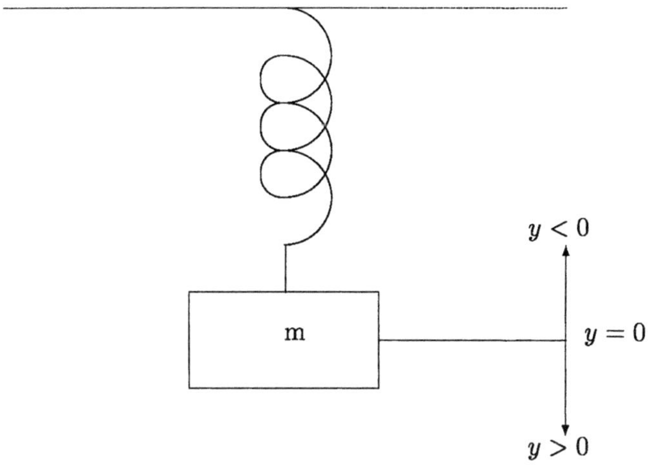

Figure 7.3: Vibrating Mass in Application 7.1.2

(7.10) solving for the c_i, $1 \leq i \leq 4$. We find

$$\frac{\partial y_1(t, \alpha)}{\partial \alpha} = e^{-t} \left(\cos(\sqrt{3}t) + \frac{2}{\sqrt{3}} \sin(\sqrt{3}t) \right) , \tag{7.52}$$

which is sometimes negative, where it should always be positive to define fuzzy number $\overline{Q}_c(t)$. The classical solution does not exist for all t in $I = [0, \infty)$.

We omit $\overline{Q}_I(t)$ since we have found $\overline{Q}_e(t)$, $t \in I$. We would use $\overline{Q}_I(t)$ to approximate $\overline{Q}_e(t)$ when it is very difficult to compute $\overline{Q}_e(t)$.

7.1.2 Vibrating Mass

Consider the vibrating mass in Figure 7.3. The mass $m = 1$ slug, the spring constant $k = 4$ lbs/ft and there is no, or negligible, damping. The forcing function is $2\cos(\beta t)$, for $0 < \beta < 2$. The differential equation of motion is

$$y'' + 4y = 2\cos(\beta t) , \tag{7.53}$$

subject to initial conditions $y(0) = \gamma_0$, $\gamma'(0) = \gamma_1$. The unique solution is

$$y = \gamma_0 \cos(2t) + \frac{\gamma_1}{2} \sin(2t) + \Phi(t) , \tag{7.54}$$

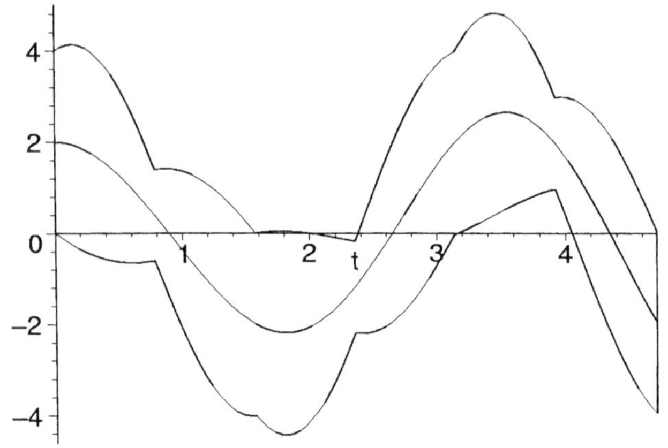

Figure 7.4: Extension Principle Solution $\overline{Y}_e(t)$ in the Vibrating Mass Application

where

$$\Phi(t) = \frac{2}{4-\beta^2}\left[\cos(\beta t) - \cos(2t)\right] . \tag{7.55}$$

The uncertain initial conditions are $\overline{\gamma}_0 = (0/2/4)$, $\overline{\gamma}_1 = (-2/0/2)$ so that $\overline{\gamma}_0[\alpha] = [2\alpha, 4-2\alpha]$, $\overline{\gamma}_1[\alpha] = [-2+2\alpha, 2-2\alpha]$. The interval condition holds so we next specify the intervals $I_k = [\delta_{k-1}, \delta_k]$ for $\delta_0 = 0$, $\delta_1 = \pi/4$, $\delta_2 = 2\pi/4$, $\delta_3 = 3\pi/4$, etc. We notice that $\partial y/\partial \gamma_0$ is positive on I_1, I_4, $I_5, \ldots$ but negative on the other I_k. Similarly, $\partial y/\partial \gamma_1 > 0$ on I_1, I_2, I_5, $I_6, \ldots$ and negative on the rest of the intervals. This allows us to find the α–cuts of $\overline{Y}_e(t)$ as follows for $y_{e1}(t,\alpha)$: (1) use $\gamma_{01}(\alpha)$ for γ_0 and $\gamma_{11}(\alpha)$ for γ_1 on I_1; (2) use $\gamma_{02}(\alpha)$ for γ_0 but $\gamma_{11}(\alpha)$ for γ_1 on I_2; (3) use $\gamma_{02}(\alpha)$ for γ_0 and $\gamma_{12}(\alpha)$ for γ_1 on I_3; and (4) use $\gamma_{01}(\alpha)$ for γ_0 and $\gamma_{12}(\alpha)$ for γ_1 on I_4. Interchange $\gamma_{01}(\alpha)$, $\gamma_{02}(\alpha)$ and interchange $\gamma_{11}(\alpha)$, $\gamma_{12}(\alpha)$ in $y_{e1}(t,\alpha)$ to get $y_{e2}(t,\alpha)$. This is the way we build α–cuts $[y_{e1}(t,\alpha), y_{e2}(t,\alpha)]$ of $\overline{Y}_e(t)$. The graph of $\overline{Y}_e(t)$, $\alpha = 0$ and $\alpha = 1$ cuts, for $\beta = 1.9$, is in Figure 7.4 (the outside curves are the $\alpha = 0$ cut and the center curve the $\alpha = 1$ cut). The fuzziness now does not disappear as $t \to \infty$.

For $\beta = 2$ we have pure resonance. But for β close to 2, say $\beta = 1.9$, we obtain "fuzzy beats". The fuzzy amplitude of the oscillations grows to a maximum and then decreases to zero, and then repeats this behavior again and again as $t \to \infty$. We did not show the "fuzzy beats" in Figure 7.4 because we would need $t \geq 500$ to get one fuzzy beat and then we would not be able to easily see the fuzziness in the solution.

We omit $\overline{Y}_I(t)$ and now consider the classical solution $\overline{Y}_c(t)$. We will change the initial conditions to $\overline{\gamma}_0 = (0/1/2)$ and $\overline{\gamma}_1 = (-1/0/1)$. We gave these fuzzy numbers a larger base in the example above so we could more easily see the fuzziness in the solution in Figure 7.4. Let $\overline{Y}_c(t)[\alpha] = [y_1(t,\alpha), y_2(t,\alpha)]$ and solve for the $y_i(t,\alpha)$ as in Example 7.1.1. Then solve for the constants c_i, $1 \leq i \leq 4$, as in equations (7.7) - (7.10). Then compute

$$\frac{\partial y_1}{\partial \alpha} = \cos(2t) + \frac{1}{2}\sin(2t) \; . \tag{7.56}$$

This means that $\overline{Y}_c(t)$ does not exist for all $t \geq 0$ since $\partial y_1/\partial\alpha$ can be negative for $t \in [0, \infty)$.

7.1.3 Dynamic Supply and Demand

This application is adopted from the example on page pages 250 – 252 in [8].

We are considering an item such that increasing its price P results in an increase in supply S but that increasing its supply S will ultimately decrease its price P. We will assume there are two factors that influence price: inflation and supply. The differential equation for price is

$$P' = \theta - k_1(S - S_0) \; , \tag{7.57}$$

where θ is the rate of inflation and S_0 is equilibrium supply, and $k_1 > 0$ is a constant. In equation (7.57) we have: (1) if $S > S_0$, the supply is too large and price is to decrease; and (2) if $S < S_0$, supply is too low and price tends to increase. This is the reason we put $-k_1(S - S_0)$, $k_1 > 0$, into the equation.

We also assume that the rate of change of S is proportional to $(P - P_0)$ for equilibrium price P_0. The equation for supply is

$$S' = k_2(P - P_0) \; , \tag{7.58}$$

for some constant $k_2 > 0$. The factor $k_2(P - P_2)$ means: (1) if $P > P_0$, price is high and supply increases; and (2) when $P < P_0$, price is low and supply decreases.

Given initial conditions $P(0) = \gamma_0$ and $S(0) = \gamma_1$ we solve equations (7.57) and (7.58) for their unique solutions. To simplify the algebra we will now use the following values for the constants: $\theta = 5\%(0.05)$, $k_1 = k_2 = 0.5$, $P_0 = 25$ and $S_0 = 1200$. The unique solution is

$$P = \gamma_0 \cos(\frac{1}{2}t) + \gamma_1 \sin(\frac{1}{2}t) + \Phi_1(t) \; , \tag{7.59}$$

$$S = \gamma_1 \cos(\frac{1}{2}t) - \gamma_0 \sin(\frac{1}{2}t) + \Phi_2(t) \; , \tag{7.60}$$

where $\Phi_i(t)$ does not contain γ_0 or γ_1, $i = 1, 2$.

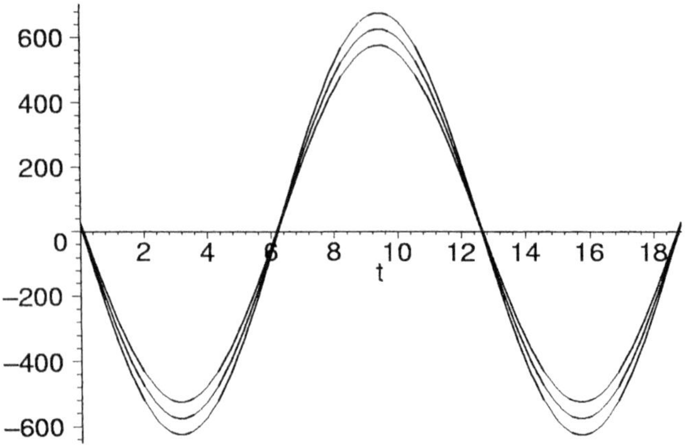

Figure 7.5: Fuzzy Price in Application 7.1.3

Now let the fuzzy initial conditions be $P(0) = \overline{\gamma}_0$, $S(0) = \overline{\gamma}_1$. Both P and S satisfy the interval condition and the intervals are the same: $I_0 = [0, \pi]$, $I_1 = [\pi, 2\pi]$, $I_2 = [2\pi, 3\pi], \ldots$. The extension principle solutions are $\overline{P}_e(t)$ and $\overline{S}_e(t)$ whose α–cuts are $[p_{e1}(t,\alpha), p_{e2}(t,\alpha)]$, $[s_{e1}(t,\alpha), s_{e2}(t,\alpha)]$, respectively. We obtain $p_{e1}(t,\alpha)$ as: (1) use $\gamma_{01}(\alpha)$ for γ_0 and $\gamma_{11}(\alpha)$ for γ_1 on I_1; (2) use $\gamma_{02}(\alpha)$ for γ_0 and $\gamma_{11}(\alpha)$ for γ_1 on I_2; (3) use $\gamma_{02}(\alpha)$ for γ_0 and $\gamma_{12}(\alpha)$ for γ_1 on I_3; etc. Interchange the γ–values in $p_{e1}(t,\alpha)$ to get $p_{e2}(t,\alpha)$. For $s_{e1}(t,\alpha)$ we: (1) use $\gamma_{11}(\alpha)$ for γ_1 and $\gamma_{02}(\alpha)$ for γ_0 on I_1; (2) use $\gamma_{12}(\alpha)$ for γ_1 and $\gamma_{02}(\alpha)$ for γ_0 on I_2; (3) use $\gamma_{12}(\alpha)$ for γ_1 and $\gamma_{01}(\alpha)$ for γ_0 on I_3; etc. The graph of $\overline{P}_e(t)$ is in Figure 7.5 and the graph of $\overline{S}_e(t)$ is in Figure 7.6, where we are using $\gamma_0 = (20/25/30)$, $\gamma_1 = (550/600/650)$. We only show the $\alpha = 0$ and $\alpha = 1$ cuts of $\overline{P}_e(t)$ and $\overline{S}_e(t)$ ($\alpha = 0$ cut is the two outside curves).

When we obtain the extension principle result we usually do not calculate the interval arithmetic solutions $\overline{P}_I(t)$ and $\overline{S}_I(t)$. So let us see if the classical solutions $\overline{P}_c(t)$ and $\overline{S}_c(t)$ exist. We show $\overline{S}_c(t)$ does not exist.

Substitute intervals $[p_1(t,\alpha), p_2(t,\alpha)]$ for $\overline{P}_c(t)$ and intervals $[s_1(t,\alpha), s_2(t,\alpha)]$ for $\overline{S}_c(t)$ into equations (7.57) and (7.58) producing

$$p_1'(t,\alpha) + 0.5s_2(t,\alpha) = 600.05\ , \tag{7.61}$$

$$p_2'(t,\alpha) + 0.5s_1(t,\alpha) = 600.05\ , \tag{7.62}$$

$$s_1'(t,\alpha) - 0.5p_1(t,\alpha) = -12.5\ , \tag{7.63}$$

$$s_2'(t,\alpha) - 0.5p_2(t,\alpha) = -12.5\ , \tag{7.64}$$

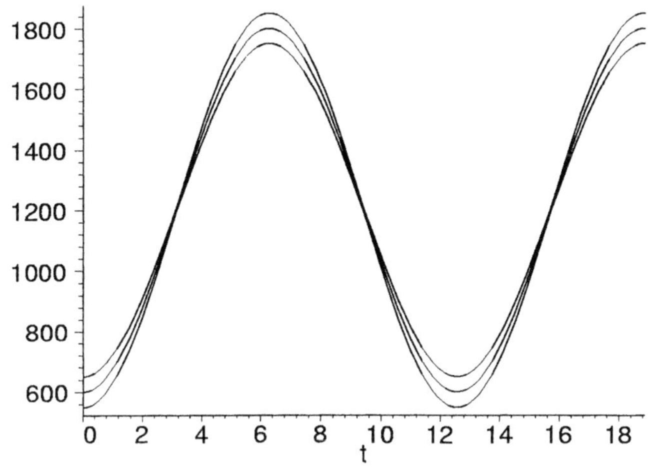

Figure 7.6: Fuzzy Supply in Application 7.1.3

where the "prime" denotes the derivative with respect to t. We eliminate p_1 and p_2 as follows: (1) differentiate equation (7.63) on t and then substitute from equation (7.61); (2) differentiate equation (7.64) on t and substitute from equation (7.62). We see that

$$s_1''(t,\alpha) + 0.25 s_2(t,\alpha) = 300.025 \; , \tag{7.65}$$

$$s_2''(t,\alpha) + 0.25 s_1(t,\alpha) = 300.025 \; , \tag{7.66}$$

Solving we obtain

$$s_1(t,\alpha) = c_1\cos(0.5t) + c_2\sin(0.5t) + c_3 e^{0.5t} + c_4 e^{-0.5t} + 1200.1 \; , \tag{7.67}$$

$$s_2(t,\alpha) = c_5\cos(0.5t) + c_6\sin(0.5t) + c_7 e^{0.5t} + c_8 e^{-0.5t} + 1200.1 \; . \tag{7.68}$$

Substituting equations (7.67) and (7.68) back into equation (7.65) we see $c_5 = c_1$, $c_6 = c_2$, $c_7 = -c_3$, $c_8 = -c_4$. Now we use the initial conditions $\overline{\gamma}[\alpha] = [20+5\alpha, 25-5\alpha]$, $\overline{\gamma}_1[\alpha] = [550+50\alpha, 650-50\alpha]$, for $\overline{S}_c(0)[\alpha] = \overline{\gamma}_1[\alpha]$, $\overline{P}_c(0)[\alpha] = \overline{\gamma}_0[\alpha]$. We use $s_1(0,\alpha) = 550+50\alpha$ and $s_2(0,\alpha) = 650-50\alpha$ giving two inconsistent equations having no solution. We conclude $\overline{S}_c(t)$ does not exist.

In the dynamic supply and demand application we only have the extension principle solution $\overline{P}_e(t)$ and $\overline{S}_e(t)$.

7.2 Other Fuzzy Parameters

Going back to equation (7.1) there are four parameters a, b, γ_0 and γ_1 that can become fuzzy. So we can consider fuzzy a, or fuzzy b, or fuzzy a and b, etc. These are uncharted waters because no general theory has been worked out for any of these cases, except fuzzy initial conditions discussed in Section 7.1.

Fuzzy b is important because: (1) in an electrical circuit b represents the resistance which may not be known exactly; and (2) in mechanical vibration b represents the retarding constant which is always difficult to measure exactly. Uncertain b can be modeled as a fuzzy number. Similar remarks can be made for fuzzy a.

To show the new difficulties in finding a solution we will only look at two cases in this section: (1) a fuzzy, $b = 0$; and (2) $a = 0$ and b fuzzy. In case (1) $\overline{a} > 0$ and $\overline{b} > 0$ in case (2).

Example 7.2.1

The differential equation is

$$y'' + ay' = 0 , \tag{7.69}$$

when $b = 0$ and we start by assuming it is homogeneous ($g(x) = 0$). The general solution is

$$y = c_1 + c_2 e^{-ax} . \tag{7.70}$$

Given initial conditions $y(0) = \gamma_0$, $y'(0) = \gamma_1$ we obtain the unique solution

$$y = \gamma_0 + \gamma_1 \left(\frac{1 - e^{-ax}}{a} \right) . \tag{7.71}$$

We first compute $\overline{Y}_e(x)$.

The term $(1 - e^{-ax})/a$ is a decreasing function of a, for $a > 0$, $x > 0$. We now choose values for the parameters $\gamma_0 = 5$, $\gamma_1 = 1$ and let $\overline{a} = (1/2/3)$. The α–cuts of $\overline{Y}_e(x)$ are $[y_{e1}(x, \alpha), y_{e2}(x, \alpha)]$ and substitute $a_2(\alpha)$ for a in equation (7.71) for $y_{e1}(x, \alpha)$ and use $a_1(\alpha)$ for a to get $y_{e2}(x, \alpha)$. We denote the α–cuts of $\overline{a}$ as $[a_1(\alpha), a_2(\alpha)]$. The resulting $\overline{Y}_e(x)$ is displayed (only the $\alpha = 0$ and $\alpha = 1$ cuts) in Figure 7.7 ($\alpha = 1$ cut is the center curve). The interval condition holds (Section 7.1) so $\overline{Y}_e(x)$ is a solution. The problem is a lot more complicated if $g(x) \neq 0$. Suppose $g(x) = 2\cos t$. The general solution is now

$$y = c_1 + c_2 e^{-ax} - \frac{2}{a^2 + 1} \cos x + \frac{2a}{a^2 + 1} \sin x . \tag{7.72}$$

Then the unique solution is

$$y = \gamma_0 + \frac{\gamma_1}{a} + \left(\frac{2}{a^2 + 1} - \frac{\gamma_1}{a} \right) e^{-ax} - \frac{2 \cos x}{a^2 + 1} + \frac{2a \sin x}{a^2 + 1} . \tag{7.73}$$

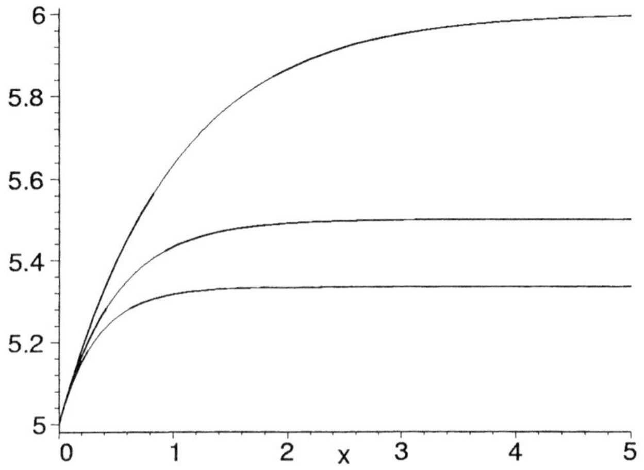

Figure 7.7: Extension Principle Solution in Example 7.2.1

To find $y_{e1}(x,\alpha)$ we need the minimum of equation (7.73) for $a \in \overline{a}[\alpha]$, for each $x \geq 0$ and each $\alpha \in [0,1]$. This is a difficult computation problem. We could use an evolutionary algorithm to estimate $[y_{e1}(x,\alpha), y_{e2}(x,\alpha)]$ for selected values of α and x. But we still have the problem of determining if $\overline{Y}_e(x)$ satisfies the original differential equation and initial conditions. For these reasons we started with $g(x) = 0$.

Next we check to see if the classical solution $\overline{Y}_c(x)$ exists. We start with

$$[y_1'(x,\alpha), y_2'(x,\alpha)] + [a_1(\alpha), a_2(\alpha)][y_1(x,\alpha), y_2(x,\alpha)] = [0,0] \tag{7.74}$$

with $\overline{a} > 0$. We solve producing

$$y_1(x,\alpha) = c_1 + c_2 e^{-a_1(\alpha)x}, \tag{7.75}$$

$$y_2(x,\alpha) = c_3 + c_4 e^{-a_2(\alpha)x}. \tag{7.76}$$

The initial conditions are $y_1(0,\alpha) = y_2(0,\alpha) = \gamma_0$, $y_1'(0,\alpha) = y_2'(0,\alpha) = \gamma_1$. That is, $\overline{Y}_c(x)$ starts off as a crisp number. Using these initial conditions we solve for the $c_1, \ldots, c_4$ in equations (7.75) and (7.76). We see

$$y_1(x,\alpha) = \gamma_1 \left(\frac{1 - e^{-a_1(\alpha)t}}{a_1(\alpha)} \right) + \gamma_0 . \tag{7.77}$$

We know for $y_1(x,\alpha)$ to define the left side of a fuzzy number we must have $\partial y_1/\partial\alpha > 0$. But the term $(1 - e^{-a_1(\alpha)t})/a_1(\alpha)$ has negative partial on α (using $a_1'(\alpha) > 0$). Hence $\overline{Y}_c(x)$ does not exist.

Example 7.2.2

We now work with

$$y'' + by = 0, \tag{7.78}$$

since $a = 0$ and we start off with $g(x) = 0$. We assume $b > 0$. The general solution is

$$y = c_1 \cos(\sqrt{b}x) + c_2 \sin(\sqrt{b}x) \ . \tag{7.79}$$

Using the initial conditions the unique solution is

$$y = \gamma_0 \cos(\sqrt{b}x) + \frac{\gamma_1}{\sqrt{b}} \sin(\sqrt{b}x) \ . \tag{7.80}$$

To simplify the calculations let $\gamma_1 = 0$ and also choose $\gamma_0 = 5$, $\bar{b} = (1/4/9)$. The unique solution is now

$$y = 5 \cos \sqrt{b}x \ . \tag{7.81}$$

We again first determine $\overline{Y}_e(x)$. The α–cuts of $\overline{Y}_e(x)$ are

$$y_{e1}(x, \alpha) = \min\left\{5\cos(\sqrt{b}x) | b \in \bar{b}[\alpha]\right\} , \tag{7.82}$$

$$y_{e2}(x, \alpha) = \max\left\{5\cos(\sqrt{b}x) | b \in \bar{b}[\alpha]\right\} . \tag{7.83}$$

We can show that the interval condition holds (Section 7.1) so $\overline{Y}_e(x)$ is a solution. We show how to get the $\alpha = 0$ cut of $\overline{Y}_e(x)$: (1) for $0 \le x \le \pi/3$ use $b_2(0) = 9$ for b to get $y_{e1}(x, 0)$ and $b_1(0) = 1$ for b in $y_{e2}(x, 0)$; (2) $\pi/3 \le x \le \pi/2$ use $b_1(0) = 1$ for b in $y_{e2}(x, 0)$ but $y_{e1}(x, 0) = -\gamma_0$; (3) $\pi/2 \le x \le 2\pi/3$, $b_2(0) = 9$ for b in $y_{e2}(x, 0)$ and $y_{e1}(x, 0) = -\gamma_0$; and (4) for $x \ge 2\pi/3$ $y_{e1}(x, 0) = -\gamma_0$, $y_{e2}(x, 0) = \gamma_0$. Of course the $\alpha = 1$ cut is

$$y_{ei}(x, 1) = \gamma_0 \cos(2x) \ . \tag{7.84}$$

In summary

$$y_{e1}(x, 0) = \begin{cases} \gamma_0 \cos(3x) & 0 \le x \le \frac{\pi}{3} , \\ -\gamma_0 & x \ge \frac{\pi}{3} , \end{cases} \tag{7.85}$$

and

$$y_{e2}(x, 0) = \begin{cases} \gamma_0 \cos x & 0 \le x \le \frac{\pi}{2} , \\ \gamma_0 \cos(3x) & \frac{\pi}{2} \le x \le \frac{2\pi}{3} , \\ \gamma_0 & x \ge \frac{2\pi}{3} . \end{cases} \tag{7.86}$$

$\overline{Y}_e(x)$ is shown in Figure 7.8 (the two outside curves are the $\alpha = 0$ cut).

If we now let $g(x) = 2\cos x$, then we find a difficult computational problem again to compute $y_{ei}(x, \alpha)$, $i = 1, 2$. We omit any more details on the $g(x) \neq 0$ case.

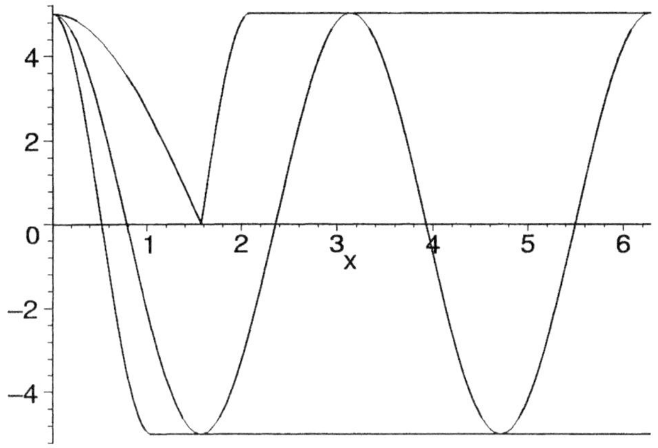

Figure 7.8: $\overline{Y}_e(x)$ Solution in Example 7.2.2

Finally, we argue that $\overline{Y}_c(x)$ does not exist. We still assume that $\gamma_0 = 5$, $\gamma_1 = 0$, $\bar{b} = (1/4/9)$ so $\bar{b}[\alpha] = [1 + 3\alpha, 9 - 5\alpha]$. We therefore find the α–cuts of $\overline{Y}_c(x)$ to be

$$y_1(x, \alpha) = \gamma_0 \cos(\sqrt{1 + 3\alpha}x) , \tag{7.87}$$

$$y_2(x, \alpha) = \gamma_0 \cos(\sqrt{9 - 5\alpha}x) . \tag{7.88}$$

as before we look at $\partial y_1 / \partial \alpha$ and see

$$\frac{\partial y_1}{\partial \alpha} = -\gamma_0 \left(\frac{3}{2\sqrt{1 + 3\alpha}} \right) \sin(\sqrt{1 + 3\alpha}x) . \tag{7.89}$$

For a fixed α in $[0, 1]$, $\sin(\sqrt{1 + 3\alpha}x)$ will be both positive and negative as $x \to \infty$. This means that there are values of x so that $\partial y_1 / \partial \alpha < 0$. Hence $\overline{Y}_c(x)$ is not defined as a fuzzy number for all x in $[0, \infty)$.

7.3 Summary and Conclusions

In this chapter we first studied solutions to

$$y'' + ay' = by = g(x) , \tag{7.90}$$

for fuzzy initial conditions. We found that the classical solution $\overline{Y}_c(x)$ may or may not exist but the extension principle solution $\overline{Y}_e(x)$ always exists (and solves the original differential equation plus the initial conditions). When $\overline{Y}_c(x)$ exists we showed that $\overline{Y}_c(x) \leq \overline{Y}_e(x)$. We did not investigate the α-cut and interval arithmetic solution $\overline{Y}_I(x)$. Three applications were presented for $\overline{Y}_e(x)$, since in each $\overline{Y}_c(x)$ failed to exist.

We also looked at fuzzifying a or b, under crisp initial conditions. It now becomes more difficult to obtain $\overline{Y}_e(x)$ and check to see if it solves the original differential equation. In the two examples investigated $\overline{Y}_c(x)$ did not exist.

For other approaches to fuzzy differential equations we refer the reader to the references in ([1], [2]) and the recent papers ([4] – [7]).

Future research will be concerned with: (1) extended to higher order equations; (2) fuzzifying the "forcing function" $g(x)$, as will be done in Chapter 9; and (3) solving systems of linear differential equations with fuzzy initial conditions. Many results discussed in this chapter have now been generalized to first order linear systems ([3]).

Bibliography

[1] J.J. Buckley and Th. Feuring: Fuzzy Differential Equations, Fuzzy Sets and Systems, 110 (1999), 43–54.

[2] J.J. Buckley and Th. Feuring: Fuzzy Initial Value Problem for N–th Order Linear Differential Equation, Fuzzy Sets and Systems, 121 (2001), pp. 247–255.

[3] J.J. Buckley, T. Feuring and Y. Hayashi: Linear Systems of First Order Ordinary Differential Equations: Fuzzy Initial Conditions, Soft Computing. To appear.

[4] P. Diamond: Time–Dependent Differential Inclusions, Cocycle Attractors, and Fuzzy Differential Equations, IEEE Trans. Fuzzy Systems, 7 (1999), pp. 734–740.

[5] E. Hüllermeier: Numerical Methods for the Fuzzy Initial Value Problem, Int. J. Uncertainty, Fuzziness and Knowledge-Based Systems, 7 (1999), pp. 439–461.

[6] J.Y. Park and H.K. Han: Fuzzy Differential Equations, Fuzzy Sets and Systems, 110 (1999), pp. 69–77.

[7] S. Song, C. Wu: Existence and Uniqueness of Solutions to Cauchy Problem of Fuzzy Differential Equations, Fuzzy Sets and Systems, 110 (1999), pp. 55–67.

[8] M.R. Spiegel: Applied Differential Equations, Third Edition, Prentice Hall, Englewood Cliffs, N. J., 1981.

Chapter 8

Fuzzy Difference Equations

In this chapter we are concerned with solutions to the second order, linear, constant coefficient difference equation

$$y(k+2) + ay(k+1) + by(k) = g(k) \ , \tag{8.1}$$

for $k = 0, 1, 2, \ldots$, where a, b are constants with $b > 0$ and $g(x)$ is continuous for $x \geq 0$. The initial conditions are $y(0) = \gamma_0$ and $y(1) = \gamma_1$. We will fuzzify the difference equation by considering fuzzy initial conditions $y(0) = \overline{\gamma}_0$, $y(1) = \overline{\gamma}_1$ for triangular fuzzy numbers $\overline{\gamma}_0$ and $\overline{\gamma}_1$.

The theory of second order, linear, constant coefficient difference equations is similar to the theory of second order, linear, constant coefficient ordinary differential equations discussed in Chapter 7. So we will consider three solution concepts called the classical solution $\overline{Y}_c(k)$, the extension principle solution $\overline{Y}_e(k)$ and the α–cut and interval arithmetic solution $\overline{Y}_I(k)$ in this chapter for the fuzzy initial value problem of difference equations. One might think the results in this chapter will then be similar to those of Chapter 7. This is not exactly true.

In both cases, difference equations and differential equations, too often $\overline{Y}_c(k)$ fails to exist. For differential equations, $\overline{Y}_e(k)$ always exists for the fuzzy initial value problem, but with difference equations many times $\overline{Y}_e(k)$ can fail to be a solution. When $\overline{Y}_e(k)$ is not a solution we will then try $\overline{Y}_I(k)$ to see if it is a solution.

However, there are enough cases where $\overline{Y}_e(k)$ is a solution for us to consider three applications at the end of the chapter: (1) an economic model of national income; (2) transmission of information in information theory; and (3) fuzzy Fibonacci numbers.

We will not allow a or b to be fuzzy in this chapter. We pointed out the difficulties in getting solutions for a or b fuzzy for differential equations in Section 7.2. We will not do this again for difference equations. This means throughout this chapter only the initial conditions can be fuzzy.

This chapter is based on paper [1]. We will be using only triangular fuzzy numbers for $\overline{\gamma}_0$ and $\overline{\gamma}_1$ and the solutions will always be triangular shaped fuzzy numbers for each $k = 0, 1, 2, \ldots$.

In the next section we present the crisp solutions to equation (8.1) followed by our solutions to the fuzzy initial value problem in Section 8.2. In Section 8.3 we look at a different method of solving difference equations called the recursive method. In this section we show $\overline{Y}_c(k)$ is never a solution to the fuzzy initial value problem for second order, linear, constant coefficient difference equations. Applications comprise Section 8.4 and 8.5 is a brief summary and our conclusions.

8.1 Difference Equations

Let $y_1(k)$ and $y_2(k)$ be any two linearly independent solutions to the homogeneous equation

$$y(k+2) + ay(k+1) + by(k) = 0 \ , \tag{8.2}$$

$k = 0, 1, 2, 3, \ldots$. If $G(k)$ is any solution to equation (8.1), then the general solution to equation (8.1) is

$$y(k) = c_1 y_1(k) + c_2 y_2(k) + G(k) \ , \tag{8.3}$$

$k = 0, 1, 2, \ldots$ for arbitrary constants c_1 and c_2. The initial conditions are used to determine values of c_1 and c_2 producing the unique solution to equation (8.1).

To find $y_1(k)$ and $y_2(k)$ substitute $y(k) = \beta^k$ into equation (8.2) giving

$$\beta^2 + a\beta + b = 0 \ . \tag{8.4}$$

Let the roots (real or complex) be r_1 and r_2. Neither root can be zero since we assumed $b \neq 0$. There are now three cases to consider.

If $r_1 \neq r_2$ and r_1, r_2 are real numbers, then $y_1(k) = (r_1)^k$ and $y_2(k) = (r_2)^k$. When $r_1 = r_2 = r$, then $y_1(k) = (r)^k$ and $y_2(k) = k(r)^k$. Now suppose $r_1 = u - vi$ and $r_2 = u + vi$ for real numbers u, v with $v > 0$. Set $\rho = \sqrt{u^2 + v^2} = \sqrt{b}$. We know that $u = -a/2$ and $v = \sqrt{4b - a^2}/2$. Switch to the polar form of a complex number

$$u + vi = \rho(\cos\varphi + i \sin\varphi) \ , \tag{8.5}$$

for $\rho\cos\varphi = u$, $\rho\sin\varphi = v$. Since $v > 0$ we see that $\varphi = \tan^{-1}(v/u)$ if $u > 0$ but $\varphi = \pi - \tan^{-1}(v/(-u))$ if $u < 0$. The result is $y_1(k) = \rho^k\cos(k\varphi)$, $y_2(k) = \rho^k \sin(k\varphi)$. For the general theory of difference equations we refer the reader to ([6] – [11]).

Next we have to find $G(k)$. If $g(k)$ is of the form d^k, $\sin(kd)$, $\cos(kd)$, $k^p, \ldots$ for constants d, p, then the method of undetermined coefficients may be used to obtain $G(k)$. We will not be using very complicated $g(k)$ so this method of undetermined coefficients will not be employed to calculate $G(k)$. As we shall see, as in Chapter 7, $G(k)$ does not effect the existence of $\overline{Y}_c(k)$, $\overline{Y}_e(k)$ or $\overline{Y}_I(k)$.

8.2 Fuzzy Initial Conditions

We first look at the classical solution $\overline{Y}_c(k)$. Then we consider the extension principle $\overline{Y}_e(k)$ and give conditions when it will be a solution. Lastly, we look at the interval arithmetic solution $\overline{Y}_I(k)$ and investigate when it is a solution. A summary is given at the end of this section.

8.2.1 Classical Solution

Let α–cuts of $\overline{Y}_c(k)$ be $[y_1(k,\alpha), y_2(k,\alpha)]$, $k = 0, 1, 2, \ldots$ and $0 \le \alpha \le 1$. Substitute these intervals into the difference equation giving

$$[y_1(k+2,\alpha), y_2(k+2,\alpha)] + a[y_1(k+1,\alpha), y_2(k+1,\alpha)] \\ +b[y_1(k,\alpha), y_2(k,\alpha)] = [g(k), g(k)] \ . \tag{8.6}$$

Simplify equation (8.6) using interval arithmetic so we will need to know if a and b are positive or negative. The simplest case is $a > 0$ and $b > 0$. Assuming a and b are positive we see we must solve, for $i = 1, 2$

$$y_i(k+2,\alpha) + ay_i(k+1,\alpha) + by_i(k,\alpha) = g(k) \ , \tag{8.7}$$

subject to

$$y_1(0,\alpha) = \gamma_{01}(\alpha) \ , \tag{8.8}$$
$$y_1(1,\alpha) = \gamma_{11}(\alpha) \ , \tag{8.9}$$
$$y_2(0,\alpha) = \gamma_{02}(\alpha) \ , \tag{8.10}$$
$$y_2(1,\alpha) = \gamma_{12}(\alpha) \ , \tag{8.11}$$

where $\overline{\gamma}_0 = [\gamma_{01}(\alpha), \gamma_{02}(\alpha)]$, $\overline{\gamma}_1 = [\gamma_{11}(\alpha), \gamma_{12}(\alpha)]$. $\overline{Y}_c(k)$ is a solution when the intervals $[y_1(k,\alpha), y_2(k,\alpha)]$ define a fuzzy number for each $k = 0, 1, 2, \ldots$. What is needed is $\partial y_1/\partial\alpha > 0$, $\partial y_2/\partial\alpha < 0$, $0 < \alpha < 1$, all k, and $y_1(k,1) = y_2(k,1)$ all k. $\overline{Y}_c(k)$ will then be a triangular shaped fuzzy number.

Example 8.2.1.1

Let us attempt to find $\overline{Y}_c(k)$ for

$$y_i(k+2,\alpha) + 3y_i(k+1,\alpha) + 2y_i(k,\alpha) = 0 \ , \tag{8.12}$$

subject to $y(0) = \overline{\gamma}_0 = (0/1/2)$, $y(1) = \overline{\gamma}_1 = (6/8/10)$. Then $\overline{\gamma}_0[\alpha] = [\alpha, 2-\alpha]$, $\overline{\gamma}_1[\alpha] = [6+2\alpha, 10-2\alpha]$. The crisp solution is

$$y = c_1(-1)^k + c_2(-2)^k \ , \tag{8.13}$$

since the roots of $\beta^2 + 3\beta + 2 = 0$ are -1 and -2. From equations (8.7) – (8.11) we are to solve

$$y_1(k+2,\alpha) + 3y_1(k+1,\alpha) + 2y_1(k,\alpha) = 0 \ , \tag{8.14}$$

$$y_1(0,\alpha) = \alpha \ , \tag{8.15}$$

$$y_1(1,\alpha) = 6 + 2\alpha \ , \tag{8.16}$$

and

$$y_2(k+2,\alpha) + 3y_2(k+1,\alpha) + 2y_2(k,\alpha) = 0 \ , \tag{8.17}$$

$$y_2(0,\alpha) = 2 - \alpha \ , \tag{8.18}$$

$$y_2(1,\alpha) = 10 - 2\alpha \ . \tag{8.19}$$

The solutions are

$$y_1(k,\alpha) = (6+4\alpha)(-1)^k - (6+3\alpha)(-2)^k \ , \tag{8.20}$$

$$y_2(k,\alpha) = (14-4\alpha)(-1)^k - (12-3\alpha)(-2)^k \ . \tag{8.21}$$

Now we check to see if these intervals $[y_1(k,\alpha), y_2(k,\alpha)]$ define a fuzzy number for all k. They do not. We see that $\partial y_1/\partial\alpha > 0$ for $k = 0,1,3,5,\ldots$ but $\partial y_1/\partial\alpha < 0$ for $k = 2,4,6,\ldots$. Also $\partial y_2/\partial\alpha < 0$ for $k = 0,1,3,5,\ldots$ but $\partial y_1/\partial\alpha > 0$ otherwise. Hence $\overline{Y}_c(k)$ exists only for $k = 0,1,3,5,\ldots$. Therefore, $\overline{Y}_c(k)$ is not a solution for all k.

Example 8.2.1.2

Let us now consider

$$y(k+2) + 2y(k+1) + 3y(k) = 0 \ , \tag{8.22}$$

with the same fuzzy initial conditions as in Example 8.2.1.1. The crisp solution is

$$y(k) = 3^{\frac{k}{2}}\left(c_1\cos(k\varphi) + c_2\sin(k\varphi)\right) \ , \tag{8.23}$$

for $\varphi = \pi - \tan^{-1}(\sqrt{2})$ since the roots of $\beta^2 + 2\beta + 3 = 0$ are $-1 \pm \sqrt{2}i$. As in Example 8.2.1.1 we now solve for $y_i(k,\alpha)$ giving

$$y_1(k,\alpha) = 3^{\frac{k}{2}}\left(\alpha\cos(k\varphi) + (3+1.5\alpha)\sin(k\varphi)\right) \ , \tag{8.24}$$

$$y_2(k,\alpha) = 3^{\frac{k}{2}}\left((2-\alpha)\cos(k\varphi) + (6-1.5\alpha)\sin(k\varphi)\right) \ . \tag{8.25}$$

Compute $\partial y_1/\partial\alpha$ to see if it is positive for all k. It is not because $\partial y_1/\partial\alpha < 0$ for $k = 2$. Hence, $\overline{Y}_c(k)$ does not exist for $k = 0,1,2,\ldots$.

Next we look for $\overline{Y}_e(k)$ to produce solutions.

8.2.2 Extension Principle Solution

There will be three cases depending on the roots of $\beta^2 + a\beta + b = 0$.

Case 1

The roots are r_1 and r_2 real with $r_1 \neq r_2$. The general solution is

$$y(k) = c_1(r_1)^k + c_2(r_2)^k + G(k) \ . \tag{8.26}$$

Using the initial conditions find values for c_1 and c_2. We obtain

$$y(k) = \left(\frac{\gamma_0 r_2 - \gamma_1}{r_2 - r_1}\right)(r_1)^k + \left(\frac{\gamma_1 - \gamma_0 r_1}{r_2 - r_1}\right)(r_2)^k + \Phi(k) \ , \tag{8.27}$$

where

$$\Phi(k) = \left(\frac{G(1) - r_2 G(0)}{r_2 - r_1}\right)(r_1)^k + \left(\frac{r_1 G(0) - G(1)}{r_2 - r_1}\right) + G(k) \ . \tag{8.28}$$

We have split the unique solution into two parts: (1) the first part in equation (8.27) contains γ_0 and γ_1; and (2) the second part $\Phi(k)$ is independent of γ_0 and γ_1.

$\overline{Y}_e(k)$ is obtained by substituting $\overline{\gamma}_0$ for γ_0 and $\overline{\gamma}_1$ for γ_1 in equation (8.27), and we evaluate using the extension principle. If we set $f_1(\gamma_0, \gamma_1)$ to be the coefficient of $(r_1)^k$ in equation (8.27) and let $f_2(\gamma_0, \gamma_1)$ be the coefficient of $(r_2)^k$, then the α–cuts of $\overline{Y}_e(k)$ are

$$y_{e1}(k, \alpha) = \tag{8.29}$$
$$\min \left\{ f_1(\gamma_0, \gamma_1)(r_1)^k + f_2(\gamma_0, \gamma_1)(r_2)^k \middle| \gamma_i \in \overline{\gamma}_i[\alpha], i = 1, 2 \right\} + \Phi(k) \ ,$$
$$y_{e2}(k, \alpha) = \tag{8.30}$$
$$\max \left\{ f_1(\gamma_0, \gamma_1)(r_1)^k + f_2(\gamma_0, \gamma_1)(r_2)^k \middle| \gamma_i \in \overline{\gamma}_i[\alpha], i = 1, 2 \right\} + \Phi(k) \ .$$

After computing $y_{ei}(k, \alpha)$, for $\overline{Y}_e(k)$ to be a solution we require the $y_{ei}(k, \alpha)$ to satisfy the difference equation and initial conditions.

The following condition is required for $\overline{Y}_e(k)$ to solve the difference equation. It is easy to see $\overline{Y}_e(k)$ satisfies the initial conditions so we only need to consider $k \geq 2$ below. For each $\alpha \in [0, 1)$ there is a $\gamma_0^* \in \overline{\gamma}[\alpha]$, $\gamma_1^* \in \overline{\gamma}_1[\alpha]$ so that

$$y_{e1}(k, \alpha) = f_1(\gamma_0^*, \gamma_1^*)(r_1)^k + f_2(\gamma_0^*, \gamma_1^*)(r_2)^k + \Phi(k) \ , \tag{8.31}$$

for all $k \geq 2$. For each $\alpha \in [0, 1)$ there is a $\gamma_0^{**} \in \overline{\gamma}_0[\alpha]$ and $\gamma_1^{**} \in \overline{\gamma}_1[\alpha]$ so that

$$y_{e2}(k, \alpha) = f_1(\gamma_0^{**}, \gamma_1^{**})(r_1)^k + f_2(\gamma_0^{**}, \gamma_1^{**})(r_2)^k + \Phi(k) \ , \tag{8.32}$$

for all $k \geq 2$. We can not have the f_i functions of k, for $k \geq 2$, $i = 1, 2$.

Now we may present conditions for $\overline{Y}_e(k)$ to be a solution. Let

$$\frac{\partial y(k)}{\partial \gamma_0} = h_1(k) = \frac{r_2 r_1^k - r_1 r_2^k}{r_2 - r_1} , \tag{8.33}$$

$$\frac{\partial y(k)}{\partial \gamma_1} = h_2(k) = \frac{r_2^k - r_1^k}{r_2 - r_1} . \tag{8.34}$$

Theorem 8.1 *$\overline{Y}_e(k)$ is a solution if and only if*

1. *$h_1(k) > 0$ or $h_1(k) < 0$, $k \geq 2$, and*
2. *$h_2(k) > 0$ or $h_2(k) < 0$, $k \geq 2$.*

Proof 8.1 *There are four cases to consider so let us look at only $h_1(k) < 0$ and $h_2(k) > 0$, $k \geq 2$. Let*

$$H(\gamma_0, \gamma_1) = f_1(\gamma_0, \gamma_1)(r_1)^k + f_2(\gamma_0, \gamma_1)(r_2)^k . \tag{8.35}$$

Under the assumptions above $H(\gamma_0, \gamma_1)$ is a decreasing function of γ_0 but an increasing function of γ_1, $k \geq 2$. This allows us to easily find $y_{ei}(k, \alpha)$: (1) use $\gamma_{02}(\alpha)$ for γ_0 and $\gamma_{11}(\alpha)$ for γ_1 to obtain $y_{e1}(k, \alpha)$ in equation (8.29) for all $k \geq 2$; and (2) substitute $\gamma_{01}(\alpha)$ for γ_0, $\gamma_{12}(\alpha)$ for γ_1 for $y_{e2}(k, \alpha)$ in equation (8.30), $k \geq 2$. Now we need to substitute equation (8.28) into equation (8.27) to get the whole expression for $y_{e1}(k, \alpha)$ and $y_{e2}(k, \alpha)$. We get

$$y_{e1}(k, \alpha) = \delta_{11} r_1^k + \delta_{12} r_2^k + G(k) , \tag{8.36}$$

$$y_{e2}(k, \alpha) = \delta_{21} r_1^k + \delta_{22} r_2^k + G(k) , \tag{8.37}$$

where $\delta_{11} = f_1(\gamma_{02}(\alpha), \gamma_{11}(\alpha)) +$ (coefficient of r_1^k in equation (8.28)), $\delta_{12} = f_2(\gamma_{02}(\alpha), \gamma_{11}(\alpha)) +$ (coefficient of r_2^k in equation (8.28)), $\delta_{21} = f_1(\gamma_{01}(\alpha), \gamma_{12}(\alpha)) +$ (coefficient of r_1^k in equation (8.28)), and $\delta_{22} = f_2(\gamma_{01}(\alpha), \gamma_{12}(\alpha)) +$ (coefficient of r_2^k in equation (8.28)). Since the δ_{ij} are just constants, $k \geq 2$, the $y_{ei}(k, \alpha)$ satisfy the difference equation (8.1), $i = 1, 2$.

Example 8.2.2.1

Consider

$$y(k+2) - 3y(k+1) + 2y(k) = 0 , \tag{8.38}$$

for $y(0) = \overline{\gamma}_0$, $y(1) = \overline{\gamma}_1$. The roots of $\beta^2 - 3\beta + 2 = 0$ are $r_1 = 1$ and $r_2 = 2$. Then $h_1(k) = 2 - 2^k < 0$, $k \geq 2$ and $h_2(k) = 2^k - 1 > 0$, $k \geq 2$. Hence, $\overline{Y}_e(k)$ is a solution for all $\overline{\gamma}_0$, $\overline{\gamma}_1$.

Example 8.2.2.2

This example shows what can go wrong if the hypotheses of Theorem 8.1 are not true. In Chapter 7 we had the "interval condition", which implies that $\overline{Y}_e(x)$ was always a solution. We do not have an "interval condition" for difference equations. The difference equation is now

$$y(k+2) + 3y(k+1) + 2y(k) = 0 \ , \tag{8.39}$$

as in Example 8.2.1.1. The zeros of $\beta^2 + 3\beta + 2 = 0$ are $r_1 = -1$ and $r_2 = -2$. Then

$$h_1(k) = 2(-1)^k - (-2)^k \ , \tag{8.40}$$
$$h_2(k) = (-1)^k - (-2)^k \ . \tag{8.41}$$

We see that $h_1(k) > 0$ for $k = 3, 5, \ldots$ and $h_1(k) < 0$ otherwise. Also, $h_2(k) > 0$ for $k = 3, 5, \ldots$ but $h_2(k) < 0$, $k = 4, 6, \ldots$.

Compute $y_{ei}(k, \alpha)$ for $k = 3, 4, 5$ and see if

$$y_{ei}(5, \alpha) + 3y_{ei}(4, \alpha) + 2y_{ei}(3, \alpha) = 0 \ , \tag{8.42}$$

$i = 1, 2$. This will not be true. To see why, define the δ_{ij} as in the proof of Theorem 8.1. The δ_{ij} will not be constant for $k \geq 2$ but will depend on k. For example, let us concentrate on $y_{e1}(k, \alpha)$. For $k = 3$ use $\gamma_{01}(\alpha)$ for γ_0 and $\gamma_{01}(\alpha)$ for γ_1 in the $f_i(\gamma_0, \gamma_1)$. For $k = 4$ we substitute $\gamma_{02}(\alpha)$ for γ_0, $\gamma_{12}(\alpha)$ for γ_1 in the $f_i(\gamma_0, \gamma_1)$. When $k = 5$ we are back to $\gamma_{01}(\alpha)$ for γ_0, $\gamma_{11}(\alpha)$ for γ_1 in the $f_i(\gamma_0, \gamma_1)$. The coefficients of r_1^k and r_2^k in $y_{e1}(k, \alpha)$ will be functions of k. $\overline{Y}_e(k)$ will not solve the difference equation.

Case 2

The roots are $r_1 = r_2 = r$. The general solution is

$$y(k) = c_1 r^k + c_2 k r^k + G(k) \ . \tag{8.43}$$

Using the initial conditions the unique solution is

$$y(k) = \gamma_0 r^k + \frac{\gamma_1 - r\gamma_0}{r} k r^k + \Phi(k) \ , \tag{8.44}$$

for

$$\Phi(k) = -G(0) r^k + \frac{rG(0) - G(1)}{r} k r^k + G(k) \ . \tag{8.45}$$

We calculate

$$\frac{\partial y(k)}{\partial \gamma_0} = (1-k) r^k \ , \tag{8.46}$$
$$\frac{\partial y(k)}{\partial \gamma_1} = k r^{k-1} \ . \tag{8.47}$$

Theorem 8.2 *$\overline{Y}_e(k)$ is a solution if and only if $r > 0$.*

Proof 8.2 *Similar to the proof of Theorem 8.1 because if $r > 0$, then $\partial y(k)/\partial\gamma_0 < 0$, $k \geq 2$ and $\partial y(k)/\partial\gamma_1 > 0$ for all $k \geq 2$.*

If $r < 0$, then $\partial y(k)/\partial\gamma_0 > 0$ for $k = 3, 5, 7, \ldots$ but it is negative when $k = 4, 6, 8, \ldots$. As in Example 8.2.2.2, $\overline{Y}_e(k)$ will not satisfy the difference equation.

Case 3

The roots are now $r_1 = u - iv$, $r_2 = u + iv$, u, v real with $v > 0$. The general solution is

$$y(k) = \rho^k \left(c_1 \cos(k\varphi) + c_2 \sin(k\varphi)\right) + G(k) \ , \tag{8.48}$$

from Section 8.1. From the initial conditions the unique solution is

$$y(k) = \rho^k \left(\gamma_0 \cos(k\varphi) + \frac{\gamma_1 - \gamma_0 u}{v} \sin(k\varphi)\right) + \Phi(k) \ , \tag{8.49}$$

$$\Phi(k) = \rho^k \left(-G(0) \cos(k\varphi) + \frac{G(0)u - G(1)}{v} \sin(k\varphi)\right) + G(k) \tag{8.50}$$

We know, from the previous two cases, that for $\overline{Y}_e(k)$ to solve the difference equation, the partials of $y(k)$ with respect to γ_0 and γ_1 must always be positive, or always be negative, for $k \geq 2$. However,

$$\frac{\partial y(k)}{\partial \gamma_1} = \frac{1}{v} \sin(k\varphi) \ , \tag{8.51}$$

which can be positive for some k and negative for other values of k. $\overline{Y}_e(k)$ is not a solution for complex roots.

8.2.3 Interval Arithmetic Solution

For this solution concept we fuzzify the unique crisp solution by substituting intervals $\overline{\gamma}_0[\alpha]$, $\overline{\gamma}_1[\alpha]$ for γ_0 and γ_1, respectively, and evaluate using interval arithmetic. This gives us α–cuts of $\overline{Y}_I(k)$ written $[y_{I1}(k,\alpha), y_{I2}(k,\alpha)]$, $k = 0, 1, 2, \ldots$, $0 \leq \alpha \leq 1$. We say $\overline{Y}_I(k)$ is a solution if the $y_{Ii}(k,\alpha)$ satisfy the difference equation plus initial conditions. As in the previous subsection on $\overline{Y}_e(k)$ there are three cases.

Case 1

Assume the zeros of $\beta^2 + a\beta + c = 0$ are $r_1 \neq r_2$, r_1, r_2 real. Substitute $[\gamma_{01}(\alpha), \gamma_{02}(\alpha)]$ for γ_0 and $[\gamma_{11}(\alpha), \gamma_{12}(\alpha)]$ for γ_1 in equation (8.27). Simplify the result using interval arithmetic. We end up with an expression like equation (8.36) for $y_{I1}(k,\alpha)$ and another like equation (8.37) for $y_{I2}(k,\alpha)$. $\overline{Y}_I(k)$ will satisfy the initial conditions but what we need for the $y_{Ii}(k,\alpha)$ to satisfy the difference equation is that coefficients δ_{ij} be independent of k (as for $\overline{Y}_e(k)$). So we have

Theorem 8.3 *$\overline{Y}_I(k)$ is a solution if and only if $r_1 > 0$ and $r_2 > 0$.*

Proof 8.3 *Assume the roots are positive. Then we get*

$$y_{I1}(k,\alpha) = \delta_{11} r_1^k + \delta_{12} r_2^k + \Phi(k) , \tag{8.52}$$

for

$$\delta_{11} = \frac{r_2\gamma_{01}(\alpha) - \gamma_{12}(\alpha)}{r_2 - r_1} , \tag{8.53}$$

$$\delta_{12} = \frac{\gamma_{11}(\alpha) - r_1\gamma_{02}(\alpha)}{r_2 - r_1} . \tag{8.54}$$

And

$$y_{I2}(k,\alpha) = \delta_{21} r_1^k + \delta_{22} r_2^k + \Phi(k) , \tag{8.55}$$

for

$$\delta_{21} = \frac{r_2\gamma_{02}(\alpha) - \gamma_{11}(\alpha)}{r_2 - r_1} , \tag{8.56}$$

$$\delta_{22} = \frac{\gamma_{12}(\alpha) - r_1\gamma_{01}(\alpha)}{r_2 - r_1} . \tag{8.57}$$

$\Phi(k)$ is given by equation (8.28). Since the δ_{ij} are constants, or they are independent of $k \geq 2$, $\overline{Y}_I(k)$ satisfies the difference equation.

If r_1 and/or r_2 are negative, we see the δ_{ij} depend on k and $\overline{Y}_I(k)$ is not a solution.

Example 8.2.3.1

The same difference equation

$$y(k+2) - 3y(k+1) + 2y(k) = 0 , \tag{8.58}$$

as in Example 8.2.2.1. The roots are $r_1 = 1$ and $r_2 = 2$. Therefore, $\overline{Y}_I(k)$ is a solution.

Example 8.2.3.2

This example shows why $\overline{Y}_I(k)$ is not a solution if r_1 or r_2 are negative. Let us look at

$$y(k+2) + 3y(k+1) + 2y(k) = 0 , \tag{8.59}$$

from Example 8.2.1.1. The roots are $r_1 = -1$ and $r_2 = -2$. Also set $\overline{\gamma}_0 = (0/1/2)$and $\overline{\gamma}_1 = (6/8/10)$ from Example 8.2.1.1. Compute the coefficient of r_1^k in equation (8.27) by substituting $[\alpha, 2-\alpha]$ for γ_0 and $[6+2\alpha, 10-2\alpha]$ for γ_1. We get

$$[6+4\alpha, 14-4\alpha]r_1^k \ . \tag{8.60}$$

Do the same for the coefficient of r_2^k in equation (8.27) producing

$$[-12+3\alpha, -6-3\alpha]r_2^k \ . \tag{8.61}$$

Now add in $\Phi(k)$ from equation (8.28). We need to do the multiplication in equations (8.60) and (8.61) to finally obtain $y_{Ii}(k, \alpha)$, $i = 1, 2$. But when we do this we will get different expressions depending if k is even or odd because the roots are negative. The coefficients of r_1^k and r_2^k in $y_{Ii}(k, \alpha)$, $i = 1, 2$ will be functions of k. Therefore, $\overline{Y}_I(k)$ is not a solution.

Case 2

The roots are $r_1 = r_2 = r$. The general and unique solution is given in Section 8.2.2.

Theorem 8.4 *$\overline{Y}_I(k)$ is a solution if and only if $r > 0$.*

Proof 8.4 *If $r > 0$, then the coefficients of r^k and $k \cdot r^k$ in the expressions for $y_{Ii}(k, \alpha)$, $i = 1, 2$, are constants (independent of k) and $\overline{Y}_I(k)$ solves the difference equation. If $r < 0$, the coefficients will depend on k.*

Case 3

The roots are complex with the unique solutions presented in Section 8.2.2. Since $\cos(k\varphi)$ and $\sin(k\varphi)$ can be positive and negative as $k = 2, 3, \ldots$ we see from equation (8.49) that the coefficients of $\cos(k\varphi)$ and $\sin(k\varphi)$ in $y_{Ii}(k, \alpha)$ will be functions of k. Hence, $\overline{Y}_I(k)$ is not a solution for complex roots.

8.2.4 Summary

We always prefer the classical solution but it never exists (proof next section). Therefore, we prefer the extension principle solution $\overline{Y}_e(k)$. We know $\overline{Y}_e(k) \leq \overline{Y}_I(k)$, for all k, when both exist. This inequality means that $\overline{Y}_e(k)$ is less fuzzy than $\overline{Y}_I(k)$. But also, $\overline{Y}_e(k)$ is more general than $\overline{Y}_I(k)$.

Both $\overline{Y}_e$ and $\overline{Y}_I$ fail to exist when there are complex roots. If the roots are equal ($r_1 = r_2 = r$), then $\overline{Y}_e$ and $\overline{Y}_I$ exist only when $r > 0$. However, when the roots are real and unequal, $\overline{Y}_e$ is more general than $\overline{Y}_I$ because $\overline{Y}_e$ can exist (Theorem 8.1) when one root is negative and the other positive.

For example, if $r_1 = -1$ and $r_2 = 2$, $\overline{Y}_e$ exists, but by Theorem 8.3, $\overline{Y}_I$ does not exist.

Before we go on to some applications let us show another method of recursively determining the α–cuts of $\overline{Y}_e(k)$. It will also be used to show $\overline{Y}_c(k)$ never exists.

8.3 Recursive Solutions

We will explain this method for $\overline{Y}_e(k)$ through the following example. We have

$$y(k+2) - y(k+1) - 2y(k) = 0 \ , \tag{8.62}$$

so the roots are $r_1 = -1$, $r_2 = 2$ and by Theorem 8.1, $\overline{Y}_e(k)$ is a solution. We may solve the homogeneous equations recursively to get $y_{ei}(k,\alpha)$, $i = 1, 2$, the α–cuts of $\overline{Y}_e(k)$. From $y(2) = y(1) + 2y(0)$ we get $y(2) = \gamma_1 + 2\gamma_0$. Substitute $\overline{\gamma}_0$ for γ_0 and $\overline{\gamma}_1$ for γ_1 and we have

$$y_{e1}(2,\alpha) = \gamma_{11}(\alpha) + 2\gamma_{01}(\alpha) \ , \tag{8.63}$$
$$y_{e2}(2,\alpha) = \gamma_{12}(\alpha) + 2\gamma_{02}(\alpha) \ . \tag{8.64}$$

Since $y(3) = y(2) + 2y(1)$ we get

$$y_{e1}(3,\alpha) = 3\gamma_{11}(\alpha) + 2\gamma_{01}(\alpha) \ , \tag{8.65}$$
$$y_{e2}(3,\alpha) = 3\gamma_{12}(\alpha) + 2\gamma_{02}(\alpha) \ . \tag{8.66}$$

Continuing this way we have

$$y_{e1}(4,\alpha) = 5\gamma_{11}(\alpha) + 6\gamma_{01}(\alpha) \ , \tag{8.67}$$
$$y_{e2}(4,\alpha) = 5\gamma_{12}(\alpha) + 6\gamma_{02}(\alpha) \ . \tag{8.68}$$

We obtain the same results as in Section 8.2.2.

Now change to the classical solution $\overline{Y}_c(k)$ and we argue that $\overline{Y}_c(2)$ is never a fuzzy number so $\overline{Y}_c(k)$ is never a solution. From equation (8.1) there are six cases: (1) $a > 0$, $b > 0$; (2)$a = 0$, $b > 0$; (3) $a < 0$, $b > 0$; (4) $a > 0$, $b < 0$; (5) $a = 0$, $b < 0$; and (6) $a < 0$, $b < 0$. We show $\partial y_1(2,\alpha)/\partial\alpha < 0$ in all cases. Let $\overline{\gamma}_0[\alpha] = [\gamma_{01}(\alpha), \gamma_{02}(\alpha)]$, $\overline{\gamma}_1[\alpha] = [\gamma_{11}(\alpha), \gamma_{12}(\alpha)]$. We use $\gamma_{01}'(\alpha) > 0$, $\gamma_{02}'(\alpha) < 0$, $\gamma_{11}'(\alpha) > 0$, $\gamma_{12}'(\alpha) < 0$, $0 < \alpha < 1$, where the "prime" denotes the derivative with respect to α. We start with

$$[y_1(2,\alpha), y_2(2,\alpha)] + a[\gamma_{11}(\alpha), \gamma_{12}(\alpha)] + b[\gamma_{01}(\alpha), \gamma_{02}(\alpha)] = 0 \ , \tag{8.69}$$

since $y(1) = \overline{\gamma}_1$, $y(0) = \overline{\gamma}_0$. Use interval arithmetic to solve for $y_1(2,\alpha)$. The results are:

$$\text{Case 1:} \quad y_1(2,\alpha) = -a\gamma_{11}(\alpha) - b\gamma_{01}(\alpha) \ , \tag{8.70}$$

$$\text{Case 2:} \quad y_1(2,\alpha) = -b\gamma_{01}(\alpha)\ , \tag{8.71}$$

$$\text{Case 3:} \quad y_1(2,\alpha) = -a\gamma_{12}(\alpha) - b\gamma_{01}(\alpha)\ , \tag{8.72}$$

$$\text{Case 4:} \quad y_1(2,\alpha) = -a\gamma_{11}(\alpha) - b\gamma_{02}(\alpha)\ , \tag{8.73}$$

$$\text{Case 5:} \quad y_1(2,\alpha) = -b\gamma_{02}(\alpha)\ , \tag{8.74}$$

$$\text{Case 6:} \quad y_1(2,\alpha) = -a\gamma_{12}(\alpha) - b\gamma_{02}(\alpha)\ . \tag{8.75}$$

One may check that in all six cases $\partial y_1(2,\alpha)/\partial\alpha < 0$.

8.4 Applications

We will consider three applications of the fuzzy initial value problem for second order, linear, difference equations. The first one is about an economic model of national income and is based on an example in ([6], pp. 153–156). See also [12]. The second application is on the transmission of information also based on an example in ([6], pp. 156–158), which was derived from [13]. The last application is about a fuzzy Fibonacci sequence.

8.4.1 National Income

In [6] the author derives the difference equation

$$y(k+2) - \Delta(1+\theta)y(k+1) - \Delta\theta y(k) = 1\ , \tag{8.76}$$

$k = 0,1,2,\ldots$ where $y(k)$ = national income for period k, Δ = the marginal propensity to consume and θ = a constant of proportionately called the "relation". It is known that $\Delta \in (0,1)$ and $\theta > 0$. The initial conditions are uncertain so let $y(0) = \overline{\gamma}_0 = (1/2/3)$ and $y(1) = \overline{\gamma}_1 = (2/3/4)$. We will look at two cases: (1) $\Delta = 0.9$ and $\theta = 2$; and (2) $\Delta = 0.9$ and $\theta = 0.5$.

Case 1:

Set $\Delta = 0.9$, $\theta = 2$ and the difference equation becomes

$$y(k+2) - 2.7y(k+1) + 1.8y(k) = 1\ . \tag{8.77}$$

The roots of $\beta^2 - 2.7\beta + 1.8 = 0$ are $r_1 = 1.2$ and $r_2 = 1.5$. We easily see that $G(k) = 10$ where $g(k) = 1$, for all k. The general solution is

$$y(k) = c_1(1.2)^k + c_2(1.5)^k + 10\ . \tag{8.78}$$

We will find $\overline{Y}_e(k)$. Using the initial conditions we see

$$c_1 = 5\gamma_0 - \frac{10}{3}\gamma_1 - \frac{50}{3}\ , \tag{8.79}$$

$$c_2 = \frac{10}{3}\gamma_1 - 4\gamma_0 + \frac{20}{3}\ , \tag{8.80}$$

k	$\alpha = 0$	$\alpha = 1$
2	[1.00,10.0]	5.5
4	[-0.53,39.2]	19.3
6	[0.30,117.6]	59.0
8	[8.34,312.6]	160.4
10	[35.3,777.8]	406.6
20	[4,124, 52,256]	28,190
100	$[0.54(10)^{18}, 0.65(10)^{19}]$	$0.35(10)^{19}$

Table 8.1: Extension Principle $\overline{Y}_e$ Solution in the National Income Application when $\Delta = 0.9$, $\theta = 2.0$

where we used the crisp initial conditions $y(0) = \gamma_0$, $y(1) = \gamma_1$. We calculate $\partial y(k)/\partial \gamma_0 < 0$ and $\partial y(k)/\partial \gamma_1 > 0$ for $k \geq 2$. Hence we obtain: (1) $y_{e1}(k, \alpha)$ by using $\gamma_{02}(\alpha)$ for γ_0, $\gamma_{11}(\alpha)$ for γ_1; and (2) $y_{e2}(k, \alpha)$ by substituting $\gamma_{01}(\alpha)$ for γ_0, $\gamma_{12}(\alpha)$ for γ_1. And then c_1 and c_2 in equations (8.79) and (8.80), respectively, are independent of k and $\overline{Y}_e(k)$ solves the difference equation. $\overline{Y}_e(k)$ is a solution. Table 8.1 gives values of $\overline{Y}_e(k)[\alpha]$ for $\alpha = 0$ and 1 for selected values of k. We notice that the fuzziness of $\overline{Y}_e(k)$ grows without bound because

$$\lim_{k \to \infty} (y_{e2}(k, 0) - y_{e1}(k, 0)) = \infty . \tag{8.81}$$

But also, the vertex of $\overline{Y}_e(k)$ grows without bound since

$$\lim_{k \to \infty} y_{e1}(k, 1) = \lim_{k \to \infty} y_{e2}(k, 1) = \infty . \tag{8.82}$$

Case 2:

Let $\Delta = 0.9$, $\theta = 0.5$. The equation becomes

$$y(k+2) - 1.35y(k+1) + 0.45y(k) = 1.0 . \tag{8.83}$$

The unique solution is

$$y(k) = c_1(0.6)^k + c_2(0.75)^k + 10 , \tag{8.84}$$

for

$$c_1 = 5\gamma_0 - \frac{20}{3}\gamma_1 + \frac{50}{3} , \tag{8.85}$$

$$c_2 = \frac{20}{3}\gamma_1 - 4\gamma_0 - \frac{80}{3} . \tag{8.86}$$

We again calculate $\partial y(k)/\partial \gamma_0 < 0$ and $\partial y(k)/\partial \gamma_1 > 0$, $k \geq 2$. This allows us to find $y_{e1}(k, \alpha)$, and $y_{e2}(k, \alpha)$ as in Case 1. $\overline{Y}_e(k)$ is a solution and its α–cuts, $\alpha = 0$, 1, for various k are in Table 8.2.

k	$\alpha = 0$	$\alpha = 1$
2	[7.75,10.00]	8.88
4	[9.34,11.82]	10.58
6	[9.85,11.68]	10.76
8	[9.99,11.18]	10.59
10	[10.02,10.75]	10.39
20	[10.00,10.05]	10.03
100	[10.00,10.00]	10.00

Table 8.2: Extension Principle $\overline{Y}_e$ Solution in the National Income Application for $\Delta = 0.9$ and $\theta = 0.5$

Now

$$\lim_{k \to \infty} \overline{Y}_e(k) = 10 , \tag{8.87}$$

the limit is crisp ten, because $r_i^k \to 0$ as $k \to \infty$, $i = 1, 2$.

8.4.2 Transmission of Information

Consider a transmission of information system that has only two signals: a "zero" or a "one". Suppose a zero requires exactly one unit of time but a one requires two units of time. Let $y(t)$ = the number of possible message signals of zero and ones that can be sent in time duration t.

First consider those message signals of duration t that end in a zero. The total number of such messages is $y(t-1)$. Similarly, the total number of messages of duration t that end in a one is $y(t-2)$. Since a message must end in a zero or one we have

$$y(t) = y(t-1) + y(t-2) . \tag{8.88}$$

Let $k = t - 2$, so the difference equation is

$$y(k+2) - y(k+1) - y(k) = 0 , \tag{8.89}$$

$k = 0, 1, 2, \ldots$. The roots of $\beta^2 - \beta - 1 = 0$ are $r_1 = (1 - \sqrt{5})/2$ and $r_2 = (1 + \sqrt{5})/2$. The general solution is

$$y(k) = c_1 r_1^k + c_2 r_2^k , \tag{8.90}$$

and for $y(0) = \gamma_0$, $y(1) = \gamma_1$, the unique solution has

$$c_1 = \frac{r_2\gamma_0 - \gamma_1}{\sqrt{5}} , \tag{8.91}$$

$$c_2 = \frac{\gamma_1 - r_1\gamma_0}{\sqrt{5}} . \tag{8.92}$$

k	$\alpha = 0$	$\alpha = 1$
2	[5.0,17.0]	11.0
4	[15.0,49.0]	32.0
6	[40.0,130.0]	85.0
8	[105.0,341.0]	223.0
10	[275.0,893.0]	584.0
20	[33,825,109,837]	71,831
100	$[1.8(10)^{21}, 5.8(10)^{21}]$	$3.8(10)^{21}$

Table 8.3: Extension Principle $\overline{Y}_e$ Solution in the Transmission of Information Application

To compute $\overline{Y}_e$ we see $\partial y(k)/\partial\gamma_0 > 0$, $\partial y(k)/\partial\gamma_1 > 0$, so that: (1) use $\gamma_{01}(\alpha)$ for γ_0, $\gamma_{11}(\alpha)$ for γ_1 to get $y_{e1}(k,\alpha)$; and (2) substitute $\gamma_{02}(\alpha)$ for γ_0, $\gamma_{12}(\alpha)$ for γ_1 for $y_{e2}(k,\alpha)$. $\overline{Y}_e(k)$ exists and is a solution for all fuzzy initial conditions. Table 8.3 shows α-cuts of $\overline{Y}_e(k)$ for $\overline{\gamma}_0 = (0/1/2)$ and $\overline{\gamma}_1 = (5/10/15)$.

The capacity of the (fuzzy) channel is defined as

$$\text{Capacity} = \lim_{k\to\infty}\left(\frac{\log_2 \overline{Y}_e(k)}{k}\right) , \tag{8.93}$$

where we define $\log_2 \overline{N}$, of fuzzy number $\overline{N}$, to be that fuzzy set with α–cuts $[\log_2 n_1(\alpha), \log_2 n_2(\alpha)]$ if $\overline{N}[\alpha] = [n_1(\alpha), n_2(\alpha)]$ and $n_1(\alpha) > 0$ for all α. Now $r_1^k \to 0$ as $k \to \infty$ because $-1 < r_1 < 0$. So for large k we get $y(k) \approx c_2 r_2^k$. Therefore, for large values of k

$$\log_2(y_{e1}(k,\alpha)) \approx \log_2\left(\frac{\gamma_{11}(\alpha) - r_1\gamma_{01}(\alpha)}{\sqrt{5}} r_2^k\right) , \tag{8.94}$$

$$\log_2(y_{e2}(k,\alpha)) \approx \log_2\left(\frac{\gamma_{12}(\alpha) - r_1\gamma_{02}(\alpha)}{\sqrt{5}} r_2^k\right) . \tag{8.95}$$

We assuming the fuzzy initial conditions $\overline{\gamma}_1$ and $\overline{\gamma}_2$ are such that the coefficients of r_2^k in equations (8.94) and (8.95) are positive. Then we see that

$$\text{Capacity} = [log_2 r_2, \log_2 r_2] , \tag{8.96}$$

a crisp number $\log_2 r_2$. This result is the same as for crisp initial conditions.

8.4.3 Fuzzy Fibonacci Numbers

The Fibonacci numbers are the numbers in the sequence

$$0, 1, 2, 3, 5, 8, 13, \dots , \tag{8.97}$$

k	$\overline{Y}_c(k)$
3	(1/3/5)
4	(2/5/8)
5	(3/8/13)
6	(5/13/21)
7	(8/21/34)
8	(13/34/55)
9	(21/55/89)
10	(34/89/144)

Table 8.4: Fuzzy Fibonacci Sequence

where each one is the sum of the two preceding numbers (starting with the fourth one). Therefore, the difference equation is

$$y(k+2) = y(k+1) + y(k) \ , \tag{8.98}$$

$k = 1, 2, \ldots$, $y(0) = 0$, $y(1) = 1$, $y(2) = 2$ and $y(k+2)$ is the $(2+k)$–th number in the sequence, $k \geq 1$.

The Fibonacci sequence is a very famous sequence in mathematics ([5] and for its relation to a population of rabbits see [6], problem 55, p. 95).

This is the same difference equation as in the previous application. But, when we fuzzify the two equations (8.89) and (8.98) we will obtain different results. In the crisp (non-fuzzy) case, the two equations (8.89) and (8.98) are the same, but in the fuzzy case these two equations are different. As an example of a fuzzy Fibonacci sequence let the fuzzy initial conditions be $\overline{\gamma}_1 = (0/1/2)$, $\overline{\gamma}_2 = (1/2/3)$. We will use equation (8.98) to find $\overline{Y}_c(k)$ the classical solution, which now exists using the difference equation in the form of equation (8.98). For example, $\overline{Y}_c(4) = \overline{\gamma}_1 + 2\overline{\gamma}_2$ and $\overline{Y}_c(10) = 21\overline{\gamma}_1 + 34\overline{\gamma}_2$. $\overline{Y}_c(k)$ is shown in Table 8.4 for $3 \leq k \leq 10$.

8.5 Summary and Conclusions

In this chapter we studied solutions to

$$y(k+2) + ay(k+1) + by(k) = g(k) \ , \tag{8.99}$$

for $k = 0, 1, 2, 3, \ldots$ where a and b were constants with $b > 0$, under fuzzy initial conditions $y(0) = \overline{\gamma}_0$, $y(1) = \overline{\gamma}_1$ for $\overline{\gamma}_0$, $\overline{\gamma}_1$ triangular fuzzy numbers. The results easily are extended to more general fuzzy numbers for $\overline{\gamma}_0$ and $\overline{\gamma}_1$. We showed that: (1) $\overline{Y}_c$ is never a solution; (2) if the roots are complex $\overline{Y}_e$ and $\overline{Y}_I$ are not solutions; (3) if the roots are equal, $\overline{Y}_e$ and $\overline{Y}_I$ are solutions if and only if the root is positive; and (4) if the roots are real and not equal, $\overline{Y}_e$ is more general than $\overline{Y}_I$. In all cases we demanded solutions to exist for all

k, and $\overline{Y}_e$ and $\overline{Y}_I$ solve the original difference equation. We also presented three applications using the $\overline{Y}_e$ solution.

For recent papers on fuzzy difference equations we refer the reader to ([2], [3], [4]).

Future research might center on also fuzzifying a, b or g in equation (8.99) but more importantly on partial solutions. If Γ is an infinite subset of $\{0, 1, 2, \ldots\}$ a partial solution exists if and only if $k \in \Gamma$. We want Γ infinite because then we may find the limiting behavior of the solution from taking the limit as $k \to \infty$ with $k \in \Gamma$. Important cases would be $\Gamma =$ positive integers or $\Gamma =$ odd integers. Then we may get $\overline{Y}_c$ a partial solution, and $\overline{Y}_e$, $\overline{Y}_I$ partial solutions for the case of complex roots. Some work on partial solutions is in [1], but more needs to be done.

Bibliography

[1] J.J. Buckley, Th. Feuring and Y. Hayashi: Fuzzy Difference Equations: The Initial Value Problem. J. Advanced Computational Intelligence. To appear.

[2] E.Y. Deeba and A. de Korvin: On a Fuzzy Difference Equation, IEEE Trans. Fuzzy Systems, 3 (1995), pp. 469–473.

[3] E.Y. Deeba and A. de Korvin: Analysis by Fuzzy Difference Equations of a Model of CO_2 Level in the Blood, Applied Math. Letters, 12 (1999), pp. 33–40.

[4] E. Deeba, A. de Korvin and S. Xie: Techniques and Applications of Fuzzy Set Theory to Difference and Functional Equations and their Utilization in Modeling Diverse Systems, in: C.T. Leondes (Ed.), Fuzzy Theory Systems, Vol. 1, Academic Press, San Diego, CA., pp. 87–110.

[5] The Fibonacci Quarterly, Fibonacci Association, Univ. Santa Clara, CA.

[6] S. Goldberg: Introduction to Difference Equations, John Wiley and Sons, N. Y., 1958.

[7] F.B. Hildebrand: Methods of Applied Mathematics, Prentice Hall, Englewood Cliffs, N.J., 1958.

[8] F.B. Hildebrand: Finite–Difference Equations and Simulations, Prentice Hall, Englewood Cliffs, N.J., 1968.

[9] W.G. Kelley and A.C. Peterson: Difference Equations: An Introduction with Applications, Second Edition, Harcourt/Academic Press, Burlington, MA, USA, 2000.

[10] V. Lakashmikanthan and D. Trigante: Theory of Difference Equations, Academic Press, San Diego, CA, 1988.

[11] K.S. Miller: Linear Difference Equations, W.A. Benjamin, N.Y., 1968.

[12] P.A. Samuelson: Interactions Between the Multiplier Analysis and the Principle of Acceleration, Review of Economic Statistics, 21 (1939), pp. 75–78.

[13] C.E. Shannon and W. Weaver: The Mathematical Theory of Communication, Univ. of Illinois Press, Urbana, Ill., 1949.

Chapter 9

Fuzzy Partial Differential Equations

We begin this chapter with discussing the type of elementary fuzzy partial differential equation we wish to solve. As in Chapter 3, 4, 7 and 8 we then consider types of solutions. If one fuzzifies the crisp partial differential equation and then solves, you are attempting to get the classical solution $\overline{Y}_c$. $\overline{Y}_c$ is the topic of the second section. When one first solves the crisp partial differential equation and then fuzzifies this crisp solution you obtain the extension principle solution $\overline{Y}_e$. We do not discuss $\overline{Y}_I$, the solution using α–cuts and interval arithmetic to evaluate the fuzzification of the crisp solution, in this chapter. $\overline{Y}_e$ is in the third section.

Most of this chapter may be extended to more general fuzzy numbers but we shall be using only triangular fuzzy numbers. Examples are presented showing $\overline{Y}_c$, $\overline{Y}_e$, how they are related and their differences. This chapter is based on [1].

In Chapters 7 and 8 we basically only fuzzified the initial conditions. There are no initial conditions for partial differential equations but instead they have boundary conditions. So we will fuzzify the boundary conditions but also, we will fuzzify the non–homogeneous part of the partial differential equation. Since we only consider very elementary (no Fourier series, etc.) fuzzy partial differential equations, topics of future research are discussed in the last section.

9.1 Elementary Partial Differential Equations

Let $I_1 = [0, M_1]$, $I_2 = [0, M_2]$ for some $M_1 > 0$, $M_2 > 0$. $F(x, y, k)$ is a continuous function for $(x, y) \in I_1 \times I_2$ and for $k = (k_1, \dots, k_n)$ a vector of constants k_i in intervals J_i, $1 \le i \le n$. The operator $\phi(D_x, D_y)$ will be a polynomial in D_x and D_y with constant coefficients where D_x (D_y) stands for the partial derivative with respect to x (y). For example

$$\phi(D_x, D_y) = aD_x^2 + bD_xD_y + cD_y^2 \,, \tag{9.1}$$

is such an operator for constants a, b, c. Also let $z = U(x,y)$ be a function so that $\phi(D_x, D_y)U(x,y)$ is continuous on $I_1 \times I_2$. The crisp partial differential equation is

$$\phi(D_x, D_y)U(x,y) = F(x,y,k) \,, \tag{9.2}$$

for F continuous on $I_1 \times I_2 \times \Pi(J_i)$ into $\mathbf{R}$, subject to certain boundary conditions. These boundary conditions can come in a variety of forms such as $U(0,y) = c_1$, $U(x,0) = c_2$, $U(M_1,y) = c_3, \ldots$, $U(0,y) = g_1(y,c_4)$, $U(x,0) = f_1(x,c_5), \ldots$, $U_x(x,0) = f_2(x,c_6)$, $U_y(0,y) = g_2(y,c_7), \ldots$. At this point we do not specify any particular boundary conditions except to say they depend on constants $c_1, \ldots, c_m$ with c_i in interval L_i, $1 \le i \le m$. Let $c = (c_1, \ldots, c_m)$ be the vector of these constants .

We assume the problem (equation (9.2)), with associated boundary conditions, has a solution $z = G(x,y,k,c)$ with $\phi(D_x, D_y)G(x,y,k,c)$ continuous on $I_1 \times I_2$, $k \in J = \Pi J_i$, $c \in L = \Pi L_i$.

Now we may explain what we mean by "elementary" partial differential equations. By "elementary" we mean that the solution $G(x,y,k,c)$ is not defined in terms of a series (Fourier, Bessel, Legendre, $\cdots$). The solution $G(x,y,k,c)$ is given in "closed form" and not in terms of infinite series. Since we will need to fuzzify $G(x,y,k,c)$ we do not wish to fuzzify an infinite series of fuzzy numbers. The theory of infinite series of fuzzy numbers has not yet been completely worked out, so we must avoid that construction. This is unfortunate because this means that we can not consider some of the basic applications to engineering such as heat conduction, vibrating beams, etc. We will not be fuzzifying the heat equation, the vibrating string equation, etc.

Some, or all, of the constants k_i and c_j may not be known exactly (obtained from measurements) and we model this uncertainty by using triangular fuzzy numbers. So we will substitute triangular fuzzy number $\overline{K}_i$ for k_i, $\overline{K}_i$ in J_i, $1 \le i \le n$, and substitute triangular fuzzy number $\overline{C}_i$ for c_i, $\overline{C}_i$ in L_i, $1 \le i \le m$. We use the extension principle to obtain $\overline{F}$ from F where $\overline{F}(x,y,\overline{K})$ has $\overline{K} = (\overline{K}_1, \ldots, \overline{K}_n)$. We know if $\overline{F}(x,y,\overline{K})[\alpha] = [F_1(x,y,\alpha), F_2(x,y,\alpha)]$, then

$$F_1(x,y,\alpha) = \min\{F(x,y,k) | k \in \overline{K}[\alpha]\} \,, \tag{9.3}$$

$$F_2(x,y,\alpha) = \max\{F(x,y,k) | k \in \overline{K}[\alpha]\} \,, \tag{9.4}$$

where $\overline{K}[\alpha] = \Pi \overline{K}_i[\alpha]$, for all $(x,y) \in I_1 \times I_2$, $\alpha \in [0,1]$.

The function $U(x,y)$ becomes $\overline{U}(x,y)$ mapping $I_1 \times I_2$ into fuzzy numbers. That is, $\overline{Z} = \overline{U}(x,y)$, $(x,y) \in I_1 \times I_2$, $\overline{Z}$ a fuzzy number. Next fuzzify the solution $z = G(x,y,k,c)$ into $\overline{Z} = \overline{G}(x,y,\overline{K},\overline{C})$ for $\overline{K} = (\overline{K}_1, \ldots, \overline{K}_n)$, $\overline{C} = (\overline{C}_1, \ldots, \overline{C}_m)$ for all the $\overline{K}_i$ triangular fuzzy numbers in J_i and all the

$\overline{C}_j$ triangular fuzzy numbers in L_j. $\overline{G}$ is also computed using the extension principle. Let $\overline{G}(x,y,\overline{K},\overline{C})[\alpha] = [g_1(x,y,\alpha), g_2(x,y,\alpha)]$, then

$$g_1(x,y,\alpha) = \min\left\{G(x,y,k,c) | k \in \overline{K}[\alpha], c \in \overline{C}[\alpha]\right\}, \tag{9.5}$$

$$g_2(x,y,\alpha) = \max\left\{G(x,y,k,c) | k \in \overline{K}[\alpha], c \in \overline{C}[\alpha]\right\}, \tag{9.6}$$

for $\overline{C}[\alpha] = \Pi \overline{C}_j[\alpha]$, for all x, y, α.

The fuzzy partial differential equation is

$$\phi(D_x, D_y)\overline{U}(x,y) = \overline{F}(x,y,\overline{K}) , \tag{9.7}$$

subject to some boundary conditions. We next fuzzify the boundary conditions, but first notice that we did not fuzzify the operator $\phi(D_x, D_y)$. The constants in $\phi(D_x, D_y)$, see equation (9.1), remain crisp and will not be fuzzified. The fuzzy boundary conditions are $\overline{U}(0,y) = \overline{C}_1$, $\overline{U}(x,0) = \overline{C}_2$, $\overline{U}(M_1,y) = \overline{C}_3, \ldots$, $\overline{U}(0,y) = \overline{g}_1(y,\overline{C}_4)$, $\overline{U}(x,0) = \overline{f}_1(x,\overline{C}_5), \ldots$, $\overline{U}_x(x,0) = \overline{f}_2(x,\overline{C}_6)$, $\overline{U}_y(0,y) = \overline{g}_2(y,\overline{C}_7), \ldots$. The $\overline{f}_i$ and $\overline{g}_i$ are extension principle extensions of f_i and g_i, respectively.

We wish to solve equation (9.7) subject to fuzzy boundary conditions. We first discuss the classical solution $\overline{Y}_c$.

9.2 Classical Solution

Let $\overline{Y}_c(x,y)$ denote the classical solution. $\overline{Y}_c(x,y)$ is a triangular shaped fuzzy number for $(x,y) \in I_1 \times I_2$. Define $\overline{Y}_c(x,y)[\alpha] = [y_1(x,y,\alpha), y_2(x,y,\alpha)]$ for all $(x,y) \in I_1 \times I_2$ and α in $[0,1]$. We assume that the $y_i(x,y,\alpha)$ have continuous partials so that $\phi(D_x, D_y)y_i(x,y,\alpha)$ is continuous for $(x,y) \in I_1 \times I_2$, α in $[0,1]$, $i = 1,2$. Substitute α–cuts of $\overline{Y}_c(x,y)$ into the fuzzy partial differential equation (9.7) producing

$$\phi(D_x, D_y)[y_1(x,y,\alpha), y_2(x,y,\alpha)] = [F_1(x,y,\alpha), F_2(x,y,\alpha)] . \tag{9.8}$$

Simplify equation (9.8) using interval arithmetic producing two equations to solve for $y_1(x,y,\alpha)$ and $y_2(x,y,\alpha)$. We append to equation (9.8) the fuzzy boundary conditions. Suppose these are $\overline{U}(0,y) = \overline{C}_1$ and $\overline{U}(M_1,y) = \overline{C}_2$. Then we add to equation (9.8)

$$y_1(0,y,\alpha) = c_{11}(\alpha) , \tag{9.9}$$

$$y_2(0,y,\alpha) = c_{12}(\alpha) , \tag{9.10}$$

$$y_1(M_1,y,\alpha) = c_{21}(\alpha) , \tag{9.11}$$

$$y_2(M_1,y,\alpha) = c_{22}(\alpha) , \tag{9.12}$$

where $\overline{C}_1[\alpha] = [c_{11}(\alpha), c_{12}(\alpha)]$, $\overline{C}_2[\alpha] = [c_{21}(\alpha), c_{22}(\alpha)]$. We solve the two partial differential equations arising from equation (9.8) plus the boundary conditions (9.9) – (9.12) for the $y_i(x,y,\alpha)$, $i = 1,2$. We say $\overline{Y}_c(x,y)$ is a solution if $[y_1(x,y,\alpha), y_2(x,y,\alpha)]$ defines a triangular shaped fuzzy number $0 \leq \alpha \leq 1$, for all $(x,y) \in I_1 \times I_2$. Assuming the $y_i(x,y,\alpha)$ are differentiable on α, what we need for these intervals to give a triangular shaped fuzzy number is $\partial y_1/\partial\alpha > 0$, $\partial y_2/\partial\alpha < 0$, $0 < \alpha < 1$, and $y_1(x,y,1) = y_2(x,y,1)$, for all x, y.

Example 9.2.1

Let $\phi(D_x, D_y) = D_y D_x$ and $F(x,y,k) = k_1 xy + k_2 e^x$ for $k_1 \in [0, M_3]$, $k_2 \in [0, M_4]$, for $M_3 > 0$, $M_4 > 0$. The boundary conditions are $U(x,0) = c_1$, $U_y(0,y) = c_2 y$, for $c_1 \in [0, M_5]$, $c_2 \in [0, M_6]$, $M_5 > 0$, $M_6 > 0$.

First we find F_1 and F_2 from equations (9.3) and (9.4), respectively. Since $\partial F/\partial k_1 > 0$ and $\partial F/\partial k_2 > 0$ we obtain

$$F_1(x,y,\alpha) = k_{11}(\alpha)xy + k_{21}(\alpha)e^x , \tag{9.13}$$

$$F_2(x,y,\alpha) = k_{12}(\alpha)xy + k_{22}(\alpha)e^x , \tag{9.14}$$

where $\overline{K}_1[\alpha] = [k_{11}(\alpha), k_{12}(\alpha)]$, $\overline{K}_2[\alpha] = [k_{21}(\alpha), k_{22}(\alpha)]$. Equation (9.8) becomes for $y_1(x,y,\alpha)$

$$\frac{\partial^2}{\partial y \partial x} y_1(x,y,\alpha) = F_1(x,y,\alpha) , \tag{9.15}$$

$$y_1(x,0,\alpha) = c_{11}(\alpha) , \tag{9.16}$$

$$\frac{\partial}{\partial y} y_1(0,y,\alpha) = c_{21}(\alpha) y . \tag{9.17}$$

A solution is

$$y_1(x,y,\alpha) = \frac{k_{11}(\alpha)}{4} x^2 y^2 + k_{21}(\alpha) y e^x + c_{11}(\alpha) + c_{21}(\alpha)\frac{y^2}{2} - k_{21}(\alpha) y . \tag{9.18}$$

Equation (9.8) for $y_2(x,y,\alpha)$ is

$$\frac{\partial^2}{\partial y \partial x} y_2(x,y,\alpha) = F_2(x,y,\alpha) , \tag{9.19}$$

$$y_2(x,0,\alpha) = c_{12}(\alpha) , \tag{9.20}$$

$$\frac{\partial}{\partial y} y_2(0,y,\alpha) = c_{22}(\alpha) y . \tag{9.21}$$

A solution is

$$y_2(x,y,\alpha) = \frac{k_{12}(\alpha)}{4} x^2 y^2 + k_{22}(\alpha) y e^x + c_{12}(\alpha) + c_{22}(\alpha)\frac{y^2}{2} - k_{22}(\alpha) y . \tag{9.22}$$

One may easily check that $\partial y_1/\partial\alpha > 0$, $\partial y_2/\partial\alpha < 0$, $0 < \alpha < 1$, $y_1(x,y,1) = y_2(x,y,1)$ so that $\overline{Y}_c(x,y)$ is a solution. We may write this as

$$\overline{Y}_c(x,y) = \frac{x^2y^2}{4}\overline{K}_1 + ye^x\overline{K}_2 + \overline{C}_1 + \frac{y^2}{2}\overline{C}_2 - y\overline{K}_2 \ , \tag{9.23}$$

for $(x,y) \in I_1 \times I_2$, $\overline{K}_i$ in J_i and $\overline{C}_i$ in L_i, $i = 1,2$. These results are true for all $M_i > 0$, $1 \le i \le 6$, or the intervals may all be $[0,\infty)$.

Example 9.2.2

Let $\phi(D_xD_y) = D_yD_x - D_x$ and $F(x,y,k) = k$, (constant) $k \geq 0$. Boundary conditions are $U(0,y) = c_1$ and $U_x(x,0) = c_2x^2$. The $c_i \in [0,M_i]$, $M_i > 0$, $i = 3,4$. Equation (9.8) is for $y_1(x,y,\alpha)$

$$\frac{\partial^2}{\partial y\partial x}y_1(x,y,\alpha) - \frac{\partial}{\partial x}y_2(x,y,\alpha) = k_1(\alpha) \ , \tag{9.24}$$

$$y_1(0,y,\alpha) = c_{11}(\alpha) \ , \tag{9.25}$$

$$\frac{\partial}{\partial x}y_1(x,0,\alpha) = c_{21}(\alpha)x^2 \ . \tag{9.26}$$

The problem for $y_2(x,y,\alpha)$ is

$$\frac{\partial^2}{\partial y\partial x}y_2(x,y,\alpha) - \frac{\partial}{\partial x}y_1(x,y,\alpha) = k_2(\alpha) \ , \tag{9.27}$$

$$y_2(0,y,\alpha) = c_{12}(\alpha) \ , \tag{9.28}$$

$$\frac{\partial}{\partial x}y_2(x,0,\alpha) = c_{22}(\alpha)x^2 \ . \tag{9.29}$$

Notice the negative D_x put y_2 into equation (9.24) and y_1 into equation (9.27). We see that the partial differential equations to solve are "coupled". Both y_1 and y_2 are in equation (9.24) and in (9.27). These equations must be solved simultaneously. This problem could have come up in Chapters 7 and 8. It did not come up explicitly in the examples in these chapters because we wanted to avoid solving systems of coupled differential (difference) equations. So, we will not consider this problem further in this chapter. We will consider only positive constants in $\phi(D_x, D_y)$. When the constants in $\phi(D_x, D_y)$ are all positive, then we get one partial differential equation containing only y_1, to solve for y_1, and another partial differential equation having only y_2, to solve for y_2. Then we may solve for y_1 and y_2 and check to see if they define a fuzzy number.

When the constants in $\phi(D_x, D_y)$ can be negative we will not consider $\overline{Y}_c$ but we may see if $\overline{Y}_e$ is a solution.

9.3 Extension Principle Solution

The extension principle solution is written $\overline{Y}_e(x,y)$ and its α–cuts are defined from fuzzification of $G(x,y,k,c)$ in equations (9.5) and (9.6). This means

$$\overline{Y}_e(x,y)[\alpha] = [g_1(x,y,\alpha), g_2(x,y,\alpha)] \ , \tag{9.30}$$

all x, y, α. For $\overline{Y}_e$ to be called a solution it must satisfy the partial differential equation plus boundary conditions. $\overline{Y}_e$ is a solution if

$$\phi(D_x, D_y)g_1(x,y,\alpha) = F_1(x,y,\alpha) \ , \tag{9.31}$$

$$\phi(D_x, D_y)g_2(x,y,\alpha) = F_2(x,y,\alpha) \ , \tag{9.32}$$

are satisfied plus the fuzzy boundary conditions are true. Equations like (9.9) – (9.12) must hold for $y_i(x,y,\alpha) = g_i(x,y,\alpha)$, $i = 1,2$, α-cuts of $\overline{Y}_e$.

Example 9.3.1

This continues Example 9.2.2. We show that $\overline{Y}_e$ is a solution. A solution $z = G(x,y,k,c)$ is

$$c_2\frac{x^3 e^y}{3} + kx(e^y - 1) + c_1 \ . \tag{9.33}$$

The partials of G with respect to k, c_1, and c_2 are all positive so we may easily evaluate equations (9.5) and (9.6) giving α–cuts of $\overline{Y}_e$

$$g_1(x,y,\alpha) = c_{21}(\alpha)\frac{x^3 e^y}{3} + k_1(\alpha)x(e^y - 1) + c_{11}(\alpha) \ , \tag{9.34}$$

$$g_2(x,y,\alpha) = c_{22}(\alpha)\frac{x^3 e^y}{3} + k_2(\alpha)x(e^y - 1) + c_{12}(\alpha) \ , \tag{9.35}$$

where $\overline{K}[\alpha] = [k_2(\alpha), k_2(\alpha)]$. We easily check that the following equations hold ($y_i = g_i$, $i = 1,2$):

$$\phi(D_x, D_y)y_1(x,y,\alpha) = k_1(\alpha) \ , \tag{9.36}$$

$$\phi(D_x, D_y)y_2(x,y,\alpha) = k_2(\alpha) \ , \tag{9.37}$$

where $\phi(D_x, D_y) = D_y D_x - D_x$. Also, the boundary conditions

$$y_1(0,y,\alpha) = c_{11}(\alpha) \ , \tag{9.38}$$

$$y_2(0,y,\alpha) = c_{12}(\alpha) \ , \tag{9.39}$$

$$\frac{\partial}{\partial x}y_1(x,0,\alpha) = c_{21}(\alpha)x^2 \ , \tag{9.40}$$

$$\frac{\partial}{\partial x}y_2(x,0,\alpha) = c_{22}(\alpha)x^2 \ , \tag{9.41}$$

hold. Hence $\overline{Y}_e$ is a solution and may be written

$$\overline{Y}_e(x,y) = \overline{C}_2 \frac{x^3 e^y}{3} + \overline{K}x(e^y - 1) + \overline{C}_1 \ . \tag{9.42}$$

This is valid for all $M_i > 0$ so all intervals can be $[0, \infty)$.

The way $\overline{Y}_e$ is defined it is always a triangular shaped fuzzy number and we expect it to satisfy the fuzzy boundary conditions. The question is whether or not it satisfies the partial differential equations (9.31) and (9.32). The following theorem clarifies when $\overline{Y}_e$ will be a solution.

Theorem 9.1 *a. $\overline{Y}_e$ is a solution if*

$$\frac{\partial G}{\partial k_i}\frac{\partial F}{\partial k_i} > 0 \ ,$$

$1 \le i \le n$, $(x,y) \in I_1 \times I_2$, $k \in \Pi J_i$.

b. If

$$\frac{\partial G}{\partial k_i}\frac{\partial F}{\partial k_i} < 0 \ ,$$

for some i, $(x,y) \in I_1 \times I_2$, $k \in \Pi J_i$, then $\overline{Y}_e$ is not a solution.

Proof 9.1 *a. For simplicity let $n = 2$ and assume that both G and F are increasing functions of k_1 and both functions are decreasing functions of k_2. From equations (9.3) – (9.4) and (9.5) – (9.6) we see that*

$$\begin{aligned} F_1(x,y,\alpha) &= F(x,y,k_{11}(\alpha),k_{22}(\alpha)) \ , & (9.43)\\ F_2(x,y,\alpha) &= F(x,y,k_{12}(\alpha),k_{21}(\alpha)) \ , & (9.44)\\ g_1(x,y,\alpha) &= G(x,y,k_{11}(\alpha),k_{22}(\alpha),c) \ , & (9.45)\\ g_2(x,y,\alpha) &= G(x,y,k_{12}(\alpha),k_{21}(\alpha),c) \ , & (9.46) \end{aligned}$$

for all α where $\overline{K}_1[\alpha] = [k_{11}(\alpha), k_{12}(\alpha)]$, $\overline{K}_2[\alpha] = [k_{21}(\alpha), k_{22}(\alpha)]$. Now G solves the partial differential equation (9.2) which means

$$\phi(D_x, D_y)G(x,y,k_1,k_2,c) = F(x,y,k_1,k_2) \ , \tag{9.47}$$

for $(x,y) \in I_1 \times I_2$, $k_1 \in J_1$, $k_2 \in J_2$, $c \in \Pi L_j$. But $k_{1j}(\alpha) \in J_1$, $k_{2j}(\alpha) \in J_2$ all α, $j = 1,2$ so

$$\begin{aligned} \phi(D_x, D_y)g_1(x,y,\alpha) &= F_1(x,y,\alpha) \ , & (9.48)\\ \phi(D_x, D_y)g_2(x,y,\alpha) &= F_2(x,y,\alpha) \ , & (9.49) \end{aligned}$$

for all x, y and α. Hence equation (9.31) and (9.32) hold and $\overline{Y}_e$ is a solution.

b. *Again let $n = 2$, and assume G is increasing in k_1, and also F is decreasing in k_1, but both G and F are decreasing in k_2. Equations (9.45) and (9.46) are still true but equations (9.43) and (9.44) become*

$$F_1(x,y,\alpha) = F(x,y,k_{12}(\alpha),k_{22}(\alpha)) \,, \tag{9.50}$$
$$F_2(x,y,\alpha) = F(x,y,k_{11}(\alpha),k_{21}(\alpha)) \,, \tag{9.51}$$

for all α. Hence equations (9.48) and (9.49) do not hold and $\overline{Y}_e$ is not a solution.

Notice that the condition (a) in Theorem 9.1 holds in Example 9.3.1 since $\partial F/\partial k > 0$ and $\partial G/\partial k > 0$.

Next let us examine the relationship between $\overline{Y}_c$ and $\overline{Y}_e$ when both are solutions. Assume the constants in $\phi(D_x, D_y)$ are all positive so we do not get a system of partial equations to solve as in Example 9.2.2. Let R_c be the subset of $I_1 \times I_2$ so that $\overline{Y}_c$ is a solution if and only if $(x,y) \in R_c$. Of course, R_c could be empty but it can also be all of $I_1 \times I_2$ as in Example 9.2.1. Similarly, let R_e be the subset of $I_1 \times I_2$ so that $\overline{Y}_e$ is a solution if and only if $(x,y) \in R_e$.

Theorem 9.2 *Assume all the constants in $\phi(D_x, D_y)$ are positive. Then $R_e \subseteq R_c$ and on R_e, $\overline{Y}_e = \overline{Y}_c$.*

Proof 9.2 *$\overline{Y}_e$ is always a triangular shaped fuzzy number and the way it was defined implies that we expect it to satisfy the fuzzy boundary conditions. So, for $\overline{Y}_e$ to be a solution equations (9.31) and (9.32) must hold.*

For $\overline{Y}_c$ to be a solution we first need to have

$$\phi(D_x,D_y)y_1(x,y,\alpha) = F_1(x,y,\alpha) \,, \tag{9.52}$$
$$\phi(D_x,D_y)y_2(x,y,\alpha) = F_2(x,y,\alpha) \,. \tag{9.53}$$

It follows that $g_1 = y_1$ and $g_2 = y_2$ for all x, y and α so that $\overline{Y}_c = \overline{Y}_e$ when $\overline{Y}_e$ is a solution.

We may have $R_c = R_e$ as in Example 9.2.1. In Example 9.2.1

$$G = k_1\frac{x^2e^y}{4} + k_2y(e^x - 1) + c_1 + \frac{c_2y^2}{2} \,, \tag{9.54}$$

so that $(\partial G/\partial k_i)(\partial F/\partial k_i) > 0$, $i = 1,2$ and Theorem 9.1 says $\overline{Y}_e$ is also a solution. The following examples show that R_e may be a proper subset of R_c (constants in $\phi(D_x, D_y)$ all positive).

Example 9.3.2

Let $\phi(D_x, D_y) = D_y^2$ and $F(x,y,k) = k_1x^2\cos(y) + k_2$. The boundary conditions are $U(x,0) = c_1$, $U(x,\pi/2) = c_2$. All the constants k_1, k_2, c_1 and c_2 are non-negative. The variable x is in $[0, M_1]$, $M_1 > 0$ and $y \in [0, \pi/2]$.

A solution G is

$$k_1 x^2(1-\cos(y)-\frac{2}{\pi}y)+k_2\frac{y}{2}(y-\frac{\pi}{2})+c_1(1-\frac{2}{\pi}y)+c_2(\frac{2}{\pi}y)\ , \tag{9.55}$$

for $(x,y) \in I_1 \times I_2$. $\overline{Y}_e$ is not a solution because the condition in part (b) of Theorem 9.1 holds for $i=1$ and $i=2$. Hence R_e is empty.

Now show that R_c is not empty. We must solve

$$\frac{\partial^2}{\partial y^2}y_1(x,y,\alpha) = k_{11}(\alpha)x^2\cos(y)+k_{21}(\alpha)\ , \tag{9.56}$$

$$\frac{\partial^2}{\partial y^2}y_2(x,y,\alpha) = k_{12}(\alpha)x^2\cos(y)+k_{22}(\alpha)\ , \tag{9.57}$$

subject to

$$y_1(x,0,\alpha) = c_{11}(\alpha)\ , \tag{9.58}$$

$$y_2(x,0,\alpha) = c_{12}(\alpha)\ , \tag{9.59}$$

$$y_1(x,\frac{\pi}{2},\alpha) = c_{21}(\alpha)\ , \tag{9.60}$$

$$y_2(x,\frac{\pi}{2},\alpha) = c_{22}(\alpha)\ . \tag{9.61}$$

Solutions are

$$y_i(x,y,\alpha)=k_{1i}(\alpha)x^2(1-\cos(y)-\frac{2}{\pi}y)+k_{2i}(\alpha)\frac{y}{2}(y-\frac{\pi}{2}) +c_{1i}(\alpha)(1-\frac{2}{\pi})+c_{2i}(\alpha)(\frac{2}{\pi}y), \tag{9.62}$$

for $i=1,2$. All we need now for $\overline{Y}_c$ to be a solution is that $\partial y_1/\partial\alpha > 0$, $\partial y_2/\partial\alpha < 0$ since $y_1(x,y,1)=y_2(x,y,1)$. R_c may not be all of $[0,\infty)\times[0,\pi/2]$ but it will be a subset of this region, depending on the fuzzy numbers $\overline{K}_i$ and $\overline{C}_i$, $i=1,2$.

Assume these triangular fuzzy numbers have base on interval $[a-1,a+1]$ with vertex at a. Then, if a "prime" denotes the derivative with respect to α, $k'_{i1}(\alpha)=1$, $k'_{i2}(\alpha)=-1$, $c'_{i1}(\alpha)=1$, $c'_{i2}(\alpha)=-1$, $i=1,2$. For equation (9.62) to define the α–cuts of triangular shaped fuzzy numbers $\overline{Y}_c$ we need

$$x^2(1-\cos(y)-\frac{2}{\pi}y)+\frac{y}{2}(y-\frac{\pi}{2})+1>0\ . \tag{9.63}$$

We get this equation from $(\partial y_1/\partial\alpha)>0$ and $(\partial y_2/\partial\alpha)<0$ under the assumptions on the $k'_{ij}(\alpha)$ and $c'_{ij}(\alpha)$ defined above. Close analysis of equation (9.63), see [1], shows that there is a number $\lambda \in [1.5, 2.0]$ so that equation (9.63) holds for $(x,y)\in[0,\lambda]\times[0,\pi/2]$. Hence $[0,\lambda]\times[0,\pi/2]\subset R_c$.

Example 9.3.3

Let $\phi(D_x, D_y) = D_y D_x$ and $F(x, y, k) = k$. The boundary conditions are $U(0, y) = c_1 \sin y$, $U_x(x, 1) = c_2 x^2$. All the k_1, c_1 and c_2 are non–negative. Assume $x \in I_1 = [0, M_1]$, $M_1 > 0$ and $y \in I_2 = [0, M_2]$, $M_2 > 1$. This is not exactly a boundary value problem since the boundary value of $y = 1$ in $U_x(x, 1)$ is inside the interval for y. However, this example will show R_e a proper subset of R_c with R_e non–empty.

A solution is G equal to

$$kx(y-1) + c_2 \frac{x^3}{3} + c_1 \sin y \ . \tag{9.64}$$

So, R_e will be $I_1 \times [1, M_2]$ since for $y > 1$, $\partial G/\partial k > 0$ and $\partial F/\partial k > 0$ (Theorem 9.1). R_e contains no part of $I_1 \times [0, 1)$ since there $\partial G/\partial k < 0$.

Now let us look for R_c. We need only look for (x, y) in $I_1 \times [0, 1)$ because $R_e \subseteq R_c$. Solving

$$\frac{\partial^2}{\partial y \partial x} y_1(x, y, \alpha) = k_1(\alpha) \ , \tag{9.65}$$

$$\frac{\partial^2}{\partial y \partial x} y_2(x, y, \alpha) = k_2(\alpha) \ , \tag{9.66}$$

subject to the fuzzy boundary conditions we obtain

$$y_i(x, y, \alpha) = k_i(\alpha) x(y-1) + c_{2i}(\alpha) \frac{x^3}{3} + c_{1i}(\alpha) \sin y \ , \tag{9.67}$$

for $i = 1, 2$. As in Example 9.3.2 let us assume $k_1'(\alpha) = c_{11}'(\alpha) = c_{21}'(\alpha) = 1$, $k_2'(\alpha) = c_{12}'(\alpha) = c_{22}'(\alpha) = -1$. So for $\partial y_1/\partial \alpha > 0$ and $\partial y_2/\partial \alpha < 0$ we need to have

$$x(y-1) + \frac{x^3}{3} + \sin y > 0 \ , \tag{9.68}$$

for $(x, y) \in [0, M_1] \times [0, 1)$. If equation (9.68) is true for an (x, y) in $[0, M_1] \times [0, 1)$ we then add it to R_c. Notice that equation (9.68) is true for $x = 2$ and $y = 0.5$. Hence $[0, M_1] \times [0, 1)$ contains points in R_c and $R_c \neq R_e$.

9.4 Summary and Conclusions

We considered two solutions, the classical and the extension principle, to elementary fuzzy partial differential equations. We were able to fuzzify both the boundary conditions and the non-homogeneous part of the equation. We showed, if all the constants in the partial differential operator are positive, that the classical solution is more general than the extension principle solution.

The theory lacks meaningful applications since we are unable, at this point, to fuzzify the classical (heat equation, wave equation) partial differential equations of physics and engineering. To accomplish this, for the extension principle solution, we first need to develop a theory of infinite series of fuzzy functions (fuzzy Fourier series).

Bibliography

[1] J.J. Buckley and Th. Feuring: Introduction to Fuzzy Partial Differential Equations, Fuzzy Sets and Systems, 105 (1999), pp. 241–248.

Chapter 10

Fuzzy Eigenvalues

The fuzzy eigenvalue problem is to solve the equation

$$\overline{A}\,\overline{X} = \overline{\lambda}\,\overline{X}\ , \tag{10.1}$$

for fuzzy eigenvalue $\overline{\lambda}$, where: (1) $\overline{A} = [\overline{a}_{ij}]$ is a $n \times n$ fuzzy matrix whose elements $\overline{a}_{ij}$ are triangular fuzzy numbers; (2) $\overline{X}^t = (\overline{x}_1, \ldots, \overline{x}_n)$ is a fuzzy eigenvector, not the zero vector, whose components $\overline{x}_i$ are triangular shaped fuzzy numbers; and (3) $\overline{\lambda}$ is a triangular shaped fuzzy number. Once we solve for $\overline{\lambda}$ we use equation (10.1) to solve for $\overline{X}$.

As in Chapters 3, 4, 7-9 we will consider three types of solutions for $\overline{\lambda}$: (1) the "classical" solution $\overline{\lambda}_c$; (2) the extension principle solution $\overline{\lambda}_e$; and (3) the α–cut and interval arithmetic solution $\overline{\lambda}_I$. These solutions are presented in the next section. We only consider symmetric fuzzy matrices because crisp (non–fuzzy) symmetric matrices only have real eigenvalues and eigenvectors, and we wish to avoid fuzzy complex numbers ([2], [4], [5]) in this book. The development in Section 10.1 is based on [8].

We will consider two applications of fuzzy eigenvalues: (1) fuzzy input–output analysis (first discussed in Chapter 3) in Section 10.2; and (2) fuzzy hierarchical analysis in Section 10.3. Section 10.2 is based on [3] while Section 10.3 is based on ([1], [6], [7]) and these references ([9], [11]) are also important for fuzzy hierarchical analysis.

10.1 Fuzzy Eigenvalue Problem

We first illustrate the solution methods for a 2×2 symmetric fuzzy matrix

$$\overline{A} = \begin{bmatrix} \overline{a} & \overline{b} \\ \overline{b} & \overline{c} \end{bmatrix}, \tag{10.2}$$

and generalize to $n \times n$ symmetric fuzzy matrices at the end of this section.

The "classical" solution method first fuzzifies the eigenvalue problem giving the fuzzy matrix equation

$$\begin{bmatrix} \overline{a} & \overline{b} \\ \overline{b} & \overline{c} \end{bmatrix} \begin{bmatrix} \overline{x}_1 \\ \overline{x}_2 \end{bmatrix} = \overline{\lambda} \begin{bmatrix} \overline{x}_1 \\ \overline{x}_2 \end{bmatrix} . \tag{10.3}$$

We substitute α–cuts for $\overline{a}$, $\overline{b}$, $\overline{c}$, $\overline{x}_1$, $\overline{x}_2$ and $\overline{\lambda}$ and evaluate using interval arithmetic. To use interval arithmetic we need to know if $\overline{a}$, $\overline{b}$, $\overline{c}$, $\overline{x}_1$, $\overline{x}_2$ and $\overline{\lambda}$ are positive or negative. Let us assume that $\overline{a} > 0$, $\overline{b} > 0$, $\overline{c} < 0$, $\overline{x}_1 \geq 0$, $\overline{x}_2 \geq 0$ and $\overline{\lambda} > 0$. We made these assumptions because the specific example we will be working with has $\overline{a} = (3/4/5)$, $\overline{b} = (2/3/4)$ and $\overline{c} = (-5/-4/-3)$. Using the vertices of the triangular fuzzy numbers we obtain the crisp symmetric matrix

$$A = \begin{bmatrix} 4 & 3 \\ 3 & -4 \end{bmatrix} , \tag{10.4}$$

with eigenvalues $\lambda_1 = 5$ and $\lambda_2 = -5$ and normalized eigenvectors $v_1^t = (3/\sqrt{10}, 1/\sqrt{10})$ for λ_1 and $v_2^t = (1/\sqrt{10}, -3/\sqrt{10})$ for λ_2. If we concentrate on the largest positive eigenvalue, then we expect the solution to equation (10.3) to give $\overline{\lambda}$ a triangular shaped fuzzy number centered at $x = 5$, $\overline{x}_1$ triangular shaped centered at $3/\sqrt{10}$, and $\overline{x}_2$ triangular shaped centered at $1/\sqrt{10}$. Hence, we assumed $\overline{x}_1 \geq 0$, $\overline{x}_2 \geq 0$ and $\overline{\lambda} \geq 0$.

Let $\overline{a}[\alpha] = [a_1(\alpha), a_2(\alpha)]$, $\overline{b}[\alpha] = [b_1(\alpha), b_2(\alpha)]$, $\overline{c}[\alpha] = [c_1(\alpha), c_2(\alpha)]$, $\overline{x}_1[\alpha] = [x_{11}(\alpha), x_{12}(\alpha)]$, $\overline{x}_2[\alpha] = [x_{21}(\alpha), x_{22}(\alpha)]$ and $\overline{\lambda}[\alpha] = [\lambda_1(\alpha), \lambda_2(\alpha)]$. We substitute these intervals into equation (10.3) and simplify, using interval arithmetic and using $a_1(\alpha) > 0$, $b_1(\alpha) > 0$, $c_2(\alpha) < 0$, $x_{11}(\alpha) \geq 0$, $x_{21}(\alpha) \geq 0$ $\lambda_1(\alpha) > 0$, producing the matrix equation

$$EW = 0 , \tag{10.5}$$

where $W^t = (x_{11}(\alpha), x_{12}(\alpha), x_{21}(\alpha), x_{22}(\alpha))$ and

$$E = \begin{bmatrix} a_1(\alpha) - \lambda_1(\alpha) & 0 & b_1(\alpha) & 0 \\ 0 & a_2(\alpha) - \lambda_2(\alpha) & 0 & b_2(\alpha) \\ b_1(\alpha) & 0 & -\lambda_1(\alpha) & c_1(\alpha) \\ 0 & b_2(\alpha) & c_2(\alpha) & -\lambda_2(\alpha) \end{bmatrix} . \tag{10.6}$$

For there to be a non-zero solution for W we need $\det(E) = 0$. The determinant of E is

$$\begin{aligned} \det(E) &= \left[(a_1(\alpha) - \lambda_1(\alpha))\lambda_1(\alpha) + b_1^2(\alpha)\right] \left[(a_2(\alpha) - \lambda_2(\alpha))\lambda_2(\alpha) + b_2^2(\alpha)\right] \\ &\quad -(a_1(\alpha) - \lambda_1(\alpha))(a_2(\alpha) - \lambda_2(\alpha))c_1(\alpha)c_2(\alpha) . \end{aligned} \tag{10.7}$$

We set the expression in equation (10.7) to zero and solve for $\lambda_1(\alpha)$ and $\lambda_2(\alpha)$ for $0 \leq \alpha < 1$. If the intervals $[\lambda_1(\alpha), \lambda_2(\alpha)]$ define a triangular shaped fuzzy number $\overline{\lambda}_c$, we say that $\overline{\lambda}_c$ is a fuzzy eigenvalue. What we need is $\partial\lambda_1(\alpha)/\partial\alpha > 0$, $\partial\lambda_2(\alpha)/\partial\alpha < 0$ and $\lambda_1(1) = \lambda_2(1)$.

Now we will look at two specific examples showing first that $\overline{\lambda}_c$ exists (is a fuzzy eigenvalue) and secondly, $\overline{\lambda}_c$ does not (in general) exist. Since the first example is rare and the second example is more likely the case, we conclude too often $\overline{\lambda}_c$ will not exist, and we then will not consider the "classical" solution further in this section.

Example 10.1.1

To simplify the solution to $\det(E) = 0$ let $\overline{c} = 0$ (real zero). Now $a_1(\alpha) = 3 + \alpha$, $a_2(\alpha) = 5 - \alpha$, $b_1(\alpha) = 2 + \alpha$ and $b_2(\alpha) = 4 - \alpha$. Of course, $c_1(\alpha) = c_2(\alpha) = 0$. Using the vertices of $\overline{a}$ and $\overline{b}$ the largest positive eigenvalue is $2 + \sqrt{13}$. Equate equation (10.7) to zero and we get two equations

$$\lambda_1^2(\alpha) - a_1(\alpha)\lambda_1(\alpha) - b_1^2(\alpha) = 0 \; , \tag{10.8}$$

$$\lambda_2^2(\alpha) - a_2(\alpha)\lambda_2(\alpha) - b_2^2(\alpha) = 0 \; . \tag{10.9}$$

Solving for $\lambda_1(\alpha)$, $\lambda_2(\alpha)$, remembering we want the largest positive (fuzzy) eigenvalue, centered at $2 + \sqrt{13}$, we obtain

$$\lambda_1(\alpha) = \frac{(3+\alpha) + \sqrt{5\alpha^2 + 22\alpha + 25}}{2} \; , \tag{10.10}$$

$$\lambda_2(\alpha) = \frac{(5-\alpha) + \sqrt{5\alpha^2 - 42\alpha + 89}}{2} \; . \tag{10.11}$$

One may easily check $\partial\lambda_1(\alpha)/\partial\alpha > 0$, $\partial\lambda_2(\alpha)/\partial\alpha < 0$, $0 \leq \alpha < 1$, and $\lambda_1(1) = \lambda_2(1) = 2 + \sqrt{13}$. Hence, $[\lambda_1(\alpha), \lambda_2(\alpha)]$ defines the α–cuts of a triangular shaped fuzzy eigenvalue $\overline{\lambda}_c$.

Example 10.1.2

Now $\overline{c} = (-5/-4/-3)$. Let us simplify equation (10.7) for $\alpha = 0$. Then $a_1(0) = 3$, $a_2(0) = 5$, $b_1(0) = 2$, etc. The resulting equation is

$$\lambda_1^2\lambda_2^2 - 3\lambda_1\lambda_2^2 - 5\lambda_1^2\lambda_2 - 16\lambda_1^2 - 4\lambda_2^2 + 65\lambda_2 + 123\lambda_1 - 177 \; , \tag{10.12}$$

using $\lambda_1 = \lambda_1(0)$, $\lambda_2 = \lambda_2(0)$. This equation does not factor, as in Example 10.1.1 when $\overline{c} = 0$. The equation in (10.12) has an infinite number of solutions for both λ_1 and λ_2. In fact, its graph gives a curve in the $\lambda_1\lambda_2$–plane. We need a solution pair (λ_1, λ_2) that goes with the largest positive eigenvalue (centered at $x = 5$). There is no obvious solution as in Example 10.1.1. More is required. We need to choose a solution pair $(\lambda_1(\alpha), \lambda_2(\alpha))$ for $0 \leq \alpha < 1$

so that $\partial\lambda_1(\alpha)/\partial\alpha > 0$, $\partial\lambda_2(\alpha)/\partial\alpha < 0$ and $\lambda_1(1) = \lambda_2(1) = 5$. We conclude that in general this is impossible, or at least much too difficult to use.

We therefore proceed on to $\overline{\lambda}_e$, the extension principle solution. This method is based on fuzzifying the crisp solution.

The crisp eigenvalue problem solves the equation $\det(A - \lambda I) = 0$ for values of λ where I is the $n \times n$ identity matrix. For the 2×2 case with

$$A = \begin{bmatrix} a & b \\ b & c \end{bmatrix}, \tag{10.13}$$

we get

$$\lambda_1 = g_1(a,b,c) = \frac{(a+c) + \sqrt{(a-c)^2 + 4b^2}}{2}, \tag{10.14}$$

$$\lambda_2 = g_2(a,b,c) = \frac{(a+c) - \sqrt{(a-c)^2 + 4b^2}}{2}. \tag{10.15}$$

We will concentrate on fuzzifying the largest positive eigenvalue (equation (10.14)), so $\overline{\lambda}_1 = g_1(\overline{a}, \overline{b}, \overline{c})$ to be evaluated using the extension principle. If $[\lambda_{11}(\alpha), \lambda_{12}(\alpha)]$ are to be α–cuts of $\overline{\lambda}_1$ we know that

$$\lambda_{11}(\alpha) = \min\left\{g_1(a,b,c) | a \in \overline{a}[\alpha], b \in \overline{b}[\alpha], c \in \overline{c}[\alpha]\right\}, \tag{10.16}$$

$$\lambda_{12}(\alpha) = \max\left\{g_1(a,b,c) | a \in \overline{a}[\alpha], b \in \overline{b}[\alpha], c \in \overline{c}[\alpha]\right\}. \tag{10.17}$$

This produces a triangular shaped fuzzy number which we call $\overline{\lambda}_e$, a fuzzy eigenvalue. In a similar manner we can fuzzify λ_2.

Next we wish to find a fuzzy eigenvector corresponding to $\overline{\lambda}_e$. We fuzzify the crisp solution. Solve

$$\begin{bmatrix} a & b \\ b & c \end{bmatrix} \begin{bmatrix} x_1 \\ x_2 \end{bmatrix} = \lambda_1 \begin{bmatrix} x_1 \\ x_2 \end{bmatrix} \tag{10.18}$$

for x_1 and x_2. We will assume the eigenvectors are normalized which means $x_1^2 + x_2^2 = 1$. The solutions are

$$x_1 = f_1(a,b,\lambda_1) = \frac{b}{\sqrt{b^2 + (\lambda_1 - a)^2}}, \tag{10.19}$$

$$x_2 = f_2(a,b,\lambda_1) = \frac{\lambda_1 - a}{\sqrt{b^2 + (\lambda_1 - a)^2}}. \tag{10.20}$$

Fuzzify the expressions for x_1 and x_2 giving $\overline{x}_1 = f_1(\overline{a}, \overline{b}, \overline{\lambda}_1)$ and $\overline{x}_2 = f_2(\overline{a}, \overline{b}, \overline{\lambda}_1)$. We obtain their α–cuts as before. For example, if $\overline{x}_1[\alpha] = [x_{11}(\alpha), x_{12}(\alpha)]$, then

$$x_{11}[\alpha] = \min\left\{f_1(a,b,\lambda_1) | a \in \overline{a}[\alpha], b \in \overline{b}[\alpha], \lambda_1 \in \overline{\lambda}_e[\alpha]\right\}, \tag{10.21}$$

$$x_{12}[\alpha] = \max\left\{f_1(a,b,\lambda_1) | a \in \overline{a}[\alpha], b \in \overline{b}[\alpha], \lambda_1 \in \overline{\lambda}_e[\alpha]\right\}. \tag{10.22}$$

Similarly, we obtain α–cuts of $\overline{x}_2$, both triangular shaped fuzzy numbers.

Let $\overline{X}^t = (\overline{x}_1, \overline{x}_2)$. Now we check to see if our solutions satisfy the original fuzzy matrix equation

$$\overline{A}\,\overline{X} = \overline{\lambda}_e\,\overline{X}\ , \tag{10.23}$$

where $\overline{\lambda}_e$ is computed from equations (10.16) and (10.17) and $\overline{x}_1$ from equations (10.21) and (10.22). How are we to evaluate equation (10.23)? Using α–cuts and interval arithmetic? Substitute the α–cuts into equation (10.23) and we obtain two interval equations. The first equation is

$$[a_1(\alpha), a_2(\alpha)][x_{11}(\alpha), x_{12}(\alpha)] + [b_1(\alpha), b_2(\alpha)][x_{21}(\alpha), x_{22}(\alpha)]$$
$$= [\lambda_{11}(\alpha), \lambda_{12}(\alpha)][x_{11}(\alpha), x_{12}(\alpha)]\ . \tag{10.24}$$

This equation (10.24) is not true in general. So let us explain in what way equations (10.24) and (10.23) are valid. First choose any $a \in \overline{a}[\alpha]$, $b \in \overline{b}[\alpha]$ and $c \in \overline{c}[\alpha]$. This determines a λ^* in $\overline{\lambda}_e[\alpha]$ from equation (10.14). Now using these a, b, c and λ^* we calculate x_1^* from equation (10.19) and x_2^* from equation (10.20). Then

$$\begin{bmatrix} a & b \\ b & c \end{bmatrix} \begin{bmatrix} x_1^* \\ x_2^* \end{bmatrix} = \lambda^* \begin{bmatrix} x_1^* \\ x_2^* \end{bmatrix}\ . \tag{10.25}$$

Also, as we allow a to vary throughout $\overline{a}[\alpha]$, b vary within $\overline{b}[\alpha]$ and c to be any value in $\overline{c}[\alpha]$, λ^* varies throughout $\overline{\lambda}_e[\alpha]$ and x_1^* (x_2^*) takes on all values in $\overline{x}_1[\alpha]$ ($\overline{x}_2[\alpha]$), $0 \le \alpha < 1$. In this sense equations (10.23) and (10.24) are true.

Lastly, we look at the α–cut and interval arithmetic solution $\overline{\lambda}_I$. We expect, as in previous chapters, $\overline{\lambda}_e \le \overline{\lambda}_I$ and the same inequality for their corresponding fuzzy eigenvectors. We usually use $\overline{\lambda}_I$ to approximate $\overline{\lambda}_e$. However, we now present an algorithm for $\overline{\lambda}_e$, for any $n \times n$ symmetric fuzzy matrix, so we will not consider $\overline{\lambda}_I$ further in this chapter.

10.1.1 Algorithm

Let $\overline{A} = [\overline{a}_{ij}]$ be a $n \times n$ symmetric fuzzy matrix whose elements $\overline{a}_{ij}$ are all triangular fuzzy numbers. If the vertices of the $\overline{a}_{ij}$ are at a_{ij}, then set $A = [a_{ij}]$. The eigenvalues of A are all real so let them be ordered $\lambda_1 > \lambda_2 > \cdots > \lambda_r$, where λ_i has multiplicity m_i, $1 \le i \le r$, and the sum of the m_i equals n.

We will focus on fuzzifying λ_1 the largest positive eigenvalue. We make this decision for two reasons: (1) the largest positive eigenvalue is important in applications (see Section 10.3); and (2) if we focus on λ_k, $1 < k \le r$, it may be difficult to keep track of λ_k during the fuzzification process. To better explain the second point mentioned above consider the following example of a 5×5 symmetric A. Let the eigenvalues be $\lambda_1 > \lambda_2 > \lambda_3 > \lambda_4 > \lambda_5$, all $m_i = 1$, and we focus on λ_3. As we allow the a_{ij} to wander throughout $\overline{a}_{ij}[\alpha]$,

$0 \leq \alpha < 1$, we may get the following eigenvalue structure: (1) for some a_{ij} values the eigenvalues are $\lambda_1 > \lambda_2 > \lambda_3 > \lambda_4$ with $m_2 = 2$ and $m_i = 1$ otherwise; and (2) for other a_{ij} values the eigenvalues are $\lambda_1 > \lambda_2 > \lambda_3$, $m_2 = 3$, $m_i = 1$ otherwise. Where does λ_3 go? So as to avoid this problem of always identifying λ_3 in the fuzzification process, we shall only fuzzify λ_1.

Let $\mathcal{A}$ be an algorithm that computes λ_1, to desired accuracy. Assume that λ_1 has multiplicity one. There are formulas for λ_1 when $n = 2, 3, 4$ but, in general, there are no formulas if $n \geq 5$. For this reason we require a numerical computation procedure to approximate λ_1, for all $n \geq 2$. Therefore, $\mathcal{A}(a_{11}, \ldots, a_{nn}) \approx \lambda_1$ for $a_{ij} \in \overline{a}_{ij}[\alpha]$.

Let v_1 be the normalized eigenvector corresponding to λ_1. Let $\mathcal{B}$ be an algorithm that will approximate v_1. That is, $\mathcal{B}(a_{11}, \ldots, a_{nn}, \lambda_1) \approx v_1$, $a_{ij} \in \overline{a}_{ij}[\alpha]$ and $\lambda_1 \in \overline{\lambda}_e[\alpha]$. Set $v_1^t = (v_{11}, v_{12}, \ldots, v_{1n})$ and $\mathcal{B} = (\mathcal{B}_1, \ldots, \mathcal{B}_n)$ so that $\mathcal{B}_i(a_{11}, \ldots, a_{nn}, \lambda_1) \approx v_{1i}$, $1 \leq i \leq n$.

Given $\mathcal{A}$ and $\mathcal{B}$ we may now fuzzify the process to compute $\overline{\lambda}_e$ and its associated, normalized, eigenvector $\overline{X}^t = (\overline{x}_1, \ldots, \overline{x}_n)$. We approximate the α-cuts of $\overline{\lambda}_e$ as follows:

$$\lambda_{11}(\alpha) = \min\{\mathcal{A}(a_{11}, \ldots, a_{nn}) | a_{ij} \in \overline{a}_{ij}[\alpha]\}, \tag{10.26}$$

$$\lambda_{12}(\alpha) = \max\{\mathcal{A}(a_{11}, \ldots, a_{nn}) | a_{ij} \in \overline{a}_{ij}[\alpha]\}, \tag{10.27}$$

for $0 \leq \alpha < 1$. We will use our evolutionary algorithm to estimate the min in equation (10.26) and the max in equation (10.27).

To estimate the corresponding fuzzy eigenvector we compute

$$x_{i1}(\alpha) = \min\left\{\mathcal{B}_i(a_{11}, \ldots, a_{nn}, \lambda) | a_{ij} \in \overline{a}_{ij}[\alpha], \lambda \in \overline{\lambda}_e[\alpha]\right\}, \tag{10.28}$$

$$x_{i2}(\alpha) = \max\left\{\mathcal{B}_i(a_{11}, \ldots, a_{nn}, \lambda) | a_{ij} \in \overline{a}_{ij}[\alpha], \lambda \in \overline{\lambda}_e[\alpha]\right\}, \tag{10.29}$$

$1 \leq i \leq n$, $0 \leq \alpha < 1$, where $\overline{x}_i[\alpha] = [x_{i1}(\alpha), x_{i2}(\alpha)]$. The evolutionary algorithm is also employed in equations (10.28) and (10.29).

Example 10.1.3

We first test our evolutionary algorithm on a 2×2 symmetric fuzzy matrix because results can be checked by hand calculations. Let $\overline{a} = (3/4/5)$, $\overline{b} = (2/3/4)$ and $\overline{c} = (-5/-4/-3)$ as were used at the beginning of this section. The algorithm $\mathcal{A}$ is simply equation (10.14) for computing λ_1. The algorithm $\mathcal{B}$ is equation (10.19) for $\mathcal{B}_1$ and equation (10.20) for $\mathcal{B}_2$.

First of all, we may easily find α-cuts $\overline{\lambda}_e$ from equations (10.16) and (10.17) since from equation (10.14), we get $\partial g_1/\partial a > 0$ $\partial g_1/\partial b > 0$, $\partial g_1/\partial c > 0$. Hence, $\lambda_{11}(\alpha) = g_1(a_1(\alpha), b_1(\alpha), c_1(\alpha))$, $\lambda_{12}(\alpha) = g_1(a_2(\alpha), b_2(\alpha), c_2(\alpha))$. However, α-cuts of x_1 (equations (10.21) and (10.22)) and x_2 are more difficult to compute because: (1) $\partial f_1/\partial b > 0$ but $\partial f_1/\partial a$ and $\partial f_1/\partial \lambda_1$ can be positive or negative; and (2) $\partial f_2/\partial a < 0$, $\partial f_2/\partial \lambda_1 > 0$ but $\partial f_2/\partial b$ may

α	Exact	Algorithm
0.0	[3.47,6.66]	[3.50,6.61]
0.25	[3.84,6.23]	[3.85,6.23]
0.5	[4.22,5.82]	[4.23,5.82]
0.75	[4.60,5.40]	[4.61,5.40]
1.0	5	5

Table 10.1: Alpha–cuts of the Fuzzy Eigenvalue $\overline{\lambda}_e$ in Example 10.1.3

α	$\overline{x}_1[\alpha]$	$\overline{x}_2[\alpha]$
0.0	[0.894,0.981]	[0.189, 0.446]
0.5	[0.923,0.979]	[0.254,0.382]
1.0	0.948	0.316

Table 10.2: Alpha–cuts of the Fuzzy Eigenvector $\overline{X}$ in Example 10.1.3

be positive or negative. Hence, we let the evolutionary algorithm find these α–cuts.

Table 10.1 shows our results for $\overline{\lambda}_e$ and Table 10.2 gives the α–cuts of the fuzzy eigenvector. Calculations were made for $\alpha = 0.0, 0.25, 0.5, 0.75$ and 1.0 in Table 10.1. We see the algorithm performs very well on finding fuzzified λ_1 so we will in the next example apply it to a 5×5 fuzzy symmetric matrix.

In Table 10.2 the fuzzy eigenvector is $\overline{X}^t = (\overline{x}_1, \overline{x}_2)$ with α–cuts of the $\overline{x}_i$ shown for α equal to zero, 0.5 and one. How can we check the results in Table 10.2 by hand calculations? First choose $\alpha \in [0, 1)$. Assume $\alpha = 0$. Next pick $a = 3.5 \in \overline{a}[0]$, $b = 2.5 \in \overline{b}[0]$ and $c = -4.5 \in \overline{c}[0]$. Compute λ_1 from $g_1(a, b, c)$ in equation (10.14) and obtain $\lambda_1 = 4.217 \in \overline{\lambda}_1[0]$ in Table 10.1. Now calculate x_1 (x_2) from equation (10.19) ((10.20)) and we find that $x_1 = 0.9612 \in \overline{x}_1[0]$ ($x_2 = 0.2757 \in \overline{x}_2[0]$) in Table 10.2.

Example 10.1.4

Let us now try our evolutionary algorithm on the 5×5 symmetric fuzzy matrix

$$\overline{A} = \begin{bmatrix} \overline{a}_{11} & \overline{a}_{12} & 0 & 0 & 0 \\ \overline{a}_{12} & \overline{a}_{22} & 0 & 0 & 0 \\ 0 & 0 & \overline{a}_{33} & 0 & 0 \\ 0 & 0 & 0 & \overline{a}_{44} & \overline{a}_{45} \\ 0 & 0 & 0 & \overline{a}_{54} & \overline{a}_{55} \end{bmatrix}, \tag{10.30}$$

where $\overline{a}_{11} = (3/4/5)$, $\overline{a}_{12} = \overline{a}_{21} = (2/3/4)$, $\overline{a}_{22} = (-5/-4/-3)$, $\overline{a}_{45} = \overline{a}_{54} = \overline{a}_{55} = (0/1/2)$, $\overline{a}_{33} = (-3/-2/-1)$ and $\overline{a}_{44} = (-2/-1/0)$. Using

α	$\overline{\lambda}_e$	$\overline{x}_1$	$\overline{x}_2$	$\overline{x}_3$	$\overline{x}_4$	$\overline{x}_5$
0.0	[3.483,6.595]	[0.894,0.980]	[0.190,0.429]	[-0.018,0.014]	[-0.019,0.027]	[-0.013,0.021]
$\frac{1}{3}$	[3.981,6.089]	[0.914,0.971]	[0.230,0.384]	[-0.012,0.009]	[-0.013,0.018]	[-0.009,0.013]
$\frac{2}{3}$	[4.760,5.538]	[0.933,0.960]	[0.273,0.354]	[-0.007,0.005]	[-0.008,0.009]	[-0.005,0.006]
1	5	0.948	0.316	0.0	0.0	0.0

Table 10.3: Alpha–cuts of Fuzzy Eigenvalue and Eigenvector for Example 10.1.4

the vertices of the $\overline{a}_{ij}$ we get a crisp symmetric matrix whose eigenvalues are $-5, -2, -\sqrt{2}, \sqrt{2}, 5$. The corresponding normalized eigenvector for $\lambda_1 = 5$ is $v^t = (1/\sqrt{10})(3, 1, 0, 0, 0)$.

The results of our evolutionary algorithm for this problem are shown in Table 10.3 for $\alpha = 0, 1/3, 2/3, 1.0$. The normalized fuzzy eigenvector corresponding to $\overline{\lambda}_e$ is $\overline{X}^t = (\overline{x}_1, \overline{x}_2, \overline{x}_3, \overline{x}_4, \overline{x}_5)$ with the α–cuts of the $\overline{x}_i$ in Table 10.3.

10.2 Fuzzy Input–Output Analysis

In this section we first discuss, and then fuzzify, Leontief's closed input–output analysis applied to a model of the world's economy. The closed model involves calculating (fuzzy) eigenvalues and eigenvectors. We have previously discussed Leontief's open fuzzy input–output model of an economy in Chapter 3. The open model involves solving systems of fuzzy linear equations.

Let us review Leontief's closed input–output analysis applied to a model of the world's economy. Let n denote the total number of countries in the world and we assume all income to any country comes only from the sale of goods internally or the sale of goods to other countries [10]. Let $a_{ij} \geq 0$ be the fraction of country j's income spent on goods from country i. If $A = [a_{ij}]$, then the column sums of A are equal to one. Let x_i denote the annual income for country i. It follows that

$$\sum_{j=1}^{n} a_{ij} x_j = x_i \ , \tag{10.31}$$

for $1 \leq i \leq n$. This is a closed system in that there are no surpluses, each country spends all of its income. In matrix notation we have

$$Ax = x \ , \tag{10.32}$$

where $x^t = (x_1, \ldots, x_n)$. For this model to be consistent A must have an eigenvalue equal to one and a corresponding eigenvector x with $x_i > 0$, for all i. If some x_i is zero, then effectively the i–th country does not exist. The column sums equal to one guarantees an eigenvalue of one [10] and if A is irreducible, it will have an eigenvector x with $x_i > 0$, for all i. We define irreducible below.

In practice the a_{ij}'s are hard to determine and are often estimated. We now propose to let an a_{ij} be a fuzzy positive fraction when its precise value is not known. We will write $\overline{a}_{ij}$ for all the elements in $\overline{A}$ even though some may be real numbers. The constraints on the $\overline{a}_{ij}$ are: (1) $0 \leq a_{ij1} < a_{ij2} \leq a_{ij3} < a_{ij4} \leq 1$ if $\overline{a}_{ij}$ is a trapezoidal fuzzy number; (2) $0 \leq a_{ij} \leq 1$ if $\overline{a}_{ij}$ equals the real number a_{ij}; and (3) $s_{j2} \leq 1 \leq s_{j3}$ if $(s_{j1}/s_{i2}, s_{j3}/s_{j4}) = \overline{S}_j = \sum_{i=1}^{n} \overline{a}_{ij}$, $1 \leq j \leq n$. The last constraint says that the column sums of $\overline{A}$ are approximately one. This fuzzy model of the world's economy is consistent if we can solve

$$\overline{A}\,\overline{x} = \overline{\lambda}\,\overline{x}\ , \tag{10.33}$$

for $\overline{\lambda} \approx (\lambda_1/\lambda_2, \lambda_3/\lambda_4) > 0$ with $\lambda_2 \leq 1 \leq \lambda_3$, and $\overline{x}_i > 0$ for all i, where $\overline{x}^t = (\overline{x}_1, \ldots, \overline{x}_n)$.

Now we discuss conditions on $\overline{A}$ for the fuzzy model to be consistent. We will evaluate equation (10.33) using α–cuts and interval arithmetic. Interval arithmetic will be quite simple since all the fuzzy numbers, and their α–cuts, will be non–negative. Let $\overline{a}_{ij}[\alpha] = [a_{ij1}(\alpha), a_{ij2}(\alpha)]$, $\overline{x}_i[\alpha] = [x_{i1}(\alpha), x_{i2}(\alpha)]$ and $\overline{\lambda}[\alpha] = [\lambda_1(\alpha), \lambda_2(\alpha)]$, $0 \leq \alpha \leq 1$. Some of the $\overline{a}_{ij}$ can be real numbers, like zero, and then its α– cut is a single number. From equation (10.33) we obtain two equations

$$\sum_{j=1}^{n} a_{ij1}(\alpha) x_{j1}(\alpha) = \lambda_1(\alpha) x_{i1}(\alpha)\ , \tag{10.34}$$

$$\sum_{j=1}^{n} a_{ij2}(\alpha) x_{j2}(\alpha) = \lambda_2(\alpha) x_{i2}(\alpha)\ , \tag{10.35}$$

for $0 \leq \alpha \leq 1$, $i = 1, \ldots, n$. Let $A_1(\alpha) = [a_{ij1}(\alpha)]$, $A_2(\alpha) = [a_{ij2}(\alpha)]$, $(x_1(\alpha))^t = (x_{11}(\alpha), \ldots, x_{n1}(\alpha))$, $(x_2(\alpha))^t = (x_{12}(\alpha), \ldots, x_{n2}(\alpha))$. In matrix form, equations (10.34) and (10.35), are

$$A_1(\alpha)x_1(\alpha) = \lambda_1(\alpha)x_1(\alpha)\ , \tag{10.36}$$

$$A_2(\alpha)x_2(\alpha) = \lambda_2(\alpha)x_2(\alpha)\ , \tag{10.37}$$

$0 \leq \alpha \leq 1$. From these last two equations we first wish to solve for $\lambda_i(\alpha)$ and then $x_i(\alpha)$, $i = 1, 2$. What we need is $\partial\lambda_1(\alpha)/\partial\alpha > 0$, $\partial\lambda_2(\alpha)/\partial\alpha < 0$. $\lambda_1(1) \leq \lambda_2(1)$ with $0 < \lambda_1(0)$ and $\lambda_1(1) \leq 1 \leq \lambda_2(1)$. For the $x_{ij}(\alpha)$ we also must have $\partial x_{i1}(\alpha)/\partial\alpha > 0$, $\partial x_{i2}(\alpha)/\partial\alpha < 0$, $x_{i1}(1) \leq x_{i2}(1)$ and $x_{i1}(0) > 0$ for all i. With one in the core of $\overline{\lambda}$ we may say that $\overline{\lambda}$ is approximately one, or $\overline{\lambda} \approx 1$.

For all of this to happen, and the fuzzy model is consistent, is that each column of $\overline{A}$ contains a fuzzy number and for $\overline{A}_1(0)$ to be irreducible. A $n \times n$ crisp matrix A is reducible (decomposable) if and only if there is a permutation matrix Π so that

$$\Pi \cdot A \cdot \Pi^{-1} = \begin{pmatrix} A_1 & A_2 \\ 0 & A_3 \end{pmatrix}\ , \tag{10.38}$$

where A_1 and A_3 are square; otherwise A is irreducible (indecomposable) ([10], [12]) A permutation matrix has a single non–zero value of one in each row and column. Notice that if $A_1(0)$ is irreducible so is $A_1(\alpha)$, $0 < \alpha \leq 1$, and $A_2(\alpha)$, $0 \leq \alpha \leq 1$. Another assumption, in place of $A_1(0)$ irreducible, is for $(A_1(0))^m = [a_{ij1}^{(m)}(0)]$ to be strictly positive, or $a_{ij1}^{(m)}(0) > 0$ all i, j, for some positive integer m. If $(A_1(0))^m$ is strictly positive, then $A_1(0)$ is irreducible ([10], [12]) and $(A_1(\alpha))^m$ is strictly positive, $0 < \alpha \leq 1$, and $(A_2(\alpha))^m$ is strictly positive for all α.

It was shown in [3] that if $\overline{A}$ has at least one fuzzy number in each column and $A_1(0)$ is irreducible (or $(A_1(0))^m$ is strictly positive for some m), then the fuzzy model is consistent.

The fuzzy model in equation (10.33) being consistent means that: (1) we can solve equations (10.36) and (10.37) for $\lambda_1(\alpha)$ and $\lambda_2(\alpha)$ so that $[\lambda_1(\alpha), \lambda_2(\alpha)]$ defines the α–cuts of fuzzy number $\overline{\lambda}$, $0 < \lambda_1(0)$, $\lambda_1(1) \leq 1 \leq \lambda_2(1)$; and (2) we then may solve for the $x_{i1}(\alpha)$, $x_{i2}(\alpha)$ so that $[x_{i1}(\alpha), x_{i2}(\alpha)]$ give the α–cuts of fuzzy numbers $\overline{x}_i$, $x_{i1}(0) > 0$, and $\overline{X}^t = (\overline{x}_1, \ldots, \overline{x}_n)$ is the strictly positive fuzzy eigenvalue corresponding to $\overline{\lambda}$. Recall there is also the constraint that the columns of $\overline{A}$ are approximately one.

Why do we need the assumption the $\overline{A}$ has at least one fuzzy number in each column? This is so that we get $a_{ij1}(\alpha) \leq a_{ij2}(\alpha)$ for all α and for all α $a_{ij1}(\alpha) < a_{ij2}(\alpha)$ for at least one i for each j. This was needed in the proof of the theorem that the fuzzy model is consistent in [3].

The assumption for the model to be consistent, in the crisp case and the fuzzy case, were essentially the same. In the crisp model one assumes that A is irreducible (or A^m is strictly positive) to guarantee a strictly positive eigenvector. In the fuzzy model the same assumptions were made on $A_1(0)$ plus each column in $\overline{A}$ has at least one fuzzy number.

In this section we argued that, under certain assumptions, the fuzzy closed input–output model exists (is consistent) using the "classical" method of solution discussed in Section 10.1.

Example 10.2.1

Here we will present a simple numerical example on finding the fuzzy vector $\overline{x}$ for a three country fuzzy input-output model. Let

$$\overline{A} = \begin{pmatrix} 0 & 0.50 & \overline{a}_{13} \\ \overline{a}_{21} & 0 & 0.25 \\ 0.33 & \overline{a}_{32} & 0 \end{pmatrix}, \tag{10.39}$$

where $\overline{a}_{21} = (0.5/0.6, 0.7/0.8)$, $\overline{a}_{32} = (0.4/0.45, 0.55/0.6)$ and $\overline{a}_{13} = (0.6/0.7, 0.8/0.9)$. Then

$$A_1(0) = \begin{pmatrix} 0 & 0.50 & 0.6 \\ 0.5 & 0 & 0.25 \\ 0.33 & 0.4 & 0 \end{pmatrix}, \tag{10.40}$$

α	$\lambda[\alpha]$	$\overline{x}_1[\alpha]$	$\overline{x}_2[\alpha]$	$\overline{x}_3[\alpha]$
0.0	[0.8579,1.1258]	[0.7212,0.7552]	[0.5800,0.6644]	[0.5478,0.5755]
0.50	[0.9002,1.0842]	[0.7275,0.7472]	[0.5967,0.6487]	[0.5484,0.5714]
1.0	[0.9423,1.0424]	[0.7341,0.7395]	[0.6134,0.6329]	[0.5501,0.5680]

Table 10.4: Fuzzy Eigenvalue and Eigenvector in Example 10.2.1

so that $A_1^2(0) > 0$ and $A_1(0)$ is irreducible. We now solve equations (10.36) and (10.37) for $x_i[\alpha]$, $\lambda_i[\alpha]$, $i = 1, 2$ for selected values of α. In $A_1(\alpha)$ we use $0.5 + 0.1\alpha$ for a_{21}, $0.4 + 0.05\alpha$ for a_{32} and $0.6 + 0.1\alpha$ for a_{13}. For $A_2(\alpha)$ we have $0.8 - 0.1\alpha$ for a_{21}, $0.6 - 0.05\alpha$ for a_{32} and $0.9 - 0.1\alpha$ for a_{13}. Solution for $\overline{\lambda}$ is shown in Table 10.4 for $\alpha = 0.0, 0.5, 1.0$. We notice that $\lambda_1(0) > 0$ and $\lambda_1(1) \leq 1 \leq \lambda_2(1)$. Solution for a corresponding fuzzy eigenvector is also in Table 10.4.

To check on the values for the fuzzy eigenvector we computed the eigenvalues and eigenvectors of the crisp matrix

$$E = \begin{pmatrix} 0 & 0.50 & 0.75 \\ 0.67 & 0 & 0.25 \\ 0.33 & 0.50 & 0 \end{pmatrix}. \tag{10.41}$$

Recall that the column sums must equal one. To get this result we chose $a_{13} = 0.75$ ($a_{32} = 0.50$) in the center of the core (where the membership value is one) of $\overline{a}_{13}$ ($\overline{a}_{32}$) but $a_{21} = 0.67$ in the core of $\overline{a}_{21}$ so that the sum of the first column was one. The matrix E has an eigenvalue equal to one with corresponding eigenvector $x^t = (0.7374, 0.6318, 0.5583)$. Notice that the x_i in x are in the core of the $\overline{x}_i$ in Table 10.4.

10.3 Fuzzy Hierarchical Analysis

In the first Section 10.3.1 we review Saaty's method of hierarchical analysis ([14] – [18]). Saaty's procedure involves finding the positive, normalized, eigenvector (called the "weights") corresponding to the largest positive eigenvalue (called $\lambda_{\max}$) of a positive, reciprocal, matrix. Saaty's method is sometimes called the $\lambda_{\max}$–method. We fuzzify Saaty's procedure in Section 10.3.2 but we employ a new method of finding the (positive, normalized) fuzzy eigenvector different from that discussed in Section 10.1. We will abbreviate hierarchical analysis as HA and fuzzy hierarchical analysis as FHA. Section 10.3.2 is based on [7]. See also [13].

Recently there have been a number of papers criticizing the methods used by Saaty in HA. See the references in [7] for a review of the literature. The criticisms include: (1) one should use the geometric means of the rows of a positive reciprocal matrix to calculate the weights and not the normalized eigenvector corresponding to $\lambda_{\max}$; (2) Saaty's measure of consistency for a positive reciprocal matrix is incorrect; (3) the method of aggregating the weights across the criteria is not correct; and (4) Saaty's procedure can produce rank reversals. The geometric mean method uses

$$w_i = \left(\prod_{j=1}^{m} a_{ij} \right)^{\frac{1}{m}} , \tag{10.42}$$

$1 \leq i \leq n$, to get the weights if $A = [a_{ij}]$ is a $m \times m$ positive reciprocal matrix. We fuzzify the geometric row mean method in Subsection 10.3.3. Even though the geometric row mean procedure does not involve eigenvalues or eigenvectors we put it into this section to compare to the alternate method in Section 10.3.2. Section 10.3.3 is based on ([1], [6]).

Acceptance of the geometric row mean method has been slow, and many researchers continue to use Saaty's $\lambda_{\max}$–procedure. So, Sections 10.3.2 and 10.3.3 cover the two basic ways of calculating fuzzy weights in FHA.

A brief summary is in the last Subsection 10.3.4.

10.3.1 The $\lambda_{\max}$–Method

In this section we review the basic computations needed to find the weights in Saaty's hierarchical analysis. In HA a person (expert, judge) is asked to give ratios a_{ij} for each pairwise comparison between issues (alternatives, candidates) $A_1, \ldots, A_m$ for each criterion (objective) in a hierarchy, and also between the criteria. For some specific criterion C_k, if a person considers A_1 more important than A_5, then a_{15} might equal 3/1, or 5/1, or 7/1. The numbers for the ratios will be taken from the set $S = \{1, 2, 3, \ldots, 9\}$ so a_{15} could be s_1/s_5 with s_1, $s_5 \in S$ and $s_1 > s_5$. The ratios a_{ij} indicate, for this expert, the strength with which A_i dominates A_j. If $a_{15} = 5/1$ then $a_{51} = 1/5$. That is, $a_{ij} = (a_{ji})^{-1}$, all i, j, with $a_{ii} = 1$, $1 \leq i \leq m$. Let $\mathcal{A}$ be the $m \times m$ matrix whose entries are the ratios ($a_{ji} = a_{ij}^{-1}$). $\mathcal{A}$ is called a positive reciprocal matrix. Since $\mathcal{A}$ is for criterion C_k we will now write $\mathcal{A}_k$ for this matrix.

Assume there are K criteria $C_1, \ldots, C_K$ with a positive reciprocal matrix $\mathcal{A}_k$ for each C_k, $1 \leq k \leq K$. Also, the judge must give pairwise comparisons of the criteria producing a positive reciprocal matrix $\mathcal{E}$. This hierarchical structure is shown in Figure 10.1. Examples, with actual fuzzy numbers in the $\mathcal{A}_k$ and $\mathcal{E}$, are presented in Subsection 10.3.2.

Next one computes weights $w_k^T = (w_{1k}, \ldots, w_{mk})$ for each $\mathcal{A}_k$ and $e^T = (e_1, \ldots, e_K)$ for $\mathcal{E}$. Given any positive reciprocal matrix $\mathcal{A}$, let the eigenvalues, counting a root of multiplicity n n-times, be $\lambda_1, \ldots, \lambda_m$. There is a dominant (real, positive) eigenvalue, let us call it $\lambda_{\max}$, so that $|\lambda_i| < \lambda_{\max}$ for all $\lambda_i \neq \lambda_{\max}$. Also, $\lambda_{\max}$ is a root of multiplicity one. Corresponding to $\lambda_{\max}$ there is a unique eigenvector $w^T = (w_1, \ldots, w_m)$ so that

$$\mathcal{A}w = \lambda_{\max} w , \tag{10.43}$$

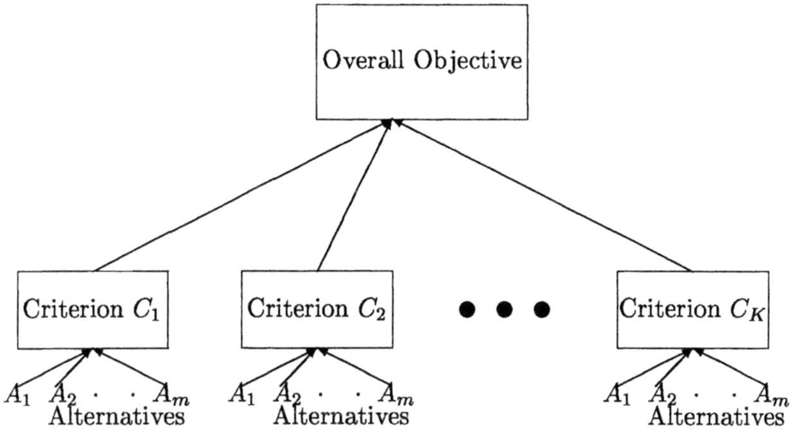

Figure 10.1: Hierarchical Structure

where $w_i > 0$ for all i and $\sum_{i=1}^{m} w_i = 1$. This positive, normalized (sum one), vector w gives the weights for $\mathcal{A}$. Then w_k is the positive, normalized, eigenvector corresponding to $\lambda_{\max}$ for $\mathcal{A}_k$, $1 \leq k \leq K$, and e is the eigenvector for $\mathcal{E}$.

The objective of HA is to rank the alternatives across all the criteria. Then, assuming that the reciprocal matrices $\mathcal{A}_k$, $1 \leq k \leq K$, and $\mathcal{E}$ are reasonably consistent ([14] – [18]), the final ranking of the alternatives is determined by the vector $r^T = (r_1, \ldots, r_m)$ where

$$r_j = \sum_{k=1}^{K} w_{jk} e_k \ , \tag{10.44}$$

$1 \leq j \leq m$. We will discuss consistency for fuzzy hierarchical analysis in Section 10.3.2. The weight for alternative A_j is r_j, $1 \leq j \leq m$. The alternatives are ranked according to the numbers r_j, $1 \leq j \leq m$. The hierarchical structure (Figure 10.1) can be expanded to more levels but we shall consider, in this book, only the three levels shown in Figure 10.1.

In FHA we will have fuzzy numbers in the $\mathcal{A}_k$ and $\mathcal{E}$. Instead of employing the method outlined in Section 10.1 for finding fuzzy $\lambda_{\max}$ and then its corresponding positive, normalized, fuzzy eigenvector for the fuzzy weights, we will directly fuzzify Saaty's method of computing the weight vector. In HA (FHA) what we really want is the weight (fuzzy weight) vector and this method gets this without first finding $\lambda_{\max}$ (fuzzy $\lambda_{\max}$).

Let $1^T = (1, 1, \ldots, 1)$, a vector of length m of all ones, let $\mathcal{A}$ be any positive reciprocal matrix with $sum(l)$ = sum of all the elements in $\mathcal{A}^l$, $l = 1, 2, 3, \ldots$.

Define

$$\lim_{l\to\infty}\left(\frac{A^l \cdot 1}{sum(l)}\right) = z \ , \tag{10.45}$$

then if

$$w = \left(\sum_{i=1}^{m} z_i\right)^{-1} z \ , \tag{10.46}$$

we know that ([14], [16] – [18]) w is the unique, positive, normalized eigenvector of $\mathcal{A}$ corresponding to $\lambda_{\max}$. This gives us a way of computing the weights w_k for $\mathcal{A}_k$ and e of $\mathcal{E}$. We evaluate equation (10.45) for $l = 2$, $l = 4$ (square), $l = 8$ (square again), ... until the vector stabilizes (changes from step-to-step are less than a given $\varepsilon > 0$). Then equation (10.46) produces the approximation to w. This is the procedure we fuzzify in the next section.

10.3.2 Fuzzy $\lambda_{\max}$–Method

The experts are allowed to use fuzzy ratios in place of exact ratios. The $\overline{a}_{ij}$, $i \neq j$, can now be fuzzy numbers in any positive reciprocal matrix. As before, $\overline{a}_{ii} = 1$ all i. The types of fuzzy numbers that can be used in paired comparisons are described by $\overline{a}_{ij} = (\alpha/\beta, \gamma/\delta)$ where α, β, γ, $\delta \in S$, $\alpha \leq \beta \leq \gamma \leq \delta$. There are eight types of fuzzy numbers, with a real number represented by $\alpha = \beta = \gamma = \delta$, and the other seven are shown in Figures 10.2 - 10.8. The judge can input these fuzzy numbers various ways including drawing them or using their verbal equivalents. For the trapezoidal (Figure 10.2) one could say approximately between β to 1 and γ to 1. The triangle (Figure 10.3) is approximately β to 1. If $\overline{a}_{ij} = (\alpha/\beta, \gamma/\delta)$, then $\overline{a}_{ij}^{-1} = \overline{a}_{ji} \approx (\delta^{-1}/\gamma^{-1}, \beta^{-1}/\alpha^{-1})$, the reciprocal of the fuzzy number $\overline{a}_{ij}$. If $\overline{a}_{ij}$ is a trapezoidal fuzzy number, then $\overline{a}_{ij}^{-1}$ is a trapezoidal shaped fuzzy number.

Now we assume the elements in the fuzzy positive reciprocal matrices $\overline{\mathcal{A}}_k$ and $\overline{\mathcal{E}}$ are $\overline{a}_{ij} = (\alpha/\beta, \gamma/\delta)$, $\overline{a}_{ii} = 1$ and $\overline{a}_{ji} = \overline{a}_{ij}^{-1}$. Some of the $\overline{a}_{ij}$ can be real numbers. We now describe how we are going to compute the fuzzy weight vectors $\overline{w}_k$ and $\overline{e}$. There are a number of other issues to be addressed in FHA, like consistency, and how do we obtain the final ranking because now the weight r_j (equation (10.44)) for alternative A_j will be a fuzzy number. These two issues will be considered below. Right now we are only concerned with finding the fuzzy weight vector for a fuzzy positive, reciprocal, matrix.

We compute the fuzzy weight vector by fuzzifying equations (10.45) and (10.46). Let $\overline{\mathcal{A}}$ be a fuzzy positive, reciprocal, matrix. Choose $\alpha \in [0,1]$. Let $\Gamma(\alpha) = \prod\{\ \overline{a}_{ij}[\alpha] \mid 1 \leq i < j \leq m\ \}$ and $v \in \Gamma(\alpha)$ we write as $v = (a_{12}, \ldots, a_{1m}, a_{23}, \ldots, a_{m-1,m})$. Define positive, reciprocal, matrix $\mathcal{A} = [e_{ij}]$ as follows: (1) $e_{ij} = a_{ij}$ if $1 \leq i < j \leq m$; (2) $e_{ii} = 1$, $1 \leq i \leq m$; and (3)

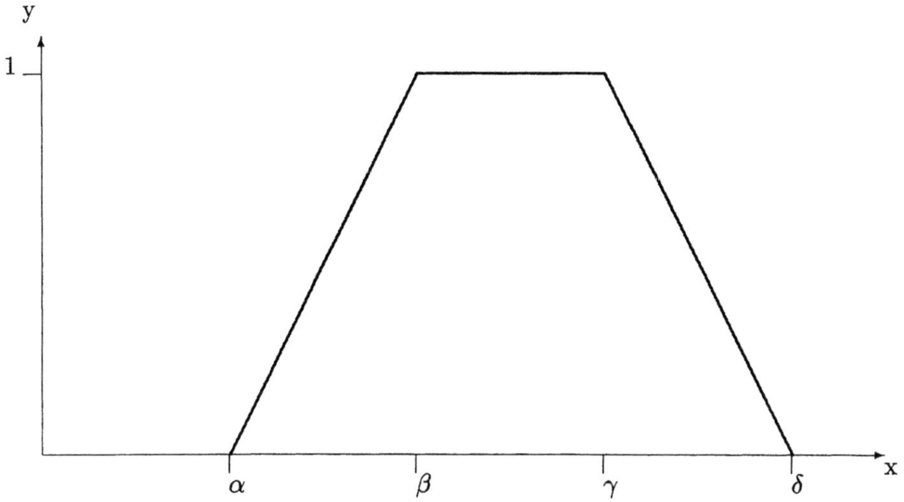

Figure 10.2: Fuzzy Numbers in FHA: Trapezoidal

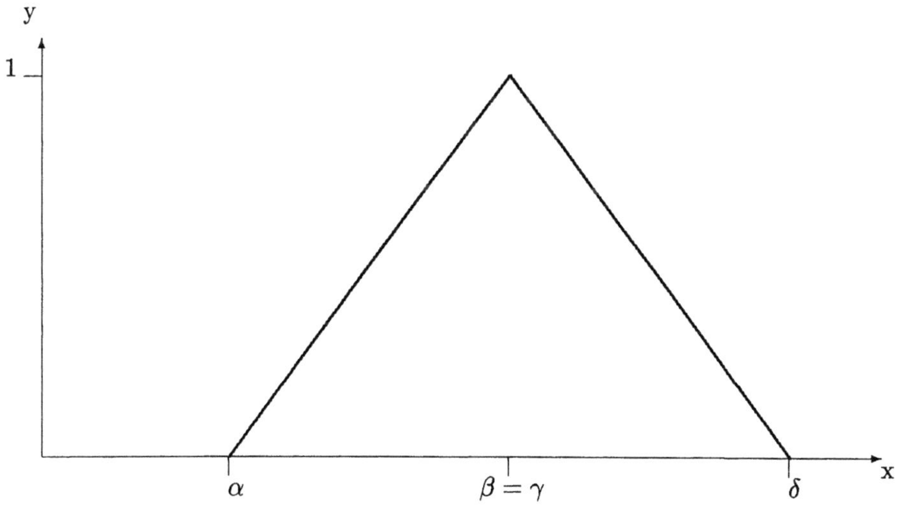

Figure 10.3: Fuzzy Numbers in FHA: Triangle

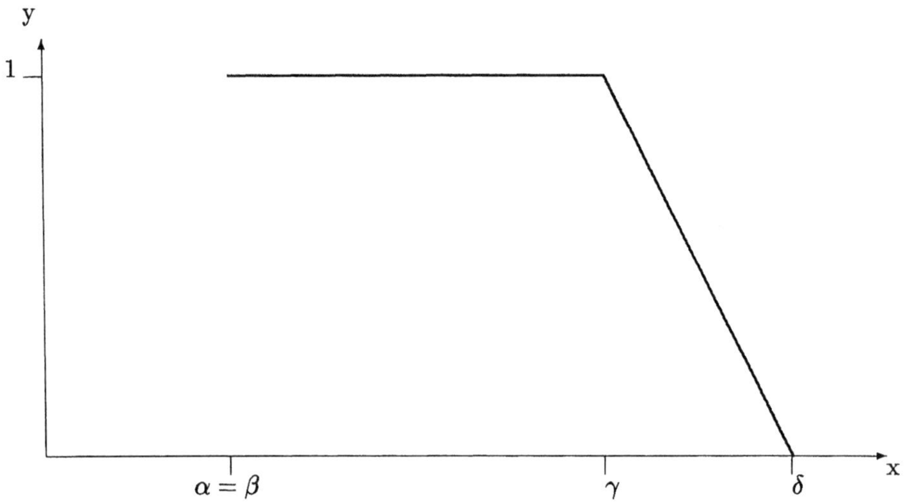

Figure 10.4: Fuzzy Numbers in FHA: More than α to 1

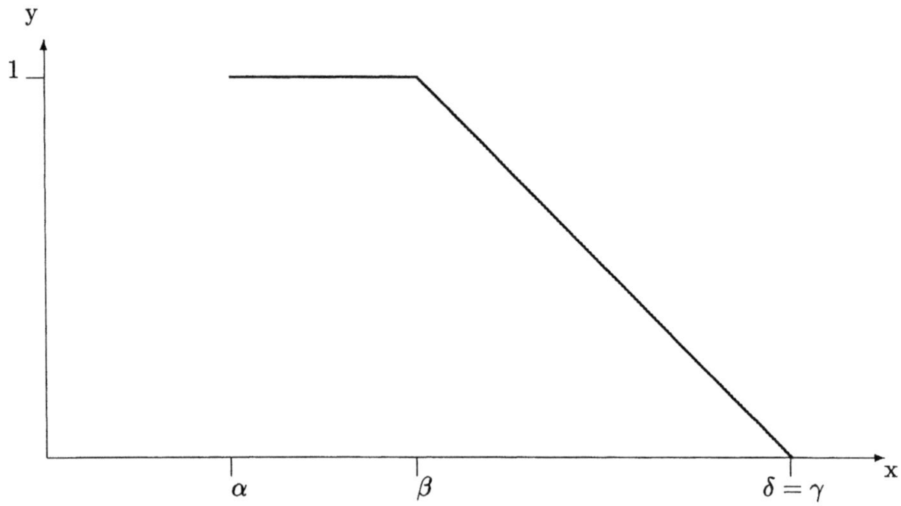

Figure 10.5: Fuzzy Numbers in FHA: Less than δ to 1

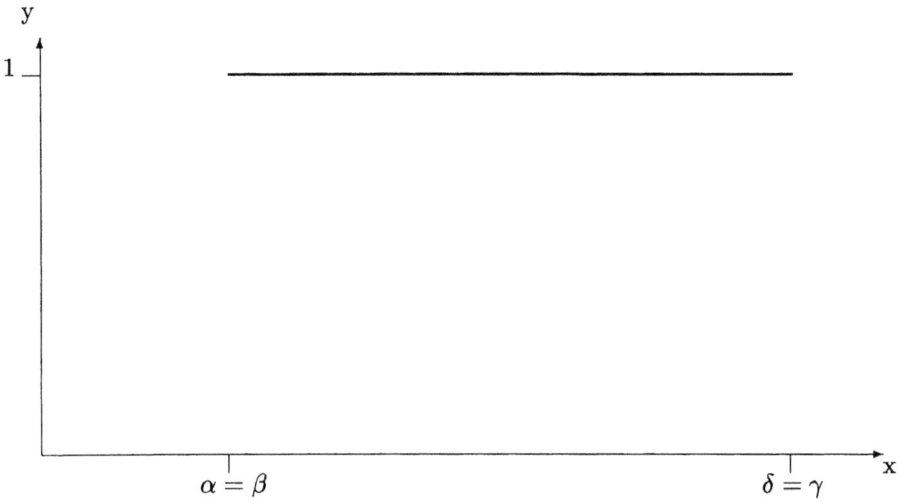

Figure 10.6: Fuzzy Numbers in FHA: Between $\alpha/1$ and $\gamma/1$

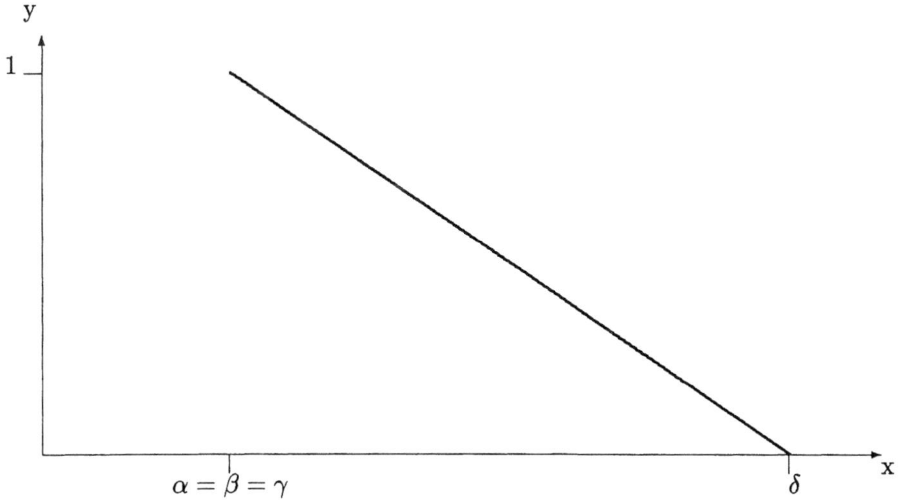

Figure 10.7: Fuzzy Numbers in FHA: At Least $\alpha/1$

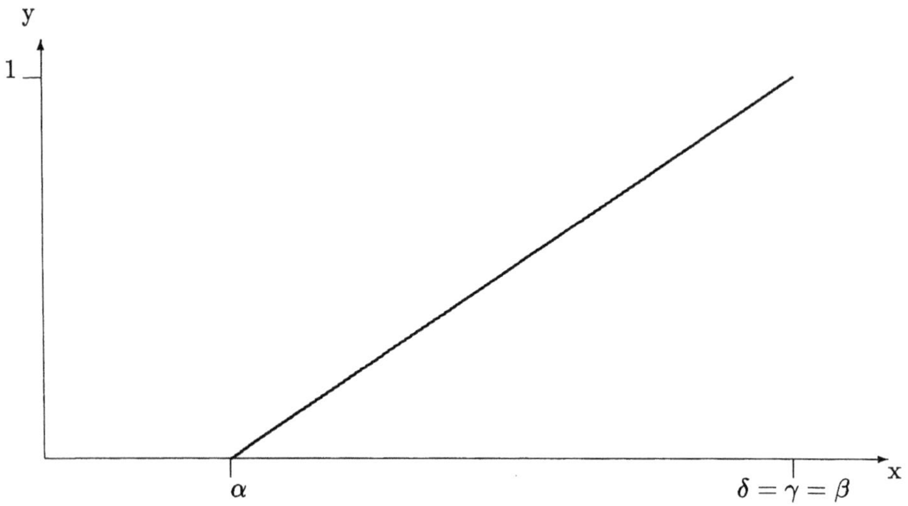

Figure 10.8: Fuzzy Numbers in FHA: At Most $\delta/1$

$e_{ji} = a_{ij}^{-1}$ for $1 \leq i < j \leq m$. Let

$$z = \lim_{l \to \infty} \left(\frac{\mathcal{A}^l \cdot 1}{sum(l)} \right) , \tag{10.47}$$

and define $w_v = (\sum_{i=1}^{m} z_i)^{-1} z$. Set $w_v^T = (w_{v1}, \ldots, w_{vm})$. We have described a continuous mapping $\Phi_i(v) = w_{vi}$, $1 \leq i \leq m$, for each α in $[0, 1]$. So, let

$$w_{i1}(\alpha) = \min\{w_{vi} \mid v \in \Gamma(\alpha)\} , \tag{10.48}$$

$$w_{i2}(\alpha) = \max\{w_{vi} \mid v \in \Gamma(\alpha)\} , \tag{10.49}$$

$1 \leq i \leq m$, for all $\alpha \in [0, 1]$. Then $[w_{i1}(\alpha), w_{i2}(\alpha)]$ becomes α–cuts of fuzzy numbers $\overline{w}_i$, $1 \leq i \leq m$, which produce the fuzzy weight vector $\overline{w}^T = (\overline{w}_1, \ldots, \overline{w}_m)$.

It is no easy job to compute the $w_{i1}(\alpha)$ and $w_{i2}(\alpha)$ so we propose an evolutionary algorithm (EA) to do this job. The basic EA is described in Chapter 13. Let us here briefly describe how the EA works to find $\overline{w}$.

Computing $w_{i1}(\alpha)$ is a complicated, non-linear, optimization problem and EAs are very good search tools for optimization. The search space is $\Gamma(\alpha)$, so members of the population will be vectors v in $\Gamma(\alpha)$. We will estimate $w_{i1}(\alpha)$ ($w_{i2}(\alpha)$) for selected values of α, say $\alpha = 0$, 0.2, 0.4, 0.6, 0.8 and 1. Then for each $i = 1, 2, \ldots, m$ and each $\alpha = 0, 0.2, \ldots, 1$ we run the EA to approximate the minimum of $\Phi_i(v) = w_{i1}(\alpha)$. Then for each $i = 1, 2, \ldots, m$ and each $\alpha = 0.0, 0.2, \ldots, 1.0$ we apply the EA to approximate the maximum of $\Phi_i(v) = w_{i2}(\alpha)$. This produces approximations to $[w_{i1}(\alpha), w_{i2}(\alpha)] = \overline{w}_i[\alpha]$ and the fuzzy weight vector $\overline{w}$ for $\overline{\mathcal{A}}$.

In order to test our procedure we will consider the special case of $m = 3$. When $m = 3$ there is a formula for the unique positive, normalized, eigenvector corresponding to $\lambda_{\max}$. See [7] for another test of our EA for the special case $m = 4$.

Testing the EA when $m = 3$

Let

$$\mathcal{A} = \begin{bmatrix} 1 & a & b \\ a^{-1} & 1 & c \\ b^{-1} & c^{-1} & 1 \end{bmatrix} , \tag{10.50}$$

be the positive, reciprocal, matrix. If $w^T = (w_1, w_2, w_3)$ is the unique positive, normalized, eigenvector corresponding to $\lambda_{\max}$, then we know that [16]

$$w_1 = \frac{a^{1/3}b^{1/3}}{T} = f_1(a,b,c) , \tag{10.51}$$

$$w_2 = \frac{a^{-1/3}c^{1/3}}{T} = f_2(a,b,c) , \tag{10.52}$$

$$w_3 = \frac{b^{-1/3}c^{-1/3}}{T} = f_3(a,b,c) , \tag{10.53}$$

where

$$T = a^{1/3}b^{1/3} + a^{-1/3}c^{1/3} + b^{-1/3}c^{-1/3} . \tag{10.54}$$

We now fuzzify w_1, w_2, w_3 by substituting fuzzy numbers $\overline{a}$ for a, $\overline{b}$ for b, and $\overline{c}$ for c, and use the extension principle to find fuzzy weights $\overline{w}_1$, $\overline{w}_2$ and $\overline{w}_3$.

Let us look more closely on how we are to get $\overline{w}_1$, $\overline{w}_2$, $\overline{w}_3$. Since the f_i are continuous we know that

$$w_{i1}(\alpha) = \min\left\{f_i(a,b,c) \;|a \in \overline{a}[\alpha], b \in \overline{b}[\alpha], c \in \overline{c}[\alpha]\right\} , \tag{10.55}$$

$$w_{i2}(\alpha) = \max\left\{f_i(a,b,c) \;|a \in \overline{a}[\alpha], b \in \overline{b}[\alpha], c \in \overline{c}[\alpha]\right\} , \tag{10.56}$$

for $i = 1,2,3$, $\alpha \in [0,1]$, where $[w_{i1}(\alpha), w_{i2}(\alpha)]$ is an α-cut of $\overline{w}_i$. It is easy to see that: (1) $\partial f_1/\partial a > 0$, $\partial f_1/\partial b > 0$; (2) $\partial f_2/\partial a < 0$, $\partial f_2/\partial c > 0$; and (3) $\partial f_3/\partial b < 0$, $\partial f_3/\partial c < 0$. However, f_1 may be increasing for some c and decreasing for other c. Similarly, for f_2 a function of b and f_3 a function of a. Hence

$$w_{11}(\alpha) = f_1(a_1(\alpha), b_1(\alpha), c^*) , \tag{10.57}$$

$$w_{12}(\alpha) = f_1(a_2(\alpha), b_2(\alpha), c^{**}) , \tag{10.58}$$

for c^*, c^{**} in $\overline{c}[\alpha]$, and

$$w_{21}(\alpha) = f_2(a_2(\alpha), b^*, c_1(\alpha)) , \tag{10.59}$$

$$w_{22}(\alpha) = f_2(a_1(\alpha), b^{**}, c_2(\alpha)) , \tag{10.60}$$

for b^*, b^{**} in $\overline{b}[\alpha]$, and

$$w_{31}(\alpha) = f_3(a^*, b_2(\alpha), c_2(\alpha)) , \tag{10.61}$$

$$w_{32}(\alpha) = f_3(a^{**}, b_1(\alpha), c_1(\alpha)) , \tag{10.62}$$

α	Method I $\overline{w}_1[\alpha]$	Method II $\overline{w}_1[\alpha]$
0	[0.5273, 0.6893]	[0.5267, 0.6908]
0.2	[0.5538, 0.6793]	[0.5535, 0.6806]
0.4	[0.5771, 0.6664]	[0.5765, 0.6695]
0.6	[0.5970, 0.6561]	[0.5965, 0.6575]
0.8	[0.6153, 0.6442]	[0.6142, 0.6445]
1.0	[0.6301, 0.6301]	[0.6301, 0.6301]

Table 10.5: Testing the Evolutionary Algorithm: $\overline{w}_1$

α	Method I $\overline{w}_2[\alpha]$	Method II $\overline{w}_2[\alpha]$
0	[0.1893, 0.2590]	[0.1888, 0.2598]
0.2	[0.1951, 0.2497]	[0.1940, 0.2503]
0.4	[0.2005, 0.2408]	[0.1996, 0.2415]
0.6	[0.2076, 0.2330]	[0.2055, 0.2332]
0.8	[0.2119, 0.2254]	[0.2118, 0.2256]
1.0	[0.2184, 0.2184]	[0.2184, 0.2184]

Table 10.6: Testing the Evolutionary Algorithm: $\overline{w}_2$

α	Method I $\overline{w}_3[\alpha]$	Method II $\overline{w}_3[\alpha]$
0	[0.1206, 0.2131]	[0.1204, 0.2136]
0.2	[0.1259, 0.1958]	[0.1254, 0.1962]
0.4	[0.1313, 0.1820]	[0.1309, 0.1821]
0.6	[0.1374, 0.1698]	[0.1370, 0.1703]
0.8	[0.1442, 0.1599]	[0.1438, 0.1602]
1.0	[0.1515, 0.1515]	[0.1515, 0.1515]

Table 10.7: Testing the Evolutionary Algorithm: $\overline{w}_3$

for a^*, a^{**} in $\overline{a}[\alpha]$, where $\overline{a}[\alpha] = [a_1(\alpha), a_2(\alpha)]$, $\overline{b}[\alpha] = [b_1(\alpha), b_2(\alpha)]$, and $\overline{c}[\alpha] = [c_1(\alpha), c_2(\alpha)]$. In this way we find the $\overline{w}_i$, $1 \leq i \leq 3$.

Let us call the procedure of calculating the weights using the expressions above Method II. Method I will use the EA and equations (10.48) and (10.49).

To test Method I we used both methods on the following positive reciprocal matrix.

$$\mathcal{A} = \begin{bmatrix} 1 & a & b \\ a^{-1} & 1 & 3 \\ b^{-1} & 1/3 & 1 \end{bmatrix}, \tag{10.63}$$

for $a = (5/6, 6/7)$, $b = (1/2, 2/3)$. The results are shown in Tables 10.5 - 10.7. We see that the EA computes a good approximation to the fuzzy weights obtained using Method II.

Application

This application has been developed from an example in ([14], [15]). A recent college graduate has be offered three jobs A_1, A_2, A_3. In order to rank these jobs he evaluates each with respect to five criteria: (1) C_1 = pay; (2) C_2 = benefits; (3) C_3 =location; (4) C_4 = colleagues (fellow workers); and (5) C_5 = potential for advancement. Using FHA he constructs the following fuzzy reciprocal matrices.

$$\overline{\mathcal{A}}_1 = \begin{bmatrix} 1 & (3/3, 5/5)^{-1} & 1/2 \\ (3/3, 5/5) & 1 & (2/3, 3/4) \\ 2 & (2/3, 3/4)^{-1} & 1 \end{bmatrix}, \tag{10.64}$$

for C_1 = pay,

$$\overline{\mathcal{A}}_2 = \begin{bmatrix} 1 & (2/3, 3/4)^{-1} & (2/3, 3/4)^{-1} \\ (2/3, 3/4) & 1 & 1 \\ (2/3, 3/4) & 1 & 1 \end{bmatrix}, \tag{10.65}$$

for C_2 = benefits,

$$\overline{\mathcal{A}}_3 = \begin{bmatrix} 1 & 1 & (7/7, 8/10) \\ 1 & 1 & (7/8, 9/10) \\ (7/7, 8/10)^{-1} & (7/8, 9/10)^{-1} & 1 \end{bmatrix}, \tag{10.66}$$

for C_3 = location;

$$\overline{\mathcal{A}}_4 = \begin{bmatrix} 1 & (1/3, 3/3)^{-1} & (2/2, 2/5) \\ (1/3, 3/3) & 1 & (6/7, 7/8) \\ (2/2, 2/5)^{-1} & (6/7, 7/8)^{-1} & 1 \end{bmatrix}, \tag{10.67}$$

for C_4 = colleagues,

$$\overline{\mathcal{A}}_5 = \begin{bmatrix} 1 & (4/4, 4/6)^{-1} & (3/4, 5/5)^{-1} \\ (4/4, 4/6) & 1 & 1 \\ (3/4, 5/5) & 1 & 1 \end{bmatrix}, \tag{10.68}$$

for C_5 = advancement, and

$$\overline{\mathcal{E}} = \begin{pmatrix} 1 & (1/2,2/3) & (3/3,5/5)^{-1} & 1 & (4/4,6/6)^{-1} \\ (1/2,2/3)^{-1} & 1 & \frac{1}{6} & (1/2,4/5)^{-1} & \frac{1}{8} \\ (3/3,5/5) & 6 & 1 & 3 & 1 \\ 1 & (1/2,4/5) & \frac{1}{3} & 1 & \frac{1}{4} \\ (4/4,6/6) & 8 & 1 & 4 & 1 \end{pmatrix} \tag{10.69}$$

for the criteria, where where the rows (top to bottom) and the columns (left to right) are labeled P = pay, B = benefits, L = location, Co = colleagues, and Av = advancement. In the $\overline{\mathcal{A}}_i$ matrices the: (1) first row/column corresponds to alternative A_1; (2) second row/column is A_2; and (3) the third row/column is for job A_3. Using our EA we compute the fuzzy weight vectors $\overline{w}_k$ for $\overline{\mathcal{A}}_k$, $1 \leq k \leq 5$, and $\overline{e}$ for $\overline{\mathcal{E}}$. Then from equation (10.44), we get

$$\overline{r}_j = \sum_{k=1}^{5} \overline{w}_{jk}\overline{e}_k \ , \tag{10.70}$$

for all j. The fuzzy weight for job A_j is $\overline{r}_j$. However, before showing the results we need to discuss consistency and the ranking of fuzzy numbers.

Consistency

Let $\mathcal{A}$ be a positive, reciprocal, matrix. $\mathcal{A}$ is said to be consistent when $a_{ik}a_{kj} = a_{ij}$ for all i, j, k. This means that if the judge states $a_{ik} = 2/1$ for A_i versus A_k and gives $a_{kj} = 3/1$ for A_k against A_j, then to be logically consistent this judge should state $6/1$ for A_i versus A_j. If $\mathcal{A}$ is consistent then $\lambda_{\max} = m$ and in general $\lambda_{\max} \geq m$. So a measure of consistency is built around the difference $(\lambda_{\max} - m)$. We would say $\mathcal{A}$ is "reasonably" consistent when $(\lambda_{\max} - m)$ is not too large (maybe $\lambda_{\max} - m \leq 1$).

We have presented a method ranking fuzzy numbers in Sections 4.4 and 6.5, but for completeness we will repeat that development again here.

To talk about consistency for fuzzy positive, reciprocal, matrices we first need to define what is meant by $\overline{M} \geq \overline{N}$, $\overline{M} > \overline{N}$ and $\overline{M} \approx \overline{N}$ for two fuzzy numbers $\overline{M}$ and $\overline{N}$. Define

$$v(\overline{M} \geq \overline{N}) = \sup_{x \geq y}(\min(\overline{M}(x), \overline{N}(y))) \ . \tag{10.71}$$

We then write $\overline{M} > \overline{N}$ if $v(\overline{M} \geq \overline{N}) = 1$ and $v(\overline{N} \geq \overline{M}) < \theta$, where θ is some fixed positive fraction less than one. Let us use $\theta = 0.8$ in this book. Next, we write $\overline{M} \approx \overline{N}$ when $\overline{M}$ is not greater than $\overline{N}$ and $\overline{N}$ is not greater than $\overline{M}$. Or, if

$$\min(v(\overline{M} \geq \overline{N}), v(\overline{N} \geq \overline{M})) \geq \theta \ , \tag{10.72}$$

then $\overline{M} \approx \overline{N}$. Finally, we say $\overline{M} \geq \overline{N}$ if $\overline{M} > \overline{N}$ or $\overline{M} \approx \overline{N}$.

A fuzzy, positive, reciprocal matrix $\overline{\mathcal{A}} = [\overline{a}_{ij}]$ is defined to be consistent when

$$\overline{a}_{ik} \cdot \overline{a}_{kj} \approx \overline{a}_{ij} \ , \tag{10.73}$$

for all i, j, k. The following theorem was proven in [1].

Theorem 10.1 *Let $\overline{\mathcal{A}} = [\overline{a}_{ij}]$ be a fuzzy positive, reciprocal, matrix with $\overline{a}_{ij} = (\alpha_{ij}/\beta_{ij}, \gamma_{ij}/\delta_{ij})$. Choose $a_{ij} \in [\beta_{ij}, \gamma_{ij}]$ and form $\mathcal{A} = [a_{ij}]$. If $\mathcal{A}$ is consistent, then $\overline{\mathcal{A}}$ is consistent.*

We shall not demand all $\overline{\mathcal{A}}_k$ and $\overline{\mathcal{E}}$ to be perfectly consistent. All we shall ask is that they be "reasonably" consistent. What this means is that each has an $\mathcal{A}$, constructed as in Theorem 10.1, which is reasonably consistent.

If we look at the $\overline{\mathcal{A}}_i$, $1 \leq i \leq 5$, and $\overline{\mathcal{E}}$ in the application, all are "reasonably" consistent. In fact $\overline{\mathcal{A}}_2$, $\overline{\mathcal{A}}_3$, and $\overline{\mathcal{A}}_5$ are consistent. Let us look at $\overline{\mathcal{A}}_1$ to see how it is "reasonably" consistent. From Theorem 10.1, for $\overline{\mathcal{A}}_1$ to be consistent we need $[\beta_{ij}, \gamma_{ij}] \subset [\beta_{ik}, \gamma_{ik}] \cdot [\beta_{kj}, \gamma_{kj}]$, for all i, j, k. Consider $i = 1$, $k = 2$ and $j = 3$. We see that $[\beta_{13}, \gamma_{13}] = 1/2$, $[\beta_{12}, \gamma_{12}] = [1/5, 1/3]$, and $[\beta_{23}, \gamma_{23}] = 3$ and $1/2 \notin [3/5, 1]$. But since $1/2$ is "reasonably" close to $3/5$ we conclude $\overline{a}_{12} \cdot \overline{a}_{23}$ is "reasonably" close to $\overline{a}_{13}$. In $\overline{\mathcal{A}}_1$ we find that $\overline{a}_{ik} \cdot \overline{a}_{kj}$ is "reasonably" close to $\overline{a}_{ij}$ for all i, k, j and we conclude that $\overline{\mathcal{A}}_1$ is "reasonably" consistent. We have no test for reasonably consistent for fuzzy positive, reciprocal, matrices as is used for crisp, positive, reciprocal matrices.

Ranking Fuzzy Numbers

We end up (equation (10.70)) with fuzzy numbers $\overline{r}_1, \ldots, \overline{r}_m$ which need to be ranked so we may obtain the final ranking of the alternatives. Let H_1 be all the undominated fuzzy numbers $\overline{r}_i$. We say $\overline{r}_i$ is undominated if no $\overline{r}_j > \overline{r}_i$, $j \neq i$. Next define H_2 to be all the undominated $\overline{r}_k$ after deleting all the fuzzy numbers in H_1. Similarly, we construct $H_3, \ldots, H_d$. Then, all the A_i corresponding to a $\overline{r}_i$ in H_1 have the highest ranking, all the A_j having $\overline{r}_j$ in H_2 have the second ranking, etc.

Results

We first found the fuzzy weights vectors $\overline{w}_k$ for A_k, $1 \leq k \leq 5$, using the formulas (10.57) – (10.62). To obtain the fuzzy weight vector for $\overline{\mathcal{E}}$ we applied our EA. All fuzzy numbers were calculated for α-cuts of $\alpha = 0$, 0.2, 0.4, 0.6, 0.8, 1.0. The final fuzzy weights $\overline{r}_1$ and $\overline{r}_2$ are given in Figure 10.9. We omitted $\overline{r}_3$ because it is very similar to $\overline{r}_1$, which can be seen from the $\alpha = 0$ and $\alpha = 1$ cuts shown in Table 10.8. From Figure 10.9 it is obvious that $H_2 = \{A_1, A_3\}$, $H_1 = \{A_2\}$ and the student selected A_2.

Alpha–cuts, for $\alpha = 0$ and $\alpha = 1$, for all the fuzzy weights are presented in [7].

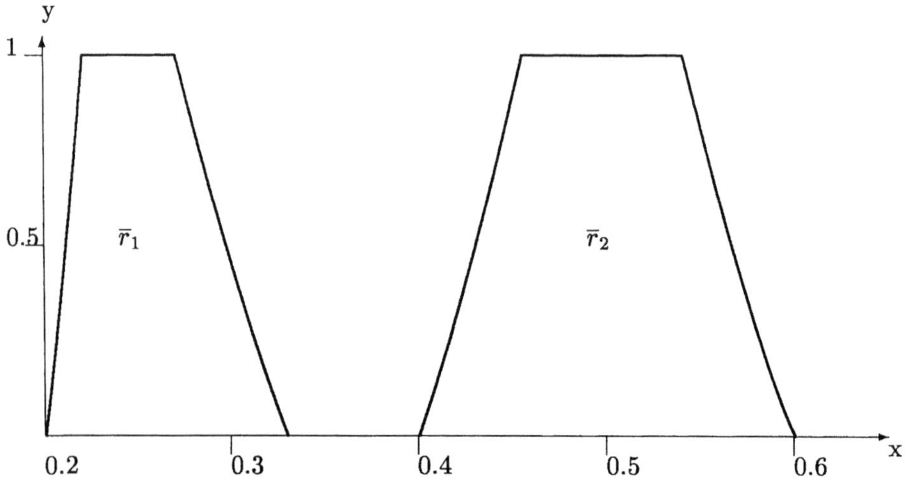

Figure 10.9: The Final Fuzzy Weights in the Application

Other Research

There has been numerous papers published on fuzzifying hierarchical analysis. See the references in [7] and ([9], [11]) for two recent papers. We believe that almost all of these publications deviate from Saaty's original method in order to obtain a more easily implementable fuzzy hierarchical analysis.

All researchers, until [9], have avoided the "direct" method of calculating the fuzzy weights. We used what we would call an "indirect" method in equations (10.47) – (10.49). The direct method computes directly (algebraically) the fuzzy weight vector $\overline{w}$ for $\overline{A}\,\overline{w} = \overline{\lambda}_{\max}\,\overline{w}$. They ([9]) applied their method to the same application and showed that using their procedure the final fuzzy weights had less "fuzziness" (less spread, smaller support).

10.3.3 Fuzzy Geometric Row Mean Method

The geometric row mean procedure is easily extended to fuzzy positive, reciprocal, matrices. First let $\mathcal{A} = [a_{ij}]$ be a positive, reciprocal, matrix, and compute the geometric row mean for each row as

$$r_i = \left(\prod_{j=1}^{m} a_{ij} \right)^{\frac{1}{m}} , \tag{10.74}$$

and then $w_i = r_i/(r_1 + \cdots + r_m)$. $w^t = (w_1, \ldots, w_m)$ is the weight vector for $\mathcal{A}$.

Now let $\overline{\mathcal{A}} = [\overline{a}_{ij}]$ be a fuzzy positive, reciprocal, matrix with $\overline{a}_{ij} = (\alpha_{ij}/\beta_{ij}, \gamma_{ij}/\delta_{ij})$ if $\overline{a}_{ij}$ is given in Figures 10.2 - 10.8, or $\overline{a}_{ij} \approx (\delta_{ij}^{-1}/\gamma_{ij}^{-1}, \beta_{ij}^{-1}/\alpha_{ij}^{-1})$ when $\overline{a}_{ji}$ is the reciprocal of a member of Figures 10.2

- 10.8. Of course, $\overline{a}_{ii} = 1$ for all i and some $\overline{a}_{ij}$ can be real numbers. We compute the fuzzy weights as

$$\overline{r}_i = \left(\prod_{j=1}^{m} \overline{a}_{ij} \right)^{\frac{1}{m}} , \tag{10.75}$$

and then $\overline{w}_i = \overline{r}_i/(\overline{r}_1 + \cdots + \overline{r}_m)$, $1 \le i \le m$.

It is not difficult to compute the membership function for each fuzzy weight $\overline{w}_i$ since all the $\overline{a}_{ij}$ are non–negative. With the y–axis vertical and the x–axis horizontal we will find the membership functions of the fuzzy weights $\overline{w}_i$ as x a function of y. For example, given the trapezoidal fuzzy number $\overline{N} = (3/5, 6/8)$, the membership function for $\overline{N}$ may be described as $x = 2y + 3$ for $0 \le y \le 1$ and $3 \le x \le 5$ and $x = -2y + 8$ for $0 \le y \le 1$, $6 \le x \le 8$. The graph of $\overline{N}$ is simply a straight line segment from $(3,0)$ to $(5,1)$ for $x \in [3,5]$ and another straight line segment for $(6,1)$ to $(8,0)$ on $[6,8]$. So let

$$f_{ij}(y) = (\beta_{ij} - \alpha_{ij})y + \alpha_{ij} , \tag{10.76}$$

$$g_{ij}(y) = (\gamma_{ij} - \delta_{ij})y + \delta_{ij} , \tag{10.77}$$

for $0 \le y \le 1$, if $\overline{a}_{ij} = (\alpha_{ij}/\beta_{ij}, \gamma_{ij}/\delta_{ij})$ with $\alpha_{ij} < \beta_{ij} \le \gamma_{ij} < \delta_{ij}$. When $\alpha_{ij} = \beta_{ij}$ as in Figure 10.4, 10.6, 10.7 we get $f_{ij}(y) = \beta_{ij}$ and if $\gamma_{ij} = \delta_{ij}$ as in Figures 10.5, 10.6, 10.8 we have $g_{ij}(y) = \gamma_{ij}$. For reciprocals $\overline{a}_{ji} = \overline{a}_{ij}^{-1}$ we see that $f_{ji}(y) = (g_{ij}(y))^{-1}$ and $g_{ji}(y) = (f_{ij}(y))^{-1}$ when $\alpha_{ij} < \beta_{ij} \le \gamma_{ij} < \delta_{ij}$, etc. Of course, $f_{ii}(y) = g_{ii}(y) = 1$ for all i and f_{ij}, g_{ij}, f_{ji} and g_{ji} are also constant if $\overline{a}_{ij}$ is a real number. In this way we construct the f_{ij} and g_{ij} for all $\overline{a}_{ij}$ in a fuzzy positive, reciprocal, matrix.

Now define

$$f_i(y) = \left(\prod_{j=1}^{m} f_{ij}(y) \right)^{\frac{1}{m}} , \tag{10.78}$$

$$g_i(y) = \left(\prod_{j=1}^{m} g_{ij}(y) \right)^{\frac{1}{m}} , \tag{10.79}$$

$0 \le y \le 1$. Also, set $f(y) = \sum_{i=1}^{m} f_i(y)$, $g(y) = \sum_{i=1}^{m} g_i(y)$, $\alpha_i = (\prod_{j=1}^{m} \alpha_{ij})^{1/m}$, $\alpha = \sum_{i=1}^{n} \alpha_i$. Similarly, define β_i, β and γ_i, γ and δ_i, δ.

The membership function for a fuzzy weight $\overline{w}_i$ is determined as follows: (1) zero to the left of α_i/δ; (2) $x = f_i(y)/g(y)$, $0 \le y \le 1$ for x in $[\alpha_i/\delta, \beta_i/\gamma]$; (3) one for $\beta_i/\gamma \le x \le \gamma_i/\beta$; (4) $x = g_i(y)/f(y)$, $0 \le y \le 1$, $\gamma_i/\beta \le x \le \delta_i/\alpha$; and (5) zero to the right of δ_i/α.

In this way we obtain the membership functions for all the fuzzy weights in each $\mathcal{A}_k$ and for $\mathcal{E}$ and combine using equation (10.70) and compute the

		Fuzzy λ_{max}	Geometric Mean
$\overline{r}_1$	$\alpha = 0$	[0.20,0.33]	[0.18,0.38]
	$\alpha = 1$	[0.22,0.27]	[0.21,0.28]
$\overline{r}_2$	$\alpha = 0$	[0.40,0.60]	[0.34,0.74]
	$\alpha = 1$	[0.45,0.54]	[0.43,0.58]
$\overline{r}_3$	$\alpha = 0$	[0.20,0.31]	[0.17,0.36]
	$\alpha = 1$	[0.24,0.28]	[0.22,0.30]

Table 10.8: Alpha-cuts of the Final Weights for Both Methods

membership functions for the final fuzzy weights $\overline{r}_j$, for all j. Further properties of FHA using this method, plus extension to multiple experts, may be found in [1]. Now let us apply this procedure to the same application studied using the fuzzy $\lambda_{\max}$–method discussed in Section 10.3.2. We find the membership functions for all the fuzzy weights as described above, and then the membership functions for the final fuzzy weights, and the $\alpha = 0$ and $\alpha = 1$ cuts are shown in Table 10.8. We can see from Table 10.8 that, for this example, the two methods produce essentially the same final fuzzy weights and the student picks A_2.

Summary

In this section we fuzzified the two basic methods of calculating the weights in hierarchical analysis. The first method was Saaty's procedure of finding the positive, normalized, eigenvector corresponding to the largest positive eigenvalue of a positive, reciprocal, matrix. The fuzzification of Saaty's procedure was computationally complicated and we used an evolutionary algorithm to achieve approximate solutions. The second method was to fuzzify the normalized geometric row means to obtain the fuzzy weights. The computations, to get the fuzzy weights, is much simpler in the second method. Both procedures were applied to the same application problem. We draw no conclusions as to which procedure is "best", but refer the reader to the discussion and references in [7].

10.4 Summary and Conclusions

In this chapter we discussed calculating fuzzy eigenvalues and their fuzzy eigenvectors, and two applications. Since there are numerous applications of eigenvalues and eigenvectors, there can be many more applications to study.

The first problem, addressed in Section 10.1, was how to fuzzify and compute eigenvalues for fuzzy symmetric matrices. We finally decided on the extension principle procedure which fuzzifies the crisp solution. We then presented an algorithm to compute the fuzzification of the largest positive eigenvalue and gave two examples.

The first application was the fuzzification of Leontief's closed input-output model. This involved finding a fuzzy eigenvector corresponding to a certain fuzzy eigenvector (having 1 in its core). The second application was to fuzzy hierarchical analysis were we needed to compute the fuzzification of the largest positive eigenvalue, and its normalized eigenvector, of a fuzzy positive, reciprocal, matrix.

Bibliography

[1] J.J. Buckley: Fuzzy Hierarchical Analysis, Fuzzy Sets and Systems, 17 (1985), pp. 233–247.

[2] J.J. Buckley: Fuzzy Complex Numbers, Fuzzy Sets and Systems, 33 (1989), pp. 333–345.

[3] J.J. Buckley: Fuzzy Eigenvalues and Input–Output Analysis, Fuzzy Sets and Systems, 34 (1990), pp. 187–195.

[4] J.J. Buckley and Y. Qu: Fuzzy Complex Analysis: Differentiation, Fuzzy Sets and Systems, 41 (1991), pp. 269–284.

[5] J.J. Buckley: Fuzzy Complex Analysis: Integration, Fuzzy Sets and Systems, 49 (1992), 171–179.

[6] J.J. Buckley and Y.R.R. Uppuluri: Fuzzy Hierarchical Analysis, in: V.T. Covello, L.B. Lave, A. Moghissi, and V.R.R. Uppuluri (eds.), Uncertainty and Risk Assessment, Risk Management and Decision Making, Plenum, N.Y., 1984, pp. 389–401.

[7] J.J. Buckley, Th. Feuring and Y. Hayashi: Fuzzy Hierarchical Analysis Revisited, European J. Operational Research, 129 (2001), pp. 48–64.

[8] J.J. Buckley, Th. Feuring and Y. Hayashi: Fuzzy Eigenvalues, Fuzzy Sets and Systems. Under revision.

[9] R. Csutora and J.J. Buckley: Hierarchical Analysis: The Lambda-Max Method, Fuzzy Sets and Systems, 120 (2001), pp. 181–195.

[10] S. Karlin: Mathematical Methods and Theory in Games, Programming and Economics, Addison–Wesley, Reading, MA, 1959.

[11] A. de Korvin and R. Kleyle: Fuzzy Analytical Hierarchical Processes, J. Intelligent and Fuzzy Systems, 7 (1999), pp. 387–400.

[12] K. Lancaster: Mathematical Economics, Macmillan, Toronto, 1968.

[13] L.C. Leung and D. Cao: On Consistency and Ranking of Alternatives in Fuzzy AHP, European J. Operational Research, 124 (2000), pp. 102–113.

[14] T.L. Saaty: A Scaling Method for Priorities in Hierarchical Structures, Journal of Mathematical Psychology, 15 (1977), pp. 234–281.

[15] T.L. Saaty: Exploring the Interface Between Hierarchies, Multiple Objectives and Fuzzy Sets, Fuzzy Sets and Systems, 1 (1978), pp. 57–68.

[16] T.L. Saaty: The Analytic Hierarchy Process, McGraw-Hill, N.Y., 1980.

[17] T.L. Saaty: Multicriteria Decision Making: The Analytic Hierarchy Process, RWS Publications, Pittsburgh, 1990.

[18] T.L. Saaty: Fundamentals of Decision Making and Priority Theory With the Analytic Hierarchy Process, RWS Publications, Pittsburgh, 1994.

Chapter 11

Fuzzy Integral Equations

Consider the Fredholm integral equation of the second kind [5]

$$f(x) = g(x) + \lambda \int_a^b K(x,y)f(y)dy \ , \tag{11.1}$$

where $K(x,y)$ is the kernel of the transformation assumed to be continuous for $a \leq x,y \leq b$, $\lambda > 0$, $g(x)$ is a known function continuous on $[a,b]$. and $f(x)$ is the unknown function assumed to be continuous for $a \leq x \leq b$. In this chapter we will allow $g(x)$ to be a fuzzy function and/or λ may be a triangular shaped fuzzy number.

Let $g(x) = \overline{A}(x)$, $a \leq x \leq b$, where $\overline{A}(x)$ is a triangular shaped fuzzy number for each x, and let $\lambda = \overline{\lambda} > 0$ be a triangular shaped fuzzy number. Since $f(x)$ must now also be a fuzzy function let $\overline{f}(x) = \overline{B}(x)$, $a \leq x \leq b$, for $\overline{B}(x)$ a triangular shaped fuzzy number for all x. We wish to consider solutions to

$$\overline{B}(x) = \overline{A}(x) + \overline{\lambda} \int_a^b K(x,y)\overline{B}(y)dy \ . \tag{11.2}$$

We will always have $K(x,y)$ a crisp (non–fuzzy) function throughout this chapter. The results below could all be generalized to $\overline{A}(x)$, $\overline{\lambda}$ and $\overline{B}(x)$ trapezoidal shaped fuzzy numbers but we shall assume them all to be triangular shaped fuzzy numbers.

Our first solution method is the "classical solution" $\overline{f}_c(x)$: (1) substitute α–cuts of $\overline{A}(x)$ and $\overline{\lambda}$, for $\overline{A}(x)$ and $\overline{\lambda}$, respectively, into equation (11.2); (2) use interval arithmetic to simplify the expression which gives two separate integral equations to solve for the α–cuts of $\overline{f}_c(x)$; and (3) if these α–cuts define a triangular shaped fuzzy number $\overline{B}(x)$ for all x, then $\overline{f}_c(x) = \overline{B}(x)$ is a solution.

Our second solution method is the "α–cut and interval arithmetic" solution $\overline{f}_I(x)$. Here we first solve equation (11.1) for f giving

$$f(x) = \Gamma(g(x), K(x,y), \lambda, a, b) \ , \tag{11.3}$$

which is the solution as a function of g, K, λ, a and b. Next we fuzzify the solution giving

$$\overline{f}_I(x) = \overline{B}(x) = \Gamma(\overline{A}(x), K(x,y), \overline{\lambda}, a, b) \ . \tag{11.4}$$

There are two ways of evaluating equation (11.4): (1) extension principle; and (2) using α–cuts and interval arithmetic. The second method produces $\overline{f}_I(x)$. So, to compute $\overline{f}_I(x)$ we substitute α–cuts of $\overline{A}(x)$ and $\overline{\lambda}$ into equation (11.4), simplify using interval arithmetic, giving α–cuts of $\overline{f}_I(x) = \overline{B}(x)$ which are $\overline{B}(x)[\alpha] = [f_{I1}(x,\alpha), f_{I2}(x,\alpha)]$. Now we substitute $\overline{B}(x)[\alpha]$ for $\overline{B}(x)$, $\overline{A}(x)[\alpha]$ for $\overline{A}(x)$ and $\overline{\lambda}[\alpha]$ for $\overline{\lambda}$ in equation (11.2), simplify using interval arithmetic, and if the resulting integral equations are valid, we say $\overline{f}_I(x)$ is a solution.

We will also consider two methods of solving the original crisp integral equation (11.1): (1) the resolvent kernel method; and (2) the symmetric and separable kernel procedure. In Section 11.1 we look at both $\overline{f}_c$ and $\overline{f}_I$ for the resolvent kernel method of solution and we will consider both $\overline{f}_c$ and $\overline{f}_I$ for the other solution procedure in Section 11.2.

In Chapters 3,4, 7-10 we also had the extension principle solution. This method is not applicable to equation (11.3). The extension principle is applicable to the following problem. Let $z = h(x,y)$ be a function mapping $a \leq x,\ y \leq b$ into real number z. Let $\overline{C}(x)$ and $\overline{D}(y)$ be fuzzy numbers for $x \in [a,b]$ and $a \leq y \leq b$, respectively. The extension principle allows us to compute $\overline{Z}(x,y) = h(\overline{C}(x), \overline{D}(y))$ for each x and y. However, as we shall see, equation (11.3) involves the integration of $\overline{A}(y)$ times another function (equations (11.8) and (11.40)), which is not the same as determining $\overline{Z}(x,y)$. That is why we only employ $\overline{f}_I$.

This chapter is based on [1]. A summary, our conclusions and suggestions for further research are in Section 11.3.

11.1 Resolvent Kernel Method

In the resolvent kernel solution method [5] we first define $K^n(x,y)$, $n = 2,3,\ldots$ as follows

$$K^2(x,y) = \int_a^b K(x,t)K(t,y)dt \ , \tag{11.5}$$

$$K^3(x,y) = \int_a^b K^2(x,t)K(t,y)dt \ , \tag{11.6}$$

etc. Define

$$R(x,y,\lambda) = \sum_{n=1}^{\infty} \lambda^{n-1} K^n(x,y) \ . \tag{11.7}$$

Let $M = \max\{K(x,y) | a \leq x,y \leq b\}$. If $0 < \lambda < (M(b-a))^{-1}$, then the series equation (11.7) converges uniformly and $R(x,y,\lambda)$ is called the resolvent kernel of K. The unique solution to (11.1) is

$$f(x) = g(x) + \lambda \int_a^b R(x,y,\lambda) g(y) dy \ . \tag{11.8}$$

Now we are ready to consider $\overline{f}_c(x)$. Let $\overline{\lambda}[\alpha] = [\lambda_1(\alpha), \lambda_2(\alpha)]$ for $\lambda_i(\alpha)$ continuous on $[0,1]$, $i = 1,2$, and $\lambda_1(0) > 0$. Also, $\overline{A}(x)[\alpha] = [g_1(x,\alpha), g_2(x,\alpha)]$ where $g_i(x,\alpha)$ is continuous on $[a,b] \times [0,1]$, $i = 1,2$. We will write $\overline{f}_c(x)[\alpha] = \overline{B}(x)[\alpha] = [f_{c1}(x,\alpha), f_{c2}(x,\alpha)]$, for $(x,\alpha) \in [a,b] \times [0,1]$, and we assume that the $\overline{f}_{ci}(x,\alpha)$ are continuous, $i = 1,2$.

11.1.1 Classical Solution

There are two cases to consider: (1) λ crisp and (2) $\lambda = \overline{\lambda}$ (fuzzy).

λ Crisp

We are to solve

$$[f_{c1}(x,\alpha), f_{c2}(x,\alpha)] = [g_1(x,\alpha), g_2(x,\alpha)] + \lambda \int_a^b K(x,y)[f_{c1}(y,\alpha), f_{c2}(y,\alpha)] dy \ , \tag{11.9}$$

for $f_{ci}(x,\alpha)$, $i = 1,2$. Simplification of equation (11.9), using interval arithmetic, depends on where $K(x,y)$ is positive and where $K(x,y)$ is negative since $\lambda > 0$.

If $K(x,y) \geq 0$ for $a \leq x,y \leq b$, then $\lambda K(x,y)[f_{c1}(y,\alpha), f_{c2}(y,\alpha)] = [\lambda K(x,y) f_{c1}(y,\alpha), \lambda K(x,y) f_{c2}(y,\alpha)]$ so that equation (11.9) produces two equations

$$f_{ci}(x,\alpha) = g_i(x,\alpha) + \lambda \int_a^b K(x,y) f_{ci}(y,\alpha) dy \ , \tag{11.10}$$

$i = 1,2$, to solve. The unique solution from equation (11.8) is

$$f_{ci}(x,\alpha) = g_i(x,\alpha) + \lambda \int_a^b R(x,y,\lambda) g_i(y,\alpha) dy \ , \tag{11.11}$$

$i = 1,2$. Now we need to check that the intervals $[f_{c1}(x,\alpha), f_{c2}(x,\alpha)]$ define a triangular shaped fuzzy number for all x. What we need is $\partial f_{c1}/\partial\alpha > 0$

and $\partial f_{c2}/\partial \alpha < 0$, $0 < \alpha < 1$, assuming they are differentiable on α, since we easily see that $f_{c1}(x,1) = f_{c2}(x,1)$ for all x. But these intervals do define a triangular fuzzy number since $\lambda > 0$, we know the $R(x,y,\lambda) \geq 0$ since we assumed $K(x,y) \geq 0$ and $\partial g_1/\partial \alpha > 0$, $\partial g_2/\partial \alpha < 0$, $0 < \alpha < 1$. Hence $\overline{f}_c(x) = \overline{B}(x)$, whose α–cuts are $[f_{c1}(x,\alpha), f_{c2}(x,\alpha)]$, is a solution.

This method gives an easy way to get $\overline{f}_c(x)$ when $K(x,y) \geq 0$ and it is not difficult to calculate $R(x,y,\lambda)$.

Example 11.1.1.1

Let $K(x,y) = xy$, $0 \leq x, y \leq 1$, and $0 < \lambda < 1$ since $M(b-a) = 1$. We find [5]

$$R(x,y,\lambda) = \frac{3xy}{3-\lambda} \ . \tag{11.12}$$

so

$$f_{ci}(x,\alpha) = g_i(x,\alpha) + \frac{3x\lambda}{3-\lambda}\int_0^1 y g_i(y,\alpha)dy \ , \tag{11.13}$$

$i = 1,2$ gives the classical solution for all $\overline{g}(x) = \overline{A}(x)$ whose α–cuts are continuous on $[0,1]^2$.

When $K(x,y)$ is not non–negative on $[a,b]^2$, because of interval arithmetic, we do not expect $\overline{f}_c(x)$ to be a solution.

Example 11.1.1.2

See references ([2], [3]). Let $\lambda = 0.1$, $K(x,y) = \sin(x/2)\sin(y)$, $0 \leq x, y \leq 2\pi$. Now $K(x,y) \geq 0$ for all x, $0 \leq y \leq \pi$, but $K(x,y) < 0$ for $0 < x < 2\pi$, $\pi < y < 2\pi$. To evaluate equation (11.9) we need to know if $f_{ci}(x,\alpha)$ is positive or negative, $i = 1,2$. Let us assume $\overline{f}_c(x) = \overline{B}(x) \geq 0$ for all x. Then we obtain from equation (11.9), via interval arithmetic, the following two equations to solve

$$f_{c1}(x,\alpha) = g_1(x,\alpha) \tag{11.14}$$

$$+(0.1)\int_0^{\pi} \sin(x/2)\sin(y) f_{c1}(y,\alpha)dy + (0.1)\int_{\pi}^{2\pi} \sin(x/2)\sin(y) f_{c2}(y,\alpha)dy,$$

$$f_{c2}(x,\alpha) = g_2(x,\alpha) \tag{11.15}$$

$$+(0.1)\int_0^{\pi} \sin(x/2)\sin(y) f_{c2}(y,\alpha)dy + (0.1)\int_{\pi}^{2\pi} \sin(x/2)\sin(y) f_{c1}(y,\alpha)dy.$$

The resolvent kernel is easily found to be $R(x,y,\lambda) = K(x,y)$ because $K^2(x,y)$ is identically zero on $[a,b]^2$. However, the resolvent kernel method is not applicable in this problem. The resolvent kernel solution (equation (11.11)) is based on equation (11.1) and does not apply to equations (11.14) and (11.15). The problem is that both f_{c1} and f_{c2} appear in both equations (11.14) and (11.15).

The resolvent kernel procedure will not apply when $K(x,y)$ can be negative over a whole interval for y, and therefore we shall assume $K(x,y) \geq 0$, except when referring to Example 11.1.1.2, for the rest of this chapter.

$\overline{\lambda}$ Fuzzy

The resolvent kernel is now formally defined as

$$R(x,y,\overline{\lambda}) = \sum_{n=1}^{\infty} (\overline{\lambda})^{n-1} K^n(x,y) \ . \tag{11.16}$$

To evaluate equation (11.16) substitute $\overline{\lambda}[\alpha] = [\lambda_1(\alpha), \lambda_2(\alpha)]$ for $\overline{\lambda}$ and get

$$R(x,y,\overline{\lambda})[\alpha] = [R_1(x,y,\alpha), R_2(x,y,\alpha)] \ , \tag{11.17}$$

where

$$R_1(x,y,\alpha) = \sum_{n=1}^{\infty} (\lambda_1(\alpha))^{n-1} K^n(x,y) \ , \tag{11.18}$$

$$R_2(x,y,\alpha) = \sum_{n=1}^{\infty} (\lambda_2(\alpha))^{n-1} K^n(x,y) \ , \tag{11.19}$$

since $[\lambda_1(\alpha), \lambda_2(\alpha)]^k = [\lambda_1^k(\alpha), \lambda_2^k(\alpha)]$, $k = 2, 3, \ldots$ because $\overline{\lambda} > 0$. Both series converge uniformly for $a \leq x, y \leq b$, $0 \leq \alpha \leq 1$, when $\lambda_2(0) < (M(b-a))^{-1}$. Notice that $R_i(x,y,\alpha) \geq 0$, $i = 1, 2$, since we assumed that $K(x,y) \geq 0$.

Now we return to equation (11.2) which becomes

$$[f_{c1}(x,\alpha), f_{c2}(x,\alpha)] \tag{11.20}$$

$$= [g_1(x,\alpha), g_2(x,\alpha)] + [\lambda_1(\alpha), \lambda_2(\alpha)] \int_a^b K(x,y)[f_{c1}(y,\alpha), f_{c2}(y,\alpha)]dy \ .$$

This simplifies to

$$[f_{c1}(x,\alpha), f_{c2}(x,\alpha)] = [g_1(x,\alpha), g_2(x,\alpha)] + \tag{11.21}$$

$$+[\lambda_1(\alpha), \lambda_2(\alpha)] \left[\int_a^b K(x,y) f_{c1}(y,\alpha)dy, \int_a^b K(x,y) f_{c2}(y,\alpha)dy \right] ,$$

because $K(x,y) \geq 0$. We cannot simplify equation (11.21) any further without knowing if the f_{ci} are positive or negative, $i = 1, 2$. Let us temporarily assume $\overline{f}_c(x) = \overline{B}(x) \geq 0$ for all x. We shall see that $\overline{g}(x) = \overline{A}(x) \geq 0$ will imply that $\overline{f}_c(x) = \overline{B}(x) \geq 0$. Using $\overline{f}_c(x) = \overline{B}(x) \geq 0$ for all x, equation (11.21) produces two equations

$$f_{ci}(x,\alpha) = g_i(x,\alpha) + \lambda_i(\alpha) \int_a^b K(x,y) f_{ci}(y,\alpha) dy \ , \tag{11.22}$$

$i = 1, 2$, to solve. Using the resolvent kernels $R_i(x,y,\alpha)$ the solutions to equation (11.22) are

$$f_{ci}(x,\alpha) = g_i(x,\alpha) + \lambda_i(\alpha) \int_a^b R_i(x,y,\alpha) g_i(y,\alpha) dy \ , \tag{11.23}$$

$i = 1, 2$. Notice that if we had assumed that $\overline{g}(x) = \overline{A}(x) \geq 0$ for all x, since $\lambda_i(\alpha) > 0$ for all α, $i = 1, 2$ and $R_i(x,y,\alpha) \geq 0$, $i = 1, 2$, equation (11.23) implies that $\overline{f}_c(x) = \overline{B}(x) \geq 0$ for all x. So we drop the assumption of $\overline{f}_c(x) \geq 0$ for all x and replace it with $\overline{g}(x) \geq 0$ for all x.

For $\overline{f}_c$ to be a solution we need the intervals $[f_{c1}(x,\alpha), f_{c2}(x,\alpha)]$ to define a triangular shaped fuzzy number for all x. But this is true because: (1) $\partial R_1/\partial\alpha > 0$ and $\partial R_2/\partial\alpha < 0$, $0 < \alpha < 1$; (2) $\partial g_1/\partial\alpha > 0$ and $\partial g_2/\partial\alpha < 0$, $0 < \alpha < 1$; (3) $\partial\lambda_1/\partial\alpha > 0$ and $\partial\lambda_2/\partial\alpha < 0$ $0 < \alpha < 1$; and (4) $f_{c2}(x,1) = f_{c2}(x,1)$ all x.

Example 11.1.1.3

This continues Example 11.1.1.1. $K(x,y) = xy$, $0 \leq x, y \leq 1$, $M = 1$, $0 < \overline{\lambda} < 1$ (or $0 < \lambda_1(0), \lambda_2(0) < 1$). Let $\overline{\lambda}[\alpha] = [(1/3)\alpha, (1/3)(2-\alpha)]$. Then

$$R_1(x,y,\alpha) = \sum_{n=1}^{\infty} \left(\frac{1}{3}\alpha\right)^{n-1} \frac{xy}{3^{n-1}} = \frac{9xy}{9-\alpha} \ , \tag{11.24}$$

$$R_2(x,y,\alpha) = \sum_{n=1}^{\infty} \left(\frac{2-\alpha}{3}\right)^{n-1} \frac{xy}{3^{n-1}} = \frac{9xy}{7+\alpha} \ , \tag{11.25}$$

since $K^n(x,y) = xy/3^{n-1}$. The classical solution is

$$f_{c1}(x,\alpha) = g_1(x,y) + \frac{3x\alpha}{9-\alpha} \int_0^1 y g_1(y,\alpha) dy \ , \tag{11.26}$$

$$f_{c2}(x,\alpha) = g_2(x,y) + \frac{3(2-\alpha)x}{7+\alpha} \int_0^1 y g_2(y,\alpha) dy \ , \tag{11.27}$$

for any $\overline{g}(x) = \overline{A}(x)$ whose α–cuts are continuous on $[0,1]^2$.

Summary

For λ crisp the assumptions were $K(x,y) \geq 0$, $0 < \lambda < (M(b-a))^{-1}$ and then $\overline{f}_c(x)$ is given by equation (11.11). When $\lambda = \overline{\lambda}$ is fuzzy we assumed that $K(x,y) \geq 0$, $0 < \overline{\lambda} < (M(b-a))^{-1}$ and $\overline{g}(x) = \overline{A}(x) \geq 0$ for all x, and then $\overline{f}_c(x)$ is given by equation (11.23). For the $\lambda = \overline{\lambda}$ the fuzzy case, one could assume $\overline{g}(x) = \overline{A}(x) \leq 0$ for all x and $\overline{f}_c(x)$ is still a solution (details left to the reader).

11.1.2 Second Solution Method

In this method we first fuzzify the solution to equation (11.1). The solution, using the resolvent kernel, is given in equation (11.8). We are assuming $K(x,y) \geq 0$, $0 < \lambda < (M(b-a))^{-1}$ so that we also have $R(x,y,\lambda) \geq 0$. We consider two cases: (1) λ crisp; and (2) $\lambda = \overline{\lambda}$ fuzzy.

λ Crisp

We fuzzify equation (11.8) by substituting α–cuts of $\overline{g}(x) = \overline{A}(x)$, which are $[g_1(x,\alpha), g_2(x,\alpha)]$ for $g(x)$ and simplify using interval arithmetic, giving α–cuts of $\overline{f}(x) = \overline{B}(x)$, which are

$$[f_{I1}(x,\alpha), f_{I2}(x,\alpha)] \tag{11.28}$$

$$= [g_1(x,\alpha), g_2(x,\alpha)] + \lambda \int_a^b R(x,y,\alpha)[g_1(y,\alpha), g_2(y,\alpha)]dy \ ,$$

or

$$f_{Ii}(x,\alpha) = g_i(x,\alpha) + \lambda \int_a^b R(x,y,\alpha) g_i(y,\alpha)dy \ , \tag{11.29}$$

$i = 1,2$, since $R_i(x,y,\alpha) \geq 0$. We say $\overline{f}_I(x)$ is a solution if the $f_{Ii}(x,\alpha)$ solve the original integral equation. But equation (11.29) gives the solution to

$$f_{Ii}(x,\alpha) = g_i(x,\alpha) + \lambda \int_a^b K(x,y) f_{Ii}(y,\alpha)dy \ , \tag{11.30}$$

$i = 1,2$ which is equation (11.2) after simplification. Hence $\overline{f}_I(x)$ is a solution. But equation (11.29) is the same as (11.11) and therefore $\overline{f}_c = \overline{f}_I$ when λ is crisp.

$\overline{\lambda}$ Fuzzy

As in Section 11.1.1 we now assume $\overline{g}(x) = \overline{A}(x) \geq 0$. Fuzzify equation (11.8) to

$$[f_{I1}(x,\alpha), f_{I2}(x,\alpha)] = [g_1(x,\alpha), g_2(x,\alpha)] + [\lambda_1(\alpha), \lambda_2(\alpha)] \int_a^b R(x,y,[\lambda_1(\alpha),\lambda_2(\alpha)])[g_1(y,\alpha), g_2(y,\alpha)]dy. \tag{11.31}$$

From Section 11.1.1 we now assume $0 < \lambda_2(0) < (M(b-a))^{-1}$. Then

$$R(x,y,[\lambda_1(\alpha),\lambda_2(\alpha)]) = \sum_{n=0}^{\infty}[\lambda_1(\alpha),\lambda_2(\alpha)]^{n-1}K^n(x,y) \ , \tag{11.32}$$

which is the same as

$$\left[\sum_{n=0}^{\infty}(\lambda_1(\alpha))^{n-1}K^n(x,y), \sum_{n=0}^{\infty}(\lambda_2(\alpha))^{n-1}K^n(x,y)\right] \ , \tag{11.33}$$

which equals

$$[R_1(x,y,\alpha), R_2(x,y,\alpha)] \ , \tag{11.34}$$

given by equations (11.18) and (11.19).

Now we may simplify equation (11.31) using $g_i(x,\alpha) \geq 0$, $\lambda_i(\alpha) \geq 0$, $R_i(x,y,\alpha) \geq 0$, $i = 1,2$, producing

$$f_{Ii}(x,\alpha) = g_i(x,\alpha) + \lambda_i(\alpha)\int_a^b R_i(x,y,\alpha)g_i(y,\alpha)dy \ , \tag{11.35}$$

$i = 1,2$. Equation (11.32) is the solution to

$$f_{Ii}(x,\alpha) = g_i(x,\alpha) + \lambda_i(\alpha)\int_a^b K(x,y)f_{Ii}(y,\alpha)dy \ , \tag{11.36}$$

$i = 1,2$, which is the simplification of equation (11.2) using interval arithmetic. Hence $\overline{f}_I(x)$ is a solution. But equation (11.32) is the same as equation (11.23) and it follows that $\overline{f}_I = \overline{f}_c$ when $\lambda = \overline{\lambda}$ is fuzzy.

Summary

Under the assumptions given in the summary just before Section 11.1.2 we see that our two solution methods produce the same solution when we employ the resolvent kernel technique of deriving a crisp solution.

11.2 Symmetric Kernel Method

In this section we assume that $K(x,y) = K(y,x)$, $a \leq x, y, \leq b$ or K is symmetric. When K is symmetric we know it has a countable number of non-zero eigenvalues θ_i. Since the solution method for separable kernels involves all the eigenvalues, in order to simplify the calculation, we will assume K has only a finite number of non-zero eigenvalues. One way to do this is to assume that K is separable and symmetric , or

$$K(x,y) = \sum_{k=1}^{p} \sigma_k(x)\sigma_k(y) \ , \tag{11.37}$$

for continuous functions σ_k, $1 \leq k \leq p$. Let $\theta_1, \ldots, \theta_m$ be the non-zero eigenvalues, or

$$\int_a^b K(x,y)\beta(y)\,dy = \theta_i\beta(x) \ , \tag{11.38}$$

for $\beta(x)$ not identically zero on $[a,b]$. The $\beta(x)$ are the associated eigenfunctions. In fact each θ_i has only a finite number of linearly independent eigenfunctions.

The spectrum of K is written

$$s_1 \leq s_2 \leq \cdots \leq s_L \ , \tag{11.39}$$

where the s_i are eigenvalues of K, repeating each eigenvalue a number of times equal to the number of linearly independent eigenfunctions associated with it. So, associated with each s_i is an eigenfunction $\beta_i(x)$. We will assume each $\beta_i(x)$ is normalized (its squared integral over $[a,b]$ equals one) and $\beta_i(x)$ and $\beta_j(x)$, $i \neq j$, are orthogonal (the integral of $\beta_i(x)\beta_j(x)$ over $[a,b]$ is zero).

Then if $(1/\lambda) \neq s_i$ for all i, the solution to equation (11.1) may be written

$$f(x) = g(x) + \sum_{k=1}^{L} \frac{\lambda s_k}{1-\lambda s_k}\left(\int_a^b g(y)\beta_k(y)\,dy\right)\beta_k(x) \ . \tag{11.40}$$

No resolvent kernel is needed in this solution method. So, this procedure is very useful when the resolvent kernel is hard to compute and L is very small (like $L = 1$). When $1/\lambda = s_i$ for some i, there is another solution [5], but we shall not use this in this chapter.

11.2.1 Classical Solution

There are two cases: (1) λ is crisp; and (2) $\lambda = \overline{\lambda}$ is fuzzy. Assume $K(x,y) \geq 0$.

λ Crisp

The solution to equation (11.10) is

$$f_{ci}(x,\alpha) = g_i(x,\alpha) + \sum_{k=1}^{L} \frac{\lambda s_k}{1-\lambda s_k} \left(\int_a^b g_i(y,\alpha)\beta_k(y)\,dy \right) \beta_k(x) \ , \quad (11.41)$$

for $1/\lambda \neq s_i$, for all i. This produces a solution to the fuzzy integral equation if $[f_{c1}(x,\alpha), f_{c2}(x,\alpha)]$ defines a fuzzy number $\overline{B}(x)$ for all x. We have a problem, because some of the s_k can be negative and the eigenfunctions $\beta_k(x)$ may also be negative for some $x \in [a,b]$, in showing we get a fuzzy number $\overline{B}(x)$ for all x. However, when there is only one eigenvalue we may easily see if equation (11.41) defines a fuzzy number.

Example 11.2.1.1

Let $K(x,y) = \sin(x)\sin(y)$, $0 \leq x, y \leq \pi$. K has one non-zero eigenvalue $s = \pi/2$ with normalized eigenfunction $\beta(x) = \sin(x)/\sqrt{\pi/2}$. Let $0 < \lambda < 2/\pi$. The solution is

$$f_{ci}(x,\alpha) = g_i(x,\alpha) + \frac{\lambda}{1-\frac{\pi}{2}\lambda} \left(\int_0^\pi g_i(y,\alpha)\sin(y)\,dy \right) \sin(x) \ , \quad (11.42)$$

$i = 1,2$, $0 \leq x \leq \pi$. This defines a fuzzy number for all x because $\partial f_{c1}/\partial\alpha > 0$, $\partial f_{c2}/\partial\alpha < 0$, $0 < \alpha < 1$, and $f_{c1}(x,1) = f_{c2}(x,1)$. In this example one may also easily find the resolvent kernel $R(x,y,\alpha)$ and obtain $\overline{f}_c(x)$ as in Section 11.1.1. We leave it to the interested reader to compare the two solutions.

$\lambda = \overline{\lambda}$ Fuzzy

Now we must simplify equation (11.21) and as in Section 11.1.1 we assume the $\overline{f}_c(x) = \overline{B}(x) \geq 0$ for all x so that we end up with equation (11.22). The solution to equation (11.22) is

$$f_{ci}(x,\alpha) = g_i(x,\alpha) + \sum_{k=1}^{L} \frac{s_k\lambda_i(\alpha)}{1-s_k\lambda_i(\alpha)} \left(\int_a^b g_i(y,\alpha)\beta_k(y)\,dy \right) \beta_k(x) \ , \ (11.43)$$

$i = 1,2$, where we need to now assume that $1/s_j \notin \overline{\lambda}[0]$, $1 \leq j \leq L$. In general, equation (11.43) may or may not define fuzzy numbers $\overline{f}_c(x) = \overline{B}(x)$ for x in $[a,b]$.

Example 11.2.1.2

This continues Example 11.1.1.1 and Example 11.1.1.3. For $K(x,y) = xy$, $0 \leq x, y \leq 1$ we calculate one eigenvalue $s = 1/3$ with corresponding (normalized) eigenfunction $\beta(x) = \sqrt{3}x$. Let $\overline{\lambda}$ be specified by its α–cuts $\overline{\lambda}[\alpha] = [\alpha, 2-\alpha]$, $0 \leq \alpha \leq 1$. Then $1/s = 3 \notin \overline{\lambda}[0] = [0,2]$. Next let $\overline{g}(x)[\alpha] = [(\alpha+1)/2, (3-\alpha)/2]x$. Then equation (11.43) becomes

$$f_{c1}(x,\alpha) = \frac{\alpha+1}{2} + \frac{\alpha}{3-\alpha}\left(\int_0^1 \frac{\alpha+1}{2} y\sqrt{3}y\,dy\right)\sqrt{3}x\ , \quad (11.44)$$

$$f_{c2}(x,\alpha) = \frac{3-\alpha}{2} + \frac{2-\alpha}{1+\alpha}\left(\int_0^1 \frac{3-\alpha}{2} y\sqrt{3}y\,dy\right)\sqrt{3}x\ . \quad (11.45)$$

We calculate

$$f_{c1}(x,\alpha) = \frac{3}{2}\left(\frac{\alpha+1}{3-\alpha}\right)x\ , \quad (11.46)$$

$$f_{c2}(x,\alpha) = \frac{3}{2}\left(\frac{3-\alpha}{\alpha+1}\right)x\ , \quad (11.47)$$

which define fuzzy numbers.

11.2.2 Second Solution Method

Again, there are two cases: (1) λ crisp; and (2) $\lambda = \overline{\lambda}$ fuzzy.

λ Crisp

We fuzzify equation (11.40) producing

$$[f_{I1}(x,\alpha), f_{I2}(x,\alpha)] \quad (11.48)$$

$$= [g_1(x,\alpha), g_2(x,\alpha)] + \sum_{k=1}^{L} \frac{\lambda s_k}{1-\lambda s_k}\left(\int_a^b [g_1(y,\alpha), g_2(y,\alpha)]\beta_k(y)\,dy\right)\beta_k(x)\ .$$

to simplify. Even if we assume the $\overline{g}(x) = \overline{B}(x) \geq 0$ we have trouble simplifying equation (11.48) without knowing if the s_k are positive or negative and without knowing if the $\beta_k(x)$ are always positive or always negative. The calculations are easy if there is only one eigenvalue.

Example 11.2.2.1

This continues Example 11.2.1.1. We obtain equation (11.42) since $\beta(x) \geq 0$ for all x. Hence $\overline{f}_I(x) = \overline{f}_c(x)$ in this example.

$\overline{\lambda}$ Fuzzy

Fuzzify equation (11.40) giving

$$[f_{I1}(x,\alpha), f_{I2}(x,\alpha)] = [g_1(x,\alpha), g_2(x,\alpha)] + \tag{11.49}$$
$$+\sum_{k=1}^{L} \frac{s_k[\lambda_1(\alpha),\lambda_2(\alpha)]}{1-s_k[\lambda_1(\alpha),\lambda_2(\alpha)]} \left(\int_0^1 [g_1(y,\alpha), g_2(y,\alpha)]\beta_k(y)\, dy \right) \beta_k(x) \, .$$

to simplify.

Example 11.2.2.2

This will continue Example 11.2.1.2. Since $\lambda_2(0) < 3$ we see that

$$\frac{\frac{1}{3}\overline{\lambda}[\alpha]}{1-\frac{1}{3}\overline{\lambda}[\alpha]} = \left[\frac{\frac{1}{3}\lambda_1(\alpha)}{1-\frac{1}{3}\lambda_1(\alpha)}, \frac{\frac{1}{3}\lambda_2(\alpha)}{1-\frac{1}{3}\lambda_2(\alpha)} \right] , \tag{11.50}$$

all α. Then, equation (11.49) reduces to equations (11.44) and (11.45). We obtain the same solution as in Example 11.2.1.2. That is, $\overline{f}_I(x) = \overline{f}_c(x)$ again in this example.

11.3 Summary and Conclusions

We observed that under the resolvent kernel method of solution for our assumptions, mainly a non–negative kernel, the two solution methods produced the same solution. The two solution methods were the classical solution and the α–cut and interval arithmetic solution, since the extension principle solution method was not applicable to this problem. Our methods of solution are not applicable to all problems, as shown in Example 11.1.1.2, where the kernel may be negative. The solution to this example may be found in ([1] - [3]). We also found that our two methods of solution give the same answer in two simple examples, when using another method of solving integral equations based on a symmetric, separable, and non–negative, kernel.

Obviously, this is just the beginning of solving fuzzy integral equations. We kept the kernel crisp (not fuzzy), so researchers might consider working with a fuzzy kernel. For recent papers and other approaches to solving fuzzy integral equations (including numerical results), we refer the reader to ([4] and [6] - [9]).

Bibliography

[1] J.J. Buckley, T. Feuring and Y. Hayashi: Fuzzy Integral Equations, Proc. FUZZ-IEEE 2000, May 7-10, San Antonio, Texas, Vol. I, pp. 319–323.

[2] M. Friedman, M. Ma and A. Kandel: Numerical Solutions to Fuzzy Differential and Integral Equations, Fuzzy Sets and Systems, 106 (1999), pp. 35–48.

[3] M. Friedman, M. Ming and A. Kandel: Solutions to Fuzzy Integral Equations with Arbitrary Kernels, Int. J. Approximate Reasoning, 20 (1999), pp. 249–262.

[4] M. Friedman, M. Ming and A. Kandel: On Fuzzy Integral Equations, Fundamenta Informaticae, 37 (1999), pp. 89–99.

[5] J. Indritz, Methods in Analysis, Macmillan, NY, 1963, Chapter 4.

[6] J. Mordeson and W. Newman, Fuzzy Integral Equations, Information Sciences, 87 (1995), pp. 215–229.

[7] J. Mordeson and W. Newman, Fuzzy Integral Equations, in: P.P. Wang (ed.), Advances in Fuzzy Theory and Technology, Vol. III, Duke University, Duham, NC, 1995, pp. 81–97.

[8] J.Y. Park and J.U. Jeong: A Note one Fuzzy Integral Equations. Fuzzy Sets and Systems, 108 (1999), pp. 193–200.

[9] J.Y. Parkand and J.U. Jeong: On the Existence and Uniqueness of Solutions of Fuzzy Volterra-Fredholm Integral Equations, Fuzzy Sets and Systems, 115 (2000), pp. 425–431.

Chapter 12

Summary and Conclusions

We first summarize Chapters 3 through 11, and then present our suggestions for future research for each Chapter 3 to 11. Our conclusions are at the end of the chapter.

12.1 Summary

12.1.1 Chapter 3: Solving Fuzzy Equations

In Chapter 3 we introduce our three solution concepts that are used throughout the book. The first solution we call the "classical" solution. In the classical solution we first fuzzify the crisp (non–fuzzy) problem and solve, using α–cuts and interval arithmetic, for the unknown variable. If the solution defines fuzzy numbers, we call it the classical solution. The classical solution, if it exists, always solves the original fuzzified problem. For the next two solutions we first solve the crisp problem to obtain the crisp solution. The second type of solution is called the extension principle solution because we obtain it from the extension principle extension (fuzzification) of the crisp solution. The extension principle solution always exists and it may, or may not, satisfy the original fuzzified problem. Sometimes we also require the extension principle extension of the crisp solution to satisfy the original fuzzified problem in order to be called a "solution". The third solution fuzzifies the crisp solution using α–cuts and interval arithmetic, so it is called the α–cut and interval arithmetic solution. We may also require it to solve the original fuzzified problem to be called a "solution". In general, we expect the classical solution (when it exists) to be less fuzzy than the extension principle solution which will be less fuzzy than the α–cut and interval arithmetic solution. Our general solution strategy is as follows: (1) choose the classical solution if it exists; (2) if the classical solution fails to exist then we prefer the extension principle solution; and (3) if the extension principle solution is difficult to compute we use the α–cut and interval arithmetic solution, which is usually easier to obtain, as our approximate (to the extension principle solution) solution.

These solution methods were first applied in solving the fuzzy linear equation $\overline{A}\,\overline{X} + \overline{B} = \overline{C}$ and then in solving systems of fuzzy linear equations. For the systems of fuzzy linear equations problem we also introduced the joint solution and its marginals, and showed a relationship between the classical, marginal, extension principle and the α–cut and interval arithmetic solutions. Four applications were given: (1) the first involved solving a fuzzy linear equation; (2) the second has a fuzzy quadratic equation to solve; (3) a 2×2 system of fuzzy linear equations needed to be solved in the third application; and (4) the final application was the fuzzification of Leontief's open input–output model for an economy.

12.1.2 Chapter 4: The Fuzzy Mathematics in Finance

In this chapter we fuzzified the following elementary concepts in finance: (1) the future value of an investment; (2) the present value of a future amount of money; and (3) the future value and the present value of a regular annuity. Cash amounts, interest rates and the number of compoundings per year may all be fuzzy. If we have to solve some equations for the value of a variable, then we apply our three solution concepts: classical solution; extension principle solution; and the α–cut and interval arithmetic solution. However, if we only have to evaluate a fuzzy expression for the answer, then we have only two solutions: the extension principle method; or the α–cut and interval arithmetic procedure. The application was to fuzzy portfolio analysis where we wish to rank future fuzzy cash flows from "best" to "worst". The two ranking methods that we fuzzified were net present value and the internal rate of return.

12.1.3 Chapter 5: Fuzzy Non–Linear Regression

We first considered univariate fuzzy non–linear regression. Let $z = f(x)$ for $x \in [a, b]$. Extend f via the extension principle to $\overline{Z} = F(\overline{X})$, for $\overline{X} \in [a, b]$. F will generate the data. Let $\overline{Z}_i = F(\overline{X}_i)$ for triangular fuzzy number $\overline{X}_i$ in $[a, b]$, $1 \leq i \leq p$, and then $(\overline{X}_i, \overline{Z}_i)$, $1 \leq i \leq p$, is the data for fuzzy regression. There is a library of fuzzy functions containing certain linear, polynomial, exponential and logarithmic fuzzy functions. The idea is to use our evolutionary algorithm (EA) to find the "best" fuzzy function in the library that "explains" the data. To test our EA we chose various F in the library and in all cases the EA essentially found the correct F. An application was suggested where the F used was not in the library, which now consists of only fuzzy polynomials. We outlined a plan that may be used to test our EA on fuzzy functions not in the library.

We then went to multivariate fuzzy regression using only two independent variables. The methods were basically the same as in the univariate case but now we started with only fuzzy polynomials, in two variables, in the library. Testing the EA was done the same way by choosing different F in the library. The application suggested was for an F not in the library, and we outlined a plan to test our EA on fuzzy functions not in the library.

12.1.4 Chapter 6: Operations Research

This chapter contained five applications of fuzzy sets to operations research: (1) linear programming; (2) PERT; (3) queuing theory; (4) inventory control; and (5) network analysis.

Fuzzy Linear Programming

We studied solutions to the fully fuzzified linear programming problem

$$\begin{aligned} &\min / \max \overline{Z} = \overline{C}_1\, \overline{X}_1 + \cdots + \overline{C}_n\, \overline{X}_n \\ &\text{subject to: } \overline{A}_{i1}\, \overline{X}_1 + \cdots + \overline{A}_{in}\, \overline{X}_n \leq \overline{B}_i \ , \ 1 \leq i \leq m \ , \\ &\qquad\qquad \overline{X}_i \geq 0 \ , \end{aligned} \tag{12.1}$$

for $\overline{A}_{ij}$, $\overline{B}_{ij}$ and $\overline{C}_j$ triangular fuzzy number. We also allowed the unknown $\overline{X}_i$ to be triangular and $\overline{Z}$ will be a triangular shaped fuzzy number. We had two problems to solve in order to discuss solutions to the fully fuzzified linear program: (1) how do you max/min fuzzy number $\overline{Z}$; and (2) what "$\leq$" shall we use between fuzzy numbers.

We first changed max/min into a multiobjective fuzzy linear program and then discussed undominated solutions. Then, using fuzzy goals the multiobjective problem became a single objective fuzzy linear program, which can be used to determine undominated solutions. We had to be very selective on choosing a method of evaluating "$\leq$" between fuzzy numbers because we required $\overline{Z}$ to be bounded (its support is in some interval $[0, M]$). Two methods for "$\leq$" were presented.

Two applications were given, with the first one the classical product mix problem (a max. problem), and the classical diet problem (a min. problem) was the second application. In both applications we used our evolutionary algorithm to generate solutions.

Fuzzy PERT

In statistics PERT job times in a project network are values of random variables but in fuzzy PERT job times are values of fuzzy variables. A fuzzy variable has its values restricted by a possibility distribution which is just a fuzzy subset of the real numbers. We first allowed the job times to all be trapezoidal fuzzy numbers. The objective was to find the possibility distribution (another trapezoidal shaped fuzzy number) of project duration.

In stochastic PERT the probability density of project duration is difficult to determine, but in fuzzy PERT it is easily obtained and we gave an algorithm for its computation. We presented an example for fuzzy trapezoidal job times.

Then we considered job times given by a discrete fuzzy set. A discrete fuzzy set for job times is a fuzzy subset of the real numbers whose membership function is not zero at only a finite number of places. An algorithm for the possibility distribution of project duration is known in this case but the details were omitted from the book.

Fuzzy Inventory Control

We looked at fuzzifying certain parameters in a single item, multiperiod (finite horizon) inventory control problem.

In the first model demand was crisp and ordering cost and holding cost were fuzzy. The second model allowed fuzzy demands and shortages with the penalty for shortages fuzzy. Examples, with computational results using our evolutionary algorithm, were presented for both of these models. Also, in both models we wanted to minimize fuzzy cost which was changed into a single crisp objective function as was done in Section 6.1.

The third model had backordering and the discounting of future money back to its present value. In this model we are selling the item so we wish to maximize fuzzy profit. Discounting was done with a fuzzy interest rate.

In all models we assumed zero delivery lag and certain variables/parameters were always crisp (not fuzzy). The decision variable of the amount to order each period was always crisp while the purchase price (we buy it and do not manufacture it ourselves) and the selling price (third model) were also always crisp.

Dynamic programming is usually used to solve the crisp multiperiod model but we did not use fuzzy dynamic programming for these fuzzy models. Instead we employed our evolutionary algorithm.

Fuzzy Queuing Theory

In this section we considered a queuing system with multiple parallel services, finite or infinite system capacity and calling source, but whose arrivals and service completions are governed by a possibility (not probability) distribution. Changes in the number of customers in the system is modeled by a transition matrix Ω. Powers of Ω are computed using the max–min product of fuzzy matrices. The limit $\lim_{k\to\infty}\Omega^k = \Pi$, if it converges, gives the steady–state possibility distribution. Powers of Ω either converge or oscillate.

Two applications are presented. The first is to determine the most economical size of a repair crew in the machine servicing problem. The objective is to minimize $\overline{Z}$, the total cost, where $\overline{Z}$ is fuzzy. We change this fuzzy objective into a single crisp objective to be minimized as was done in fuzzy inventory control. The second application is the fuzzy queuing decision problem of deciding on the optimal number of servers for a system with finite capacity but infinite calling source. Again the fuzzy objective to be minimized is changed into a single crisp objective.

Fuzzy Network Analysis

Here we looked at only two fuzzy network problems: (1) the fuzzy shortest route problem; and (2) the fuzzy minimum cost capacitated flow problem. In both problems our evolutionary algorithm was used to generate solutions.

In the fuzzy shortest route problem we have a network of n nodes connected by arcs and the distance (cost) of going from node i to node j is given by a triangular (or other) fuzzy number. The problem is to find the "shortest" paths from node 1 to node n. Now the distance from node 1 to node n will be a fuzzy number $\overline{D}$. To determine the "shortest" distance (cost) from node 1 to node n we define $\leq$, $<$ and $\approx$ between fuzzy numbers and specify the hierarchy H_i. H_1 will contain all the undominated distances from node 1 to node n, H_2 will be all the undominated distances not in H_1, etc. Our solution is H_1. A numerical example is given showing our evolutionary algorithm computation of H_1.

The minimum cost capacitated flow problem generalizes the transportation, transshipment, assignment and max–flow problems. Again we have a network of n nodes connected by arcs with some nodes as input nodes and some other nodes as output nodes. Arcs have capacities and unit costs associated with them. Now the costs and capacities will be triangular fuzzy numbers. The objective is to minimize the fuzzy total cost of a flow, subject to the fuzzy capacity constraints and the fuzzy conservation of flow constraints. The flow from node i to node j will be a triangular fuzzy number. There is also the fuzzy feasibility constraint that says the sum of all fuzzy inputs is approximately equal to the sum of all the fuzzy outputs. The objective, the total fuzzy cost, is changed to a single crisp objective as in previous sections. A numerical example is given showing the evolutionary algorithm solution.

12.1.5 Chapter 7: Fuzzy Differential Equations

We concentrated on the second order, linear, constant coefficient ordinary differential equation

$$y'' + ay' + by = g(x) , \tag{12.2}$$

for initial conditions $y(0) = \gamma_0$, $y'(0) = \gamma_1$.

First we fuzzified the initial conditions with γ_0 and γ_1 triangular fuzzy numbers. We showed (also in the three applications) that the "classical" solution usually does not exist. However, the extension principle solution always exists. We added the constraint that the extension principle extension of the crisp solution must also solve the original differential equation for it to be called a "solution", and still it is always a solution. With the extension principle method always producing a solution we did not bother with the third type of solution called the α–cut and interval arithmetic solution. We present three applications: (1) an electrical circuit; (2) a vibrating mass; and (3) a dynamic supply and demand model.

Then we investigated two special cases of a or b (equation (12.2)) fuzzy. We looked mostly at the homogeneous case ($g(x)$ is zero) and crisp initial conditions. Not much is known about solutions for a and b fuzzy. What we did find in two examples was: (1) the extension principle solution existed for $g(x) \equiv 0$, but is difficult to compute and work with when $g(x)$ was not identically zero; and (2) the "classical" solution failed to exist.

12.1.6 Chapter 8: The Fuzzy Difference Equations

In this chapter we looked at solutions to the second order, linear, constant coefficient difference equation

$$y(k+2) + ay(k+1) + by(k) = g(k) \ , \tag{12.3}$$

for $k = 0, 1, 2, \ldots$ for fuzzy initial conditions $y(0) = \overline{\gamma}_0$, $y(1) = \overline{\gamma}_1$ with $\overline{\gamma}_0$, $\overline{\gamma}_1$ triangular fuzzy numbers. We showed that the "classical" solution never exists while the extension principle solution sometimes exists, and other times fails to exist. We also required the extension principle extension of the crisp solution to solve the original difference equation in order for it to be called a "solution". We also considered the α–cut and interval arithmetic solution, and required it to satisfy the original difference equation. We argued that the extension principle solution is more general (exists more often) than the α–cut and interval arithmetic solution. The extension principle solution was used in three applications: (1) an economic model for national income; (2) transmission of information; and (3) fuzzy Fibonacci numbers.

12.1.7 Chapter 9: Fuzzy Partial Differential Equations

We only considered simple partial differential equations whose crisp solutions would be given in "closed form", or the crisp solution did not involve series (Fourier, Bessel, Legendre, ...). The extension principle solution method fuzzifies the crisp solution and we are unable, at this time, to fuzzify series of functions. We fuzzified more of the original equation than in Chapters 7 and 8 because now the non–homogeneous part and the boundary conditions both get fuzzified. The partial differential operator was not fuzzified.

We considered only two types of solutions: (1) the classical solution; and (2) the extension principle solution (required to solve the original equation). We showed through one theorem and examples that the classical solution is more general than the extension principle solution if all the constants in the partial differential operator are positive.

12.1.8 Chapter 10: Fuzzy Eigenvalues

The fuzzy eigenvalue problem is to solve

$$\overline{A}\,\overline{X} = \overline{\lambda}\,\overline{X}\ , \tag{12.4}$$

for fuzzy eigenvalue $\overline{\lambda}$ where: (1) $\overline{A}$ is a $n \times n$ matrix whose elements are triangular fuzzy numbers; (2) $\overline{X}$ is a $n \times 1$ vector whose members are triangular shaped fuzzy numbers; and (3) $\overline{\lambda}$ is a triangular shaped fuzzy number. Once we obtain $\overline{\lambda}$ we then solve for fuzzy eigenvector $\overline{X}$.

We argued that too often the classical solution will fail to exist so we then concentrated on the extension principle solution. We did not consider the third solution method called the α–cut and interval arithmetic solution. In order to avoid fuzzy complex numbers $\overline{A}$ is always a symmetric fuzzy matrix.

We presented an algorithm for computing the largest positive fuzzy eigenvalue and its corresponding eigenvector. We tested our algorithm on a 2×2 matrix $\overline{A}$ where we may compute the correct results through hand calculations. Then we applied the algorithm to a 5×5 symmetric fuzzy matrix $\overline{A}$.

Two applications are given: (1) fuzzy input–output analysis, and (2) fuzzy hierarchical analysis. Fuzzy input–output analysis fuzzifies Leontief's closed input–output model (the "open" model is in Chapter 3). Fuzzy hierarchical analysis first fuzzified Saaty's $\lambda_{\max}$–method and then we also fuzzified the geometric row mean procedure in hierarchical analysis.

12.1.9 Chapter 11: Fuzzy Integral Equations

This chapter is about solutions to the Fredholm integral equation of the second kind

$$f(x) = g(x) + \lambda \int_a^b K(x,y) f(y)dy\ , \tag{12.5}$$

for kernel K, $\lambda > 0$, known function g when λ and/or g are fuzzy. When λ and/or g are fuzzy we wish to solve for the unknown fuzzy function f.

We considered the classical solution and the α–cut and interval arithmetic solution because the extension principle method is not directly applicable. Since the α–cut and interval arithmetic method fuzzifies the crisp solution we look at two procedures for obtaining a crisp solution: (1) the resolvent kernel method; and (2) the symmetric and separable kernel procedure.

For the resolvent kernel method and $K \geq 0$ it is shown that: (1) both solutions exist and are equal if λ is crisp; and (2) if λ is fuzzy and fuzzy g is non–negative, then both solutions exist and are equal. In the symmetric and separable kernel procedure simple examples are presented showing solutions exist and are equal.

12.2 Research Agenda

Here we will summarize what we think are the major areas of research with respect to the topics covered in this book.

12.2.1 Chapter 3: Solving Fuzzy Equations

This chapter is about fuzzifying some simple equations in elementary algebra and linear algebra. Much more can be done. What is obviously missing is a theory of fuzzy trigonometry.

It would be nice to know when the extension principle method and the α–cut and interval arithmetic procedure, for fuzzifying functions and algebraic expressions, will produce the same result.

12.2.2 Chapter 4: The Fuzzy Mathematics of Finance

This seems a natural application of fuzzy sets since future interest rates and future cash flows are uncertain and may be modeled by fuzzy numbers. We only looked at some simple calculations such as future value, present value and regular annuities. This can be expanded into other types of annuities, etc.

We also presented only two methods of fuzzy portfolio analysis: (1) net present value; and (2) the internal rate of return. Other procedures for comparing future cash flows may be fuzzified. Much of financial analysis seems ready for the application of fuzzy logic.

12.2.3 Chapter 5: Fuzzy Non–Linear Regression

Further testing of our evolutionary algorithm is needed on fuzzy functions not in the library. In order to make the algorithm run faster we need to find ways to reduce the size of the search space. Also, we need to be able to employ fuzzy numbers that are not always positive. Lastly, we need to find methods of constructing the right library to fit the regression problem. We do not want to always fit fuzzy polynomials to fuzzy data.

12.2.4 Chapter 6: Operations Research

Fuzzy Linear Programming

We assumed all our fuzzy numbers were non–negative triangular (shaped) fuzzy numbers. So, we need to allow for other types of fuzzy numbers (some could be negative, etc.). We would like to find some sufficient conditions on the fully fuzzified linear program so that the undominated set will be non–empty. Also, we need to be able to incorporate other methods of evaluating "$\leq$" between fuzzy numbers.

Fuzzy PERT

The model needs to be expanded to include the computation of the possibility distributions of floats, earliest start time, latest start time, etc. Resource allocation to the jobs can be added and then one could consider the concept of "crashing".

Fuzzy Inventory Control

This seems a natural application of fuzzy sets since many of the parameters are uncertain and could be modeled with fuzzy numbers.

In this section we only considered the single item multiperiod (finite horizon) model. Within this model much more can be added such as: (1) having fuzzy marginal purchasing costs; (2) having uncertain termination dates or infinite planning horizon; (3) allowing delivery lags; and (4) producing the product ourselves instead of buying it. We also need to find ways to speed up our evolutionary algorithm since it was quite slow in a few cases.

The fuzzy EOQ (economic order quantity) model has been widely studied but other areas of inventory control seem ripe for the application of fuzzy sets.

Fuzzy Queuing Theory

It seems that a good application area for possibility theory would be in the study of queuing systems. In this section we only considered the simplest queuing systems to model using possibility theory. The theory should be expanded to more complex queuing systems. With this comes the expansion of possibility theory into queuing decision models.

Fuzzy Network Analysis

Our evolutionary algorithm worked quickly on the fuzzy shortest path problem but was very slow on the fuzzy min–cost capacitated network problem (spent too much time checking all the constraints). For larger fuzzy network problems we need to find ways to speed up the algorithm. Also, fuzzy sets may be applied to other network problems in operations research.

12.2.5 Chapter 7: Fuzzy Differential Equations

In equation (12.2) we need to consider fuzzifying the non–homogeneous part ($g(x)$)as was done in Chapter 9. More study is needed on making a and/or b in equation (12.2) fuzzy.

Also, these results should be extended to: (1) higher order differential equations; and (2) certain systems of linear differential equations under fuzzy initial conditions.

12.2.6 Chapter 8: Fuzzy Difference Equations

As for fuzzy differential equations we should consider fuzzifying a, b and $g(k)$ in equation (12.3). Since many times solutions do not exist more research should be concerned with "partial" solutions. A solution is partial if it exists for k in an infinite subset of $\{0, 1, 2, \ldots\}$. We want it to be an infinite subset since then we can find the limiting solution as k grows without bound.

12.2.7 Chapter 9: Fuzzy Partial Differential Equations

In order to have some meaningful applications we will have to develop fuzzy Fourier series. Also, it was assumed that the constants in the partial differential operator were always crisp and positive. We should look at when these constants can be crisp and negative, or fuzzy.

12.2.8 Chapter 10: Fuzzy Eigenvalues

There is the general unsolved problem of computing fuzzy eigenvalues. This is a difficult computational problem because a fuzzy eigenvalue can be a (real) fuzzy number or a fuzzy complex number. Also, in the crisp case eigenvalues can be distinct but in the fuzzy case they can overlap (their supports overlap). After computing the fuzzy eigenvalues we then want to obtain their corresponding fuzzy eigenvectors. Since there are numerous applications of eigenvalues and eigenvectors in economics and engineering we need to solve the fuzzy eigenvector problem.

12.2.9 Chapter 11: Fuzzy Integral Equations

In equation (12.5) we fuzzified λ and g to solve for unknown fuzzy function f. All that is left is to fuzzify the kernel $K(x, y)$. No applications were presented so we need some good applications for fuzzy integral equations.

12.3 Conclusions

We see this book as only the beginning because in the future we envision separate books on: (1) applications of fuzzy sets/logic to finance; (2) applications of fuzzy sets/logic to economics; (3) applications of fuzzy sets/logic to operations research; and (4) applications of fuzzy sets/logic to (chemical, electrical, ...) engineering. We look forward in participating in this project.

Chapter 13

Evolutionary Algorithms

13.1 Introduction

Our general purpose evolutionary algorithm is described in the next section. We have not included a disk containing our evolutionary algorithm in this book since you can download your own software (genetic, evolutionary) from the internet. Use the term "genetic algorithm", or "evolutionary algorithm" in your search engine to locate their web pages and then you can solve your own fuzzy optimization problems.

13.2 General Purpose Algorithm

In this Section we briefly discuss the design of our general purpose evolutionary algorithm used for obtaining (good) approximate solutions to fuzzy optimization problems. More details may be obtained from the book [1]. Evolutionary algorithms perform a directed search. Therefore they are useful for fuzzy optimization problems since the usual methods of solving crisp optimization problems (calculus, ..) usually do not apply to fuzzy optimization problems.

Fuzzy optimization problems usually have constraints , and these are usually fuzzy constraints, which we shall refer to as "the constraints" in this Section. Also, the fuzzy function, or fuzzy expression, we wish to maximize, or minimize, has been changed into a single crisp objective function (as in Chapters 5,6 and 10) which we will refer to as the "fitness function" in this Section. The fuzzy shortest path problem in Section 6.5 is different in that there was no fitness function as described above. The "fittest individuals" in a population means: (1) in a max. problem those with the greatest values of the fitness function; and (2) in a min. problem those with the smallest values of the fitness function.

The overall theory of evolutionary algorithms is adopted from nature where the fittest members of a population have better chances to survive than other population members. The main operators of evolution are selection, which selects the fittest members of the population, and the recombination and mutation operators which are needed to create a new generation out of the fittest population members selected by the selection operator. These operators are then applied to the new generation. The process continues until a predefined number of generations is reached or some fitness limit is obtained.

Since our evolutionary algorithm is a general purpose algorithm, it has to be adapted for any specific fuzzy optimization problem. How it was adapted is discussed in the text within the section it was applied. So now we can describe the general procedure.

In any fuzzy optimization problem there are decision variables. Let us call these decision variables x_i, $1 \leq i \leq n$, which are all real numbers. We wish to find values of the x_i to max/min the fitness function. The meaning of the x_i are problem dependent. For example, in the fuzzy shortest route problem in Section 6.5, the x_i represent a path through a network. In Chapter 5 they represent the fuzzy parameters in a fuzzy function. In the fully fuzzified linear programming problem in Section 6.1 they represent triangular fuzzy numbers which are the fuzzy decision variables.

For each evolutionary algorithm we first create an initial population consisting of P elements. The size of P may be small like 50, or P can be 10,000 depending on the fuzzy optimization problem. Each element of the population is represented as a vector $v = (x_1, \cdots, x_n)$. The x_i are real numbers, there is no need to code the decision variable in binary notation (zeros and ones). To each member of the population we add σ which represents its mutation rate. So, members of a population look like

$$v = (x_1, \cdots, x_n, \sigma), \tag{13.1}$$

with σ usually set initially at 0.3. We randomly generate the initial population subject to the constraints. That is, we randomly generate a v , $\sigma = 0.3$, and if it satisfies the constraints we take it into the initial population. Otherwise, it is discarded and we randomly generate another v.

After producing the initial population the evolution starts with the selection process. The algorithm selects the Q fittest individuals to produce the next generation (population). Selection is a deterministic operation and the Q fittest individuals are chosen to build the offspring of the next generation. A typical value of Q would be 10% of P, assuming $Q \geq 1000$, or it could be as high as 50 % of P.

The recombination process constructs a temporary population of size P by applying a crossover operator to the Q-population, and then mutation gives the final population which is the next generation. The crossover operator may be a one point crossover, or a two point crossover, or a multipoint crossover operator. Our general purpose evolutionary algorithm employs a multipoint crossover operator. However, let us only describe the single point crossover operator in this Section.

Randomly choose two members of the Q-population. Assume they are q_a and q_b, where

$$q_a = (x_{a1}, \cdots, x_{an}), \tag{13.2}$$

and

$$q_b = (x_{b1}, \cdots, x_{bn}). \tag{13.3}$$

We delete the σ term since it is not effected by crossover. Now randomly choose an integer k in $\{1, 2, \cdots, n\}$. For example, suppose $k = 7$. Then form q_a^{temp} and q_b^{temp}, two possible members of the temporary population, as follows

$$q_a^{temp} = (x_{a1}, \cdots, x_{a6}, x_{b7}, \cdots, x_{bn}), \tag{13.4}$$

and

$$q_b^{temp} = (x_{b1}, \cdots, x_{b6}, x_{a7}, \cdots, x_{an}). \tag{13.5}$$

This is the one point crossover operation. We interchange $(x_{a7}, \cdots, x_{an})$ and $(x_{b7}, \cdots, x_{bn})$. However, we are not finished. We test q_a^{temp} and q_b^{temp} to see if they satisfy the constraints. If q_a^{temp} satisfies the constraints, then we take it into the temporary population and discard q_b^{temp}. If q_a^{temp} fails to satisfy the constraints but q_b^{temp} satisfies them, then we discard q_a^{temp} and take q_b^{temp} into the temporary population. When both fail to satisfy the constraints, both are discarded and we randomly choose two more individuals from the Q-population and perform the crossover operation again. Continue picking two from the Q-population until we have P members in the temporary population. The reader may now generalize to a multipoint crossover operator. In our multipoint crossover operator, we first randomly choose the number of crossover points. The crossover operator may need to be modified in certain fuzzy optimization problems. Next, the mutation operator transforms the temporary population into the next generation.

There are many types of mutation operators. A simple mutation operation is to randomly change a position x_i, $1 \leq i \leq n$, for some individuals in the temporary population. First randomly choose s members from the temporary population , put them in a set S and rename them $r_1, \cdots, r_s$. The size of S could be 10% to 50% of the total temporary population. For each $r_i = (r_{i1}, \cdots, r_{in})$ in S (omit the σ) randomly generate an integer $k \in \{1, \cdots, n\}$. Suppose $k = 4$. Then randomly generate a real number r in some interval and replace r_{i4} with r giving

$$r_i^* = (r_{i1}, r_{i2}, r_{i3}, r, r_{i5}, \cdots, r_{in}). \tag{13.6}$$

If r_i^* satisfies the constraints, then place it into the next generation. Otherwise, discard it and randomly generate positions k and replacements r until it satisfies the constraints. Of course, more than one r_{ij} may be replaced this way. This simple mutation operator ignored the mutation rate σ.

Mutation is the driving force of evolutionary algorithms. It is to randomly spread the population out over the whole search space in search of the optimum solution. Now let us consider a more complicated mutation operation that uses the mutation rate σ. First choose s members from the temporary population and put them in set S as before and rename them $r_1, \cdots, r_s$ with $r_i = (r_{i1}, \cdots, r_{in})$, $1 \leq i \leq s$. Calculate, for each $1 \leq i \leq s$

$$r^*_{ij} = r_{ij} + \sigma N(0,1), \tag{13.7}$$

for $1 \leq j \leq n$, where $N(0,1)$ stands for a normally distributed random variable having zero mean and standard deviation one. We then check to see if a r^*_i satisfies the constraints. If it does, then take it into the next generation. Continue the mutation operation until we get s new r^*_i, $1 \leq i \leq s$, individuals to place in the next generation of size P. We also mutate the σ for each of these new members. For each i, $1 \leq i \leq s$, replace σ with a slightly altered σ^*_i and place this new mutation rate σ^*_i in the last position in r^*_i. Then a new r^*_i is

$$r^*_i = (r^*_{i1}, \cdots, r^*_{in}, \sigma^*_i), \tag{13.8}$$

$1 \leq i \leq s$. We place all of these r^*_i, $1 \leq i \leq s$, into the next generation. The mutation operator may have to be changed to meet the fuzzy optimization problem. For example, in the fuzzy min-cost capacitated network problem in Section 6.5, mutation was to simply interchange two adjacent elements in the vector $v = (x_1, \cdots, x_n, \sigma)$, not including σ.

Now we have our next generation : (1) $P - s$ individuals created by crossover from the previous generation; and (2) s individuals first altered by crossover and then mutation. The selection operation now selects the fittest Q individuals for crossover and mutation. We continue this way from generation to generation until we see the results stabilizing (converging) and we stop the algorithm and record the results. Run the evolutionary algorithm, with new random initial populations, a number of times to check and see if we always end up in the same place. By repeating the experiment again and again we gain confidence in the results as being good approximate answers to the fuzzy optimization problem.

There are, of course, many variations of the above description. If you download a genetic, or evolutionary, algorithm it will probably differ from that discussed above. However, the whole procedure appears quite robust and many different algorithms will give good approximate solutions to fuzzy optimization problems.

Bibliography

[1] J.J. Buckley and T. Feuring: Fuzzy and Neural: Interactions and Applications, Physica-Verlag, Heidelburg, 1999.

Index

List of Figures

List of Tables

Printing: Strauss GmbH, Mörlenbach
Binding: Schäffer, Grünstadt

Batch number: 10725802

Printed by Printforce, the Netherlands